HANDBOOK OF ROCKY MOUNTAIN PLANTS

by Ruth Ashton Nelson 1896–1987
revised by Roger L. Williams
drawings by Dorothy V. Leake

Text copyright © 1992 by Roger L. Williams
Drawings copyright © 1969 by Dorothy V. Leake

Published by Roberts Rinehart Publishers
Post Office Box 666, Niwot, Colorado 80544

Published in the United Kingdom and Europe
by Roberts Rinehart Publishers,
3 Bayview Terrace, Schull, West Cork, Republic of Ireland

Published in Canada by Key Porter Books
70 The Esplanade, Toronto, Ontario M5E 1R2

International Standard Book Number 0-911797-96-3
Library of Congress Catalog Card Number 91-66684

Cover Design: Hugh Anderson
Graphic Production: Archetype, Inc.
Interior Design: Gretchen Kingsley

This book is presented to
plant lovers of the Rocky Mountain region
as a memorial to

Aven Nelson

Inspiring Teacher
Faithful Friend
Beloved Companion

CONTENTS

ILLUSTRATIONS

LINE DRAWINGS AND CHARACTERS ILLUSTRATING CHARACTERISTICS OF PLANT AND FLOWER

LINE DRAWINGS OF INDIVIDUAL PLANTS

FERNS AND FERN ALLIES

PINE FAMILY

WATERLILY FAMILY

BUTTERCUP FAMILY

BARBERRY FAMILY

HONEYSUCKLE FAMILY

MOSCHATEL FAMILY

VALERIAN FAMILY

COMPOSITE FAMILY

WATER-PLANTAIN FAMILY

Key to national parks and monuments

A. Glacier National Park

B. Yellowstone National Park

C. Grand Teton National Park

D. Craters of the Moon National Monument

E. Dinosaur National Monument

F. Rocky Mountain National Park

G. Black Canyon of the Gunnison National Monument

H. Colorado National Monument

I. Arches and Canyonlands National Parks

J. Bryce Canyon National Park

K. Zion National Park

L. Great Sand Dunes National Monument

M. Mesa Verde National Park

N. Grand Canyon National Park

O. Canyon de Chelly National Monument

P. Walnut Canyon National Monument

Q. Bandelier National Monument

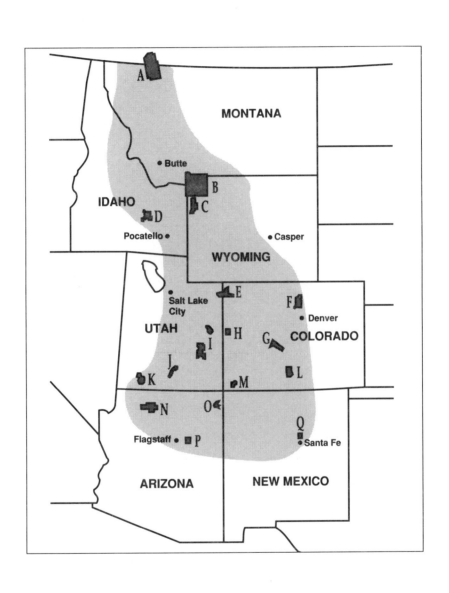

PREFACE

Nearly a quarter of a century ago, Ruth Ashton Nelson addressed the first edition of this flora to outdoor people who are neither botanists nor natural scientists, but who would like to identify the wild flowers they see on their mountain excursions, and who might, along the way, be encouraged to acquire information about plants in general and the places where plants live. Some of that information was drawn from historical, ecological, geological, ethno-botanical, economic, and conservation sources. Much of it came from her personal observation and experience.

The book was designed to give residents and visitors to the region the means of identifying plants on a simplified botanical basis. The popularization of any scientific subject is difficult and provokes controversy. Generalizations are dangerous, but the uninitiated need them. Yet, living organisms vary and never perfectly fit our schemes to classify them. The keys, whether to the plant families, genera, or species, are based whenever possible on the most obvious characters. Even the best of keys may mislead on occasion. The need to simplify them in a popular work compounds the peril.

Mrs. Nelson's original keys were remarkably simple; but to achieve that simplicity, she omitted keys to a number of plant families, genera, and even species that she treated in her text. It may be that the casual user of the book was never inconvenienced by the omissions—or never discovered them. We have reason to believe, in fact, that beginners often shun the keys, fearing a technical vocabulary and preferring to rely on the drawings for identification. The drawings are, no doubt, helpful, but the keys in any flora are fundamental, and the beginner is encouraged to become familiar with how they work. Besides, as Ruth Nelson pointed out, the use of keys is a valuable exercise in observation.

The keys in this revised edition are more extensive than in earlier editions, but the emphasis remains on simplicity and clarity. The expansion permits inclusion, not only of all groups and species mentioned in the text, but of numerous additional species likely to be found within our range that cannot be treated in the limited space of an introductory text. Measurements are now given in the metric system (traditional in botany), on the assumption that Americans are more familiar with it than they were twenty-five years ago. The use of keys is explained in Part II of the Introduction.

It has become increasingly common in popular floras to arrange the sequence of plant families alphabetically for the convenience of the user in the field. Ruth Nelson, evidently believing that even the beginner in botany ought to be made aware of the evolutionary relationships (phylogeny) in the plant world, elected to follow the phylogenetic sequence of plant families established by Adolph Engler at the end of the nineteenth century. For this edition, we have adopted the newer and preferable phylogenetic sequence for the flowering plants published by Arthur Cronquist in 1968.

As for plant names, Ruth Nelson gave precedence to popular names to assist the popular reader. The scientific names followed. We accept that order for a practical reason: the line drawings of individual plants were labelled with the popular names. The conscientious user of the keys will soon discover, however, why precedence must be given to the scientific names in the keys. Popular names are unstable, varying from region to region. What is more (as is detailed in Part II of the Introduction), only the scientific names reveal how species are grouped into genera, and how genera are grouped into plant families. Botanical nomenclature, in short, is a precise and elegant medium, popular names the gentle anarchy of the uninformed.

In preparing the first edition, Ruth Nelson acknowledged the extensive use she had made of two books: H. D. Harrington, *Manual of Plants of Colorado*; and W. A. Weber, *Rocky Mountain Flora*, both of which have continued to serve me. The staff of the Rocky Mountain Herbarium at the University of Wyoming (which contains the preeminent collection of Rocky Mountain species) was as cordial to Ruth Nelson as it has been to me. Regional authorities read parts of the original manuscript and are here gratefully acknowledged again: C. W. T. Penland of Colorado College; George J. Goodman of the University of Oklahoma; and Charles T. Mason, Jr., of the University of Arizona. By the time the revised edition was undertaken, I had the use of additional distinguished regional flora: C. L. Hitchcock et al., *Vascular Plants of the Pacific Northwest*; Robert D. Dorn, *Vascular Plants of Wyoming*; and the several volumes of the incomplete Arthur Cronquist et al., *Intermountain Flora*. Several dated titles have proved to be useful on occasion because of their comprehensive coverage of the region: Coulter and Nelson, *New Manual of Botany of the Central Rocky Mountains* (1909), and P. A. Rydberg, *Flora of the Rocky Mountains and Adjacent Plains* (1922).

The original line drawings of individual plants were expertly made by Dorothy Van Dyke Leake, who has graciously consented to their republication in this revised edition in memory of her long friendship with Ruth Nelson. On behalf of the late Ruth Ashton Nelson, who died on 4 July 1987, I thank all those who encouraged this revision so that her work might live on. I cannot claim that she would approve every alteration I have made; but she knew full well, when entrusting the revision to me, that progress in botanical knowledge in the past several decades dictated commensurate change in the book.

Roger L. Williams

INTRODUCTION

T he Rocky Mountain region provides a distinct set of living conditions for plants. These are different from those of other geographic regions such as New England, the Mississippi Valley, the Pacific Coast, or the semi-desert Southwest. Because this region is so different, its plants are worthy of special attention. In the montane regions, at least 75 percent of the species belong exclusively to the mountains. For the subalpine regions the figure is even higher. The mountains have such a variety of subsidiary climatic conditions that the number of kinds of plants found here is very great. More than nine hundred are described in this book, but that is only a portion of the whole number.

When you start learning about plants, you will soon find that environment (that is, living conditions) has much to do in determining where certain kinds of plants grow. The study of living organisms (plants or animals) in relation to their environment is called *ecology*. There are several factors that determine environment. The most important are climate, soil structure, and, in our region, the effects of elevation.

Climbing a mountain is a bit like travelling northward. Students of geography and plant or animal life have recognized distinct latitudinal zones of climate, each characterized by its own type of vegetation and animal life. In western North America, from south to north, one finds the Sonoran (Mexican), Transition (central United States), Canadian (northern United States and southern Canada), Hudsonian (Hudson Bay Region), and the Arctic zones. Botanists prefer a different set of names for mountain zones. In central Colorado, for example, they recognize the Plains Zone, below 5,400 feet; Foothill Zone, 5,400 to 7,000 feet; Montane Zone, 7,000 to 9,500 feet; Subalpine Zone, 9,500 to 11,500 feet; and Alpine Zone, above 11,500 feet. Because of the extensive area covered by this book, these elevations can only be used to give an approximate idea of the altitudes, as they apply to the central part of our range.

A climb of 1,000 feet is roughly equal to a trip of 600 miles northward. Average temperature is decreased approximately 3° F. for every 1,000 feet of elevation. Within our range, the elevation of timberline decreases northward at the rate of about 360 feet per degree of latitude. In New Mexico, accordingly, the corresponding plant zones will be at a higher elevation than those given above, but considerably lower in Montana. To a certain extent, blooming seasons are correlated with elevation or latitude. The old saying, "To find the season just past, go either up a mountain or up north," is true of the three lower zones. It does not apply to the Alpine and Arctic zones, because these are more directly affected by light intensity and day length, and because there is a greater accumulation of snow below timberline. Summer comes in the Alpine earlier than in the next lower zone.

The corresponding plant communities in different areas may contain different species, but the appearance of the vegetation will be similar. The boundaries of these zones are not marked by continuous, definite lines. They vary up and down with the topography, soil, and exposure, with much overlapping and interfingering. Along water courses and on north-facing slopes, there is always a downward extension of the next higher zone; on dry, sunny slopes, an upward extension of the lower zone. For reasons not easily understood, one finds islands of shrub or tree communities surrounded by grassland. Consequently, the best way to recognize a zone is by the plant community *typical* of it.

FOOTHILLS OR SHRUB ZONE

This region extends from where the plains break at the foot of the mountains up to the region where large trees are dominant. Here the grassland that is typical of the plains begins to give way to a shrub-type vegetation. At first, the grassland continues up the south-facing slopes in fingerlike extensions, and shrubs appear only in the breaks and on the north sides of mesas and ridges. A little higher, shrubs become more conspicuous and appear on the northwest and northeast slopes. Fingers of grassland run up between them on the south- and southwest-facing slopes. Still higher, a few small trees begin to appear in the ravines and on the north exposures. Gradually, as we proceed upward, the trees become more numerous until we pass from the shrub zone into the montane zone, which is characterized by forest. In an area as large as this, we find considerable variation in the composition of the shrub belt. But brushland of some kind exists as a transition between the grassland and forest belts throughout this great region.

Along the east face of the Rockies, a division point in central Colorado is formed by a ridge of hills extending eastward from the Front Range: the divide between the basins of the Arkansas and the Platte rivers. Located about 20 miles north of Pikes Peak, it is sometimes called the Palmer Lake Divide; and on the highway, it is called Monument Hill. It constitutes a natural barrier to weather conditions as well as to plants. Several species have their northern or southern limits in this vicinity, and the aspect of the foothill slopes north and south of this dividing place is quite different.

The intermountain valleys have a distinct type of shrub formation, and the western border of our range a still different and complex form of brushland: semi-desert species in the southwest, and species with high moisture requirements in the northwest.

The most common and widely distributed shrubs throughout the Rocky Mountain foothill regions are species of juniper, sagebrush, and mountain mahogany. There are several species of each, and each species occupies its own portion of the whole area, with some overlapping. Yet, they are similar in general appearance and occupy corresponding habitats. One exception: the upright juniper is replaced by a creeping form in the northeastern section.

In northern New Mexico and Arizona, and in southern Colorado and Utah, much of the foothill region is covered by an open, shrubby forest of pinon and juniper, interrupted here and there by deciduous shrub communities. Scrub-oak also occurs, usually at a higher level than the pinon-juniper, and interspersed with groups of other shrubs. In extreme northern New Mexico, apache plume (132) is characteristic of the transition from shrub land to ponderosa pine forest.

Farther south it is much more widely distributed. In central Colorado, pinon and scrub-oak disappear, the pinon at the Arkansas-Platte divide, the oak just north of Denver and Boulder. Farther north, the mountain mahongany (137), antelopebrush (133), buckbrush (269), and sagebrush (317) cover the foothill slopes and merge into the lower edge of the montane forest.

The intermountain valleys are ringed with sagebrush, and the small hills within the valleys are usually covered with it. On the slightly higher slopes are thickets of tall shrubs, mainly chokecherry (142) and serviceberry (131), which often extend upward into the montane forest. In western Colorado and Utah, the lower part of this foothill shrub zone features almost pure stands of sagebrush with thickets of taller shrubs: scrub-oak, serviceberry, and chokecherry. In the mountains of north-central Utah, and occasionally northward through Idaho into Montana, some moist canyon sides are covered with thickets of the wasatch maple (191). In drier areas, scrub-oak is abundant, but only as far north as Idaho.

No matter the side from which you approach the mountains, as you ascend the foothills you will pass from grassland (or farmland) through a zone of brushland into a zone of forest. The brushland may consist of small, gray-leaved shrubs, of pinon-juniper scrub forest, of scrub-oak or maple thickets, or other medium or tall shrubs; but your approach to the true forest will always cross this vegetation zone of brushland.

MONTANE OR PONDEROSA PINE—DOUGLAS-FIR ZONE

The montane zone is a forested region but has as much variation in its vegetation as we find in the foothills zone. In northern Montana, the montane begins almost at the edge of the plains at an elevation of about 4,000 feet. In New Mexico and Arizona, it is above the pinon-juniper belt and almost entirely above 8,000 feet.

From northern new Mexico into southern Wyoming (along the east side of the mountains), across southern Colorado, and from northern Arizona into Utah, the ponderosa pine (13) is the dominant and characteristic tree of this zone. At the lower edge of its elevational range, it may appear stunted, deformed, often shrubby; but at about 8,000 feet it becomes a tall, spreading tree with a beautiful orange-brown trunk. It grows in open formation so that plenty of light reaches the ground, permitting the growth of grasses and many wild flowers in the vicinity. On north-facing slopes, in ravines, and elsewhere in the upper portion of this zone, the douglas fir (15) occurs mingled with ponderosa. On the western side of the Continental Divide in northern Colorado, Utah, and Wyoming, douglas-fir is the characteristic tree of the region, and the ponderosa is rarely seen.

Occasionally we find the limber pine (14) on barren ridges and rock outcrops throughout the montane zone. The early stage of succession on rock includes seedlings of hardy grasses and different perennial wildflowers such as sedums, saxifrages, geraniums, potentillas, and penstemons in addition to the shrubs and limber pines which often find footholds in rock crevices. These may be followed by either one or both of two trees which cover large areas in this region and often extend into the subalpine region: lodgepole pine and aspen.

Lodgepole pine and aspen often occupy areas where the natural plant succession has been interrupted, preventing the development of the climax vegetation, that is, the ponderosa or douglas-fir forest. Fire is the usual cause of such a change, but logging or excessive erosion can have the same effect. Aspen usually appears on moist soil and lodgepole on drier areas. The loldgepoles grow in such dense stands that practically nothing can live beneath them; but aspens are more open, permitting enough light to reach the ground so that lovely flower gardens are found among them and in the small meadow openings between groves. In the northwestern and northern part of our range, the montane forests are almost entirely of lodgepole pine, notably in Yellowstone and Glacier National Parks.

The montane zone includes many open valleys where mountain streams flow between flowery meadows and beautiful trees in natural groupings. The early settlers thought them to have a planted appearance and called them parks. Many place names still include the word *park*: Sandia Park in New Mexico, South Park, Estes Park, and others in Colorado. Here are found a greater number of different kinds of trees and shrubs than in any other part of the mountain region. Along the streams are narrowleaf and balsam popular, alders (43), and water birch.

Two of the most beautiful evergreens of North America are found in the protected canyons of the montane. In our southern section, from northern New Mexico as far north as Pikes Peak and westward, is found the white fir. In favorable situations, it develops into a tall, handsome, silver-green conical tree. It is confined to stream banks and the lower canyon slopes, where it is found in groves or as scattered individuals. In the canyons west of Denver and northward into southern Wyoming, the Colorado blue spruce (16) occupies the same habitat and is equally beautiful. It is more rigid in pattern than the fir, and many individual trees, if not all, are distinctly blue in color. In the Pikes Peak region, this spruce forms a poor-looking forest at middle elevations, never seeming to attain the symmetry of those trees in the stream-side groves. Farther north, it occurs only in such groves.

Along its upper edge the montane forest becomes more dense and, with its overlapping areas of aspen and lodgepole, merges with the subalpine forest.

SUBALPINE OR ENGELMANN SPRUCE-SUBALPINE FIR ZONE

This region has a more homogeneous tree growth than the zone below it. It begins in the south between 11,000 and 11,500 feet, in the north at about 6,500 feet, and is composed of an almost continuous forest of engelmann spruce with some mixture of subalpine fir. This spruce-fir forest covers the whole area, north, south, and west, except as it is sometimes interrupted by lodgepole pine and aspen intrusions. But the latter are not nearly as conspicuous as in the montane zone. The only breaks in the dense forests are made by steep water courses, rock outcrops, or fire scars. The water courses are usually marked by the lighter green of aspens and bordering, flower-filled meadows.

The rock outcrops are places where rough, often gnarled five-needled pines find footholds in crevices. Three such pines may be found on windswept slopes and crags. The limber pine (14) is the most widely distributed. The bristlecone pine, also called the foxtail pine, is found on Mount Evans and south and westward into the mountains of Arizona, Nevada, and California. From northern Wyoming north and westward, there is white-barked pine. All three are very rugged, often twisted and distorted by the constant winds into interesting and picturesque shapes. Many are several hundreds of years old.

The ground under the spruce forest is brown and soft with accumulations of rotting needles and the rotted wood of long-fallen tree trunks. This is the elevation where moisture is most abundant. Much snow falls here, accumulating all winter and remaining late into the summer, because the trees protect it from melting and from blowing away, creating an environment in which moisture-loving plants can thrive. Many dainty woodland plants grow and bloom here such as fairy slipper (357), woodnymph (101), twin flower (264), and dotted saxifrage (126).

At the upper edge of this zone we come to the timberline region, for many people the most interesting and delightful part of the entire mountain territory. There is a sufficient moisture to supply a luxuriant growth of wildflowers. The season is short and the nights cold, but the days are long in June and July and the sunlight brilliant. Because of the severe winter winds and the weight of snow, the trees here are twisted, becoming shorter and shorter until they finally spread out over the ground or form little canopies 3 or 4 feet high, closed on the windy side by a dense growth of twigs and needles.

This timberline forest, called elfin timber or *Krummholz*, seems to send out bands of scouts that advance a short distance beyond the main frontier and stand as isolated wind-beaten sentinels. Timberline varies in elevation with the topography. On southeast and southwest slopes, or in sheltered draws, the tree line extends higher. On north and northwest exposures, and on wind-swept ridges, it is lower.

In the little protected meadows formed by this irregular line and watered by slowly melting snowdrifts, there are luxuriant flower gardens. Here we find some of the many species common at much lower altitudes in northern latitudes. The white-flowered marsh marigold (29) is one, rose crown (119) another. Great beds of the yellow-flowered snow-lily (350) spring into bloom almost before the snow is gone and follow as it retreats up the open slopes. The dainty leaves and pale sky-blue flowers of jacobs ladder (224) are always under the elfin timber. Parry primrose (108), with clusters of brilliant rose flowers, stands with its roots in the icy water of little streams. Many more are here to delight the adventurous nature lover.

ALPINE ZONE OR TUNDRA

Above timberline we find a region of grassland and rock fields. This is called *tundra*, being very similar to the great treeless regions of the Arctic known by that term. Spring comes earlier in this very high region than it does in the sub-alpine zone, because the snow disappears more rapidly from the exposed alpine slopes than it does in the tree-shaded areas below. Much of the alpine region is so windswept that it is bare of snow most of the winter. But great permanent snow banks form where the drifting snow is trapped in draws and depressions. Flowers bloom throughout the short summer season around the melting edges of these perpetual drifts.

The great alpine flower display of the mountain tops comes in late June and early July, about the same date whether you visit them in northern New Mexico, in Utah, or in Montana. The date is influenced more by the snow accumulation of the previous winter than by the relative distance north. The alpine flowers must bloom during the longest days in order to mature seed before winter-like storms bring sustained subfreezing temperatures to these heights. Most of the alpine plants are able to withstand considerable frost.

The climax vegetation of arctic regions and of mountain tundra is a tight, short, sedge turf. When well established, this formation excludes other plants; but on gravel and rock chip pavements, around rock outcrops, or wherever the soil is more or less disturbed by frost or other action, there are pioneer communities: natural gardens of flowering plants.

Most of these are short-stemmed; many of them grow in a cushion or carpet form. Most of them have brilliant and comparatively large blossoms. The rose, mustard, saxifrage, and composite families are each represented by several species; and all colors are in evidence. A few of the commonest and most conspicuous species in order of their blooming times are: fairy primrose (107), alpine forget-me-not (232), moss campion (55), alpine sandwort (59), alpine avens (147), rydbergia (292), and arctic gentian (212).

Flowering plants disappear completely somewhere between 12,000 and 14,000 feet, and the only evident plant life on the very highest mountain tops are members of that pioneer group called lichens.

ROCKY MOUNTAIN PLANTS

PLANT NAMES AND CLASSIFICATION

Each language has its own common names for plants. Very often, the same plant has several different names in different parts of our own country. The same common name is frequently applied to different kinds of plants in different regions. Even travelers within the Rocky Mountain region are apt to find that the common names in general use in one area for a given plant are not always the same as those used in another area for the same plant. But all plants have scientific names, and Latin the universal language of botany. Its use makes it possible for two botanists of different countries, who speak different languages, to communicate with each other about plants and know that they are talking about the same kind of plant. If botanists doing work on plants had to depend on the common, or vernacular, plant names, it would make for endless confusion. By using the universally accepted scientific names, they are able to be exact in their work.

Most plant names, whether scientific or common, are intended to be descriptive. In scientific nomenclature, each *genus* is given a name that is a Latin noun. Each *species* is given an additional descriptive name that is a Latin adjective. *Rosa*, for instance, is the name for a genus that includes all the various true roses. Included in the genus is *Rosa acicularis*, which means in English "the rose with the needlelike prickles." *Rosa woodsii* was named in honor of a man whose name was Joseph Wood, and *Rosa arkansana* is the "rose which grows near the Arkansas River." Some other descriptive names are *Fragaria americana*, "American wild strawberry," and *Saxifraga chrysantha*, the "yellow-flowered saxifrage."

The system of rules for botanical nomenclature was established by an international congress of botanists. Sometimes later research finds that a scientific name that had come into general use for a certain plant does not accord with the rules. In that case, the name has to be superseded by the correct one. Botanical research may also reveal that a plant, which had been known under a distinct name, is the same as one of another region that has a much older name. In that case, the older name must be accepted. These two provisions account for many of the recent name changes of Rocky Mountain plants.

During recent years, extensive botanical study in our area has revealed that some of our plants, at first believed to be distinct species, are actually identical to Arctic, Siberian, or European species. Consequently, the names by which we have known our plants become *synonyms*, and the older names take their place. In some instances, the Rocky Mountain form of a European or Arctic species has been found to be sufficiently distinct to warrant designation as a *variety* or *subspecies*. Such plants are indicated by the abbreviation var. or ssp., depending on the form used by the botanists who described them. All this causes some confusion, but not nearly as much as would be involved if we used common names.

Because this region has been comparatively recently settled, and because of the great number of plants that are different from those in older parts of the country, many of our plants do not even have common names. In fact, it is not nearly as difficult to learn the scientific names as most people think. Many of them are already in general use without the user realizing that they are Latin: *Geranium, Yucca, Viburnum, Phacelia,* and *Phlox* are examples. In this book, an attempt has been made to include the most widely used and most appropriate common names. In some cases where there is no well-known common name in use, the generic name is used as a common name. The common names are given first, followed by the scientific name in italics. When there is a number in parentheses, it designates the illustration of this particular plant. The same number appears with the illustration and in the text of Part I when reference is made to this plant. But readers should understand by now why they are encouraged to learn and use the generic names.

Plants are classified by *species, genus,* and *family,* but the unit of classification is the species. All the plants of one kind belong to one species. A better definition is that the species is a reproductive community. A group of closely related species constitutes a *genus,* and the plural of this Latin word is *genera.* All the plants of one genus have some characteristics in common. All violets have five separate petals forming an irregular flower, and all maples have opposite leaves. Closely related *genera* are grouped into plant families. The members of a particular plant family have certain fundamental distinguishing features in common. Flowers of plants in the Mustard Family, for instance, have four separate petals and a two-celled seed pod, whereas flowers of the Lily Family have six petals and a three-celled seed pod. Flowers of the Figwort Family have petals united into an irregular, five-lobed corolla and a two-celled seed pod.

Such family and generic characters are used in the keys for identification. Certain statements, therefore, can be made that are true in general in regard to the plants that make up a family, genus, or species. But all individuals vary to some extent, occasionally varying beyond the limits of a general description; this variability accounts for the use of qualifying adjectives in keys and descriptions. In botany, one can never safely make unqualified statements.

THE FIRST THING TO LEARN ABOUT PLANTS
(Characteristics of the Plant and Flower Described and Illustrated)

THE PLANT

The flowering plants comprise only one of the several large groups that make up the plant kingdom and are referred to as *seed plants.* Mosses, ferns, and mushrooms reproduce by spores. But the flowering plants are the most conspicuous in our environment and are the plants with which this book is mainly concerned. Many of the descriptive terms used in the text are briefly explained in the next few pages and are illustrated in Plates A through G. The glossary also provides definitions, and much botanical terminology will be defined in a good dictionary.

Plants are *annual* if they complete their life cycle of growth and fruition within one year. They are *perennial* if they live for an indefinite number of years. Plants are *herbs* if their leaves and stems die back to the ground at the end of each growing season. They are *woody* if their stems above ground remain alive through the winter, sending out new growth in the spring.

Woody plants are either *vines, shrubs,* or *trees,* and the woodiness of a stem is usually obvious. The Rocky Mountain region has very few woody vines, as vines in general are not adapted to severe climates. *Shrubs* may be anywhere from a few centimeters to five meters tall, but are bushy in habit with several main stems. *Trees* are usually five meters or more in height and have one main stem, called a *trunk,* which is eight centimeters or more in diameter. In the Rocky Mountains, several woody species are on the borderline between shrubs and trees. In favorable locations, they may grow to five meters or more in height but have several trunks and a bushy habit. In other places, the same species will be considerably under five meters tall. Western red birch, alder (43), and Rocky Mountain maple (190) fall into this class. Trees and shrubs are *deciduous* or *evergreen* according to whether they drop their leaves in the fall or hold them through the winter. Leaves that are evergreen are firmer and tougher than those that are deciduous.

A typical flowering plant consists of *stem, roots, leaves, flowers,* and *fruit.*

STEMS (PLATE A)

The stem is the axis of the plant. All the other parts arise from it. A stem may be long and intricately branched, as in tall trees, or it may be short and compact, as in the base of an onion bulb. It may even be green and take over the food-making process, as in the cacti. Buds and roots originate in the stem.

The Latin term *acaulescent,* meaning *stemless,* applies to plants that have only short stems at ground level (Plate A2). In such cases, leaves and flower stalks arise directly from this consolidated stem, which is often called a crown or *caudex.* These plants may appear to be tufted. If the *caudex* is branched, the plants are said to be *caespitose* and appear mat-like or cushion-like (the moss campion [55] is an example). Stems may rise strictly *erect* from the crown, or they may lean outward and be described as *ascending.* If they are horizontal, they are said to be *decumbent,* and if they start out horizontally and then bend so as to become more or less erect, they are described as "decumbent at base." A spreading underground stem is called a *rootstock* or *rhizome.* In a strict botanical sense, a leafless stalk of a flower is not a stem. Stems bear leaves and buds. This distinction is important in using keys.

Stems have joints called *nodes* (Plate A7). The spaces between the nodes are the *internodes* (A7). Leaves and branches grow from buds at the nodes, and in some cases roots also grow from the nodes. If there is only one leaf at each node, the *leaf arrangement* is *alternate* (A3). If there are two leaves at each node, it is *opposite* (A4). If there are more than 2 leaves at each node, the leaves are in *whorls* (A6). The angle formed by the stem and the leaf is the *axil* (A3). Buds formed in the axils are called *axillary buds* (A3). The bud at the tip of a stem is called a *terminal bud* (A7). If leaves are alternate, these axillary buds produce an *alternate branching* pattern (A3); if they are opposite, the buds produce an *opposite* branching pattern (A4). Flowers are sometimes axillary.

ROOTS

The roots of plants serve to anchor them to the ground and are organs through which water and nutrient materials in solution are absorbed. The professional botanist needs to determine the type of root when classifying an unknown plant. In a popular flora such as this, the root characters are used infrequently in the keys for the convenience of the beginner, but they are sometimes unavoidable. We include only a brief statement in their regard.

Taproots are those which have a main axis extending downward, with smaller side branches. This main axis is sometimes thickened by the storage of food, as in the carrot and several related wild plants. Many annuals and biennials and some perennials have tap roots. Annual plants usually have less extensive root systems than perennials, a characteristic providing an easy distinction between the two.

Other plants have an irregularly branched root system, which is said to be *fibrous*. Stems may also give rise to roots, particularly underground stems. When this happens, the plant spreads out in a mat or carpet form, or it trails along the ground and is described as *rooting at the nodes*.

LEAVES (PLATES B AND C)

A typical leaf consists of *blade, petiole,* and *stipules* (B8). Often stipules are absent, or very tiny, or they soon drop off. If the petiole is absent, the leaf is *sessile*. Leaves are *simple* if the blade consists of a single piece. *Compound* leaves are made up of *leaflets* (B9, 10). Simple leaves may be distinguished from the leaflets of compound leaves by the presence of axillary buds. That is, leaflets never have axillary buds, but there is always a bud at the base of a compound leaf. Simple leaves and the leaflets of compound leaves may be variously *toothed* or *lobed* (B11, 12). Compound leaves may be *palmate* or *pinnate* (B11, 12). Pinnate leaves may have an even or odd number of leaflets.

Leaf forms (Plate C) may be *linear, lanceolate, oblanceolate, ovate, obovate, oblong, spatulate, heart-shaped, arrow-shaped,* or *round*. Leaf tips may be *acute, acuminate,* or *obtuse*. The bases may be *heart-shaped, arrow-shaped, wedge-shaped,* or *auriculate*. Leaf margins may be *entire, toothed,* or *scalloped*. Leaves at ground level are called *basal* leaves. They often form what is called a *rosette*.

Leaves have distinctive *venation*. If a leaf has a midrib running from the base to the apex with lateral veins extending from the midrib toward the margin, it is said to be feather-veined or *pinnately-veined*. If it has more than one main vein and all originate at the base and spread as a fan (or as the fingers of a hand spread from the palm), it is said to be *palmately-veined*. Both types of leaves occur on plants that belong to the group of Dicots. Their veins branch, and the veinlets form a netted pattern ending at the leaf margin. Such leaves are said to be *netted-veined*.

If the leaf has several veins extending the length of the leaf parallel to each other and parallel to the margins, it is *parallel-veined*, (B14). Parallel-veined leaves are usually long and narrow (for example, grass leaves), but they may be broad, as are the leaves of Solomon's plume (340). Leaves of Monocots are always parallel-veined, even though this is not always evident; and leaves of Dicots are always netted-veined, though occasionally their principal veins may appear to be parallel.

FLOWERS AND INFLORESCENCES (PLATES D, E, F)

The flowering part of a plant is called the *inflorescence*. Some plants have *solitary* flowers, or flower clusters, on unbranched stalks that rise from the caudex. Such a stalk is called a *scape*, and the plant is said to be *scapose* (A2). If the plant has a branched stem and the flower stalks rise from the axils, the inflorescence is *axillary*, and the main stalk of the flower cluster is a *peduncle*. The stalk of an individual flower in a cluster is a *pedicel* (D29).

If the inflorescence is branched, it follows a pattern that will usually agree with one of the patterns illustrated on Plate F. If the axis is unbranched and the individual flowers are without pedicels, it is a *spike*. If the axis is unbranched and the individual flowers are pedicellate, it is a *raceme*, one of the most common types of inflorescence. If the axis is alternately and irregularly branched, the inflorescence is a *panicle*. If it is oppositely and regularly branched, with the terminal flower blooming first, it is a *cyme*. If all the pedicels arise from one point and are of equal length (like the rays of an umbrella), it is an *umbel* (F52). Umbels may be compounded, as in cow parsnip (203) or sulphur flower (66). If the pedicels arise from different positions on the axis and are of different lengths, but come to one level, the inflorescence is a *corymb* (F 53). A spike may be modified into a *catkin*, as in the aspen (82), or it may be compacted into a *head* (F 58), as in parry clover (156).

The essential parts of a flower are the *pistil* and the *stamen*, the female and male organs respectively. If both female and male organs are present and functional, the flower is called *perfect* and *bisexual*. Some flowers contain pistils only or stamens only, called *pistillate* or *staminate* flowers, and are *unisexual* and *imperfect*. Both perfect and imperfect flowers may be present on the same plant or limited to different plants.

Some plants depend upon the wind for pollination, others upon insects. Most flowers that depend upon insects secrete nectar and advertise its presence by scent or showy structures attached to the flower, such as brightly colored petals, sepals, or petal-like bracts. The different parts of the flower are in series, and they are attached to the enlarged top of a stalk called the *receptacle* in a definite arrangement.

The easiest part of the arrangement to recognize is the numerical pattern, which is a guide to family relation. For instance, the Lily Family has a pattern based on the numbers 3 and 6; the Mustard Family has flower parts based on the number 4; most members of the Rose Family have 5 petals. The numbers of pistils and stamens are not always the same as those of petals, but they are usually constant for the group (see the definition of *-merous* in the Glossary).

Diagrams showing some of these patterns are given in Plate D. Learning these patterns helps one to recognize at first sight the family to which a plant belongs.

A *complete* flower consists of *sepals, petals, stamens,* and *pistil.* The outermost series is formed by the sepals, the next by the petals, the next by the stamens, and the innermost by one or more pistils. The parts may all be separate, as in a buttercup (Plate E40, 44), or they may be joined to each other or to parts of a different series. The sepals collectively constitute the *calyx*; the petals collectively are the *corolla.* These two terms are generally used when the sepals or petals are *united* (Plate E46, 47). When sepals and petals are very similar, as in many species of the monocot group, they are collectively called the *perianth.*

In the dicot group, if there is only one series outside the stamens, it is usually made up of sepals—even though they are brightly colored and petal-like. One example is the pasque flower (31). Sepals may be entirely separate and early deciduous, as in the buttercup (E40), or they may be united at base, as in the potentilla (E41). They may be united into a cup, as in the rose (E42), or a tube, as in the milk vetches (E43).

Flowers may be *regular* or *irregular.* If a flower is regular, it is radially symmetrical, as is the case in potentilla (E44) and phlox (E46). If a flower is irregular, it is bilaterally symmetrical, as in the pea flower (E49) and the penstemon (E51). *Papilionaceous* (butterfly-like) is a term used to describe irregular, separate-petal corollas such as those in the Pea Family (E49). United corollas are of many forms, some of which are illustrated in Plate E.

Corollas of regular flowers may be *rotate* (E45), *bell-shaped* (or *campanulate*) (E47), *funnel-shaped* (E48), or *salver-shaped* (E46). Corollas of irregular flowers are usually 2-lipped (*bilabiate*) (E51). When corollas are united, the stamens are often inserted on the inside of the corolla tube.

The stamen is composed of *anther* and *filament* (D29). Anthers contain *pollen.* The pistil may be simple or compound; it comprises *stigma, style,* and *ovary* (D29). The stigma is the surface that receives the pollen. The ovary contains the *ovules,* which, when fertilized, develop into seeds. The style, which connects stigma and ovary, may be short or long. Sometimes the style is absent, making the stigma *sessile* on top of the ovary. A *carpel* is the structural unit of the pistil. Thus, if the pistil is compound, the ovary is composed of two or more united carpels.

The ovary is *superior* if the stamens, petals, and sepals are attached to the receptacle at its base (thus, at the base of the ovary), as in *Caltha* (29), or if the petals and stamens are inserted on the calyx cup as in the plum (D31). The ovary is *inferior* if the stamens are attached above it (D32), as in the gooseberry (117) and the harebell (260). There are intermediate stages where the lower part of the ovary is embedded in the receptacle, as in some saxifrages.

FRUITS

After fertilization takes place, the ovary (and its associated parts) ripens into a *fruit.* Fruits are of many kinds. They may be one-seeded and enclosed in a hard covering, as in the grains of grasses or the *achenes* of the composites. They may be dry *capsules,* as in the seed-pods of poppies, or *follicles,* as in columbines; or they may be fleshy.

The fleshy parts of fruits are derived from various structures associated with the flower. Serviceberries are like apples, where the calyx enlarges and encloses the ovary. Strawberries are formed when the receptacle enlarges and each individual ovary becomes an achene attached to it. Plums and cherries, technically called *drupes*, are formed when the outer seed coat becomes thick and juicy. Raspberries and blackberries, which are really clusters of small drupes called *drupelets*, are attached to a receptacle from which the whole cluster separates when ripe. True berries, such as huckleberries, are soft fruits with seeds embedded in a juicy pulp. These developments help to insure wide distribution of seeds.

HOW TO USE THE KEYS

Nearly one thousand different plants are described in this book. Over one-third of them are illustrated by line drawings. Useful as the drawings are in identification, discrepancies will illustrate the fundamental need to learn to use keys. To cope with the technical vocabulary in the keys, the reader is provided, first of all, with the section called "The First Things to Learn About Plants," along with Plates A through G. Secondly, there is a glossary.

The dichotomous key, so-named because the author always gives the reader a choice between two contrasted groups of characters, was an ingenious device conceived by Lamarck in the late eighteenth century. The unknown plant will fit into one or other of the two groups. By making a choice in accord with your plant's characters at each successive pair of statements, you will eventually come to the name of your plant. You are given, in fact, a series of keys, starting with the largest plant groupings and concluding with a key that distinguishes between species:

Key 1: The Major Plant Groups (including definitions)
Key 2: Division Pteridophyta: Ferns and Fern Allies
Key 3: Division Spermatophyta: Gymnosperms
Key 4: Division Spermatophyta: Angiosperms; Apetalous Dicots
Key 5: Division Spermatophyta: Angiosperms; Free-petal Dicots
Key 6: Division Spermatophyta: Angiosperms; United-petal Dicots
Key 7: Division Spermatophyta: Angiosperms; Monocots

Once you establish in which major group your plant belongs, the key within that group will lead you to the correct plant family. And each plant family has a key to its genera, each genus a key to its species. It takes patience and practice to learn to use keys effectively, and the necessary brevity of these keys increases the chance for errors. If the description of your plant in the text does not fit the plant you identified by progressing through the keys, you must go back through the keys to find the point where you may have made an incorrect choice. Remember that it is difficult to construct a key that will fit every individual within a species. After you have followed the process through several times, you will begin to recognize the groups and families to which plants belong. You will then be able to save time by going directly to the family or the genus key.

The Plant

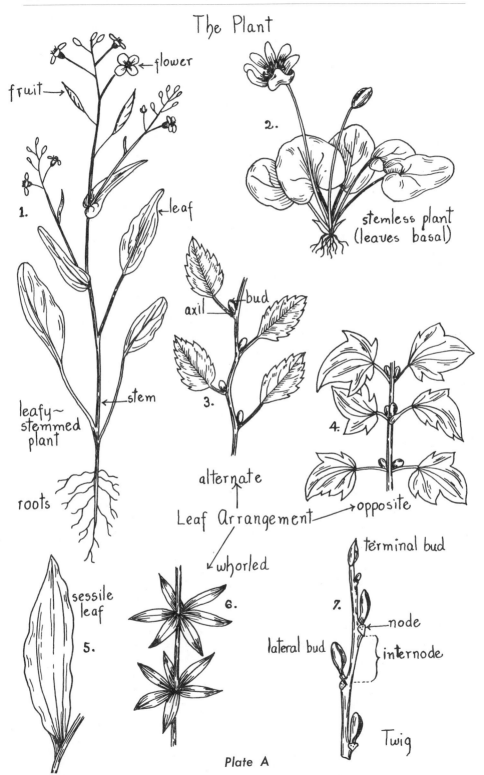

fruit

flower

1.

leaf

2.

stemless plant
(leaves basal)

axil

bud

3.

4.

leafy-
stemmed
plant

stem

roots

alternate

Leaf Arrangement

opposite

whorled

sessile
leaf

5.

6.

7.

terminal bud

node

lateral bud

internode

Twig

Plate A

Leaves

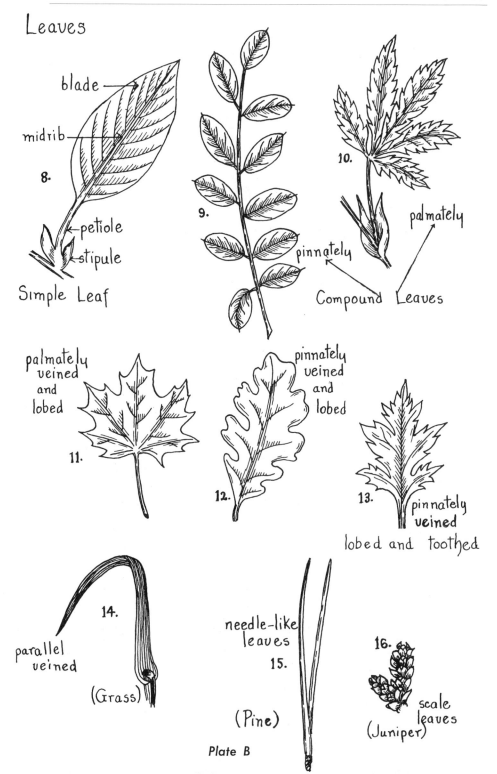

blade

midrib

8.

petiole

stipule

Simple Leaf

10.

palmately

pinnately

Compound   Leaves

9.

palmately
veined
and
lobed

11.

pinnately
veined
and
lobed

12.

13.   pinnately
veined

lobed and toothed

14.

parallel
veined

(Grass)

needle-like
leaves

15.

(Pine)

16.

scale
leaves
(Juniper)

Plate B

Leaf Forms

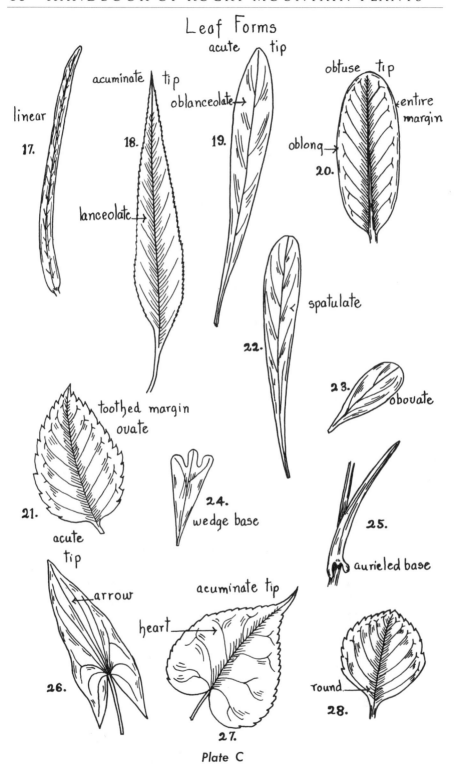

acute tip

acuminate tip

oblanceolate→

linear

obtuse tip

entire margin

oblong→

17.

18.

19.

20.

lanceolate→

spatulate

22.

23. obovate

toothed margin
ovate

24.
wedge base

25.

21.

aurieled base

acute tip

arrow

acuminate tip

heart

round

26.

27.

28.

Plate C

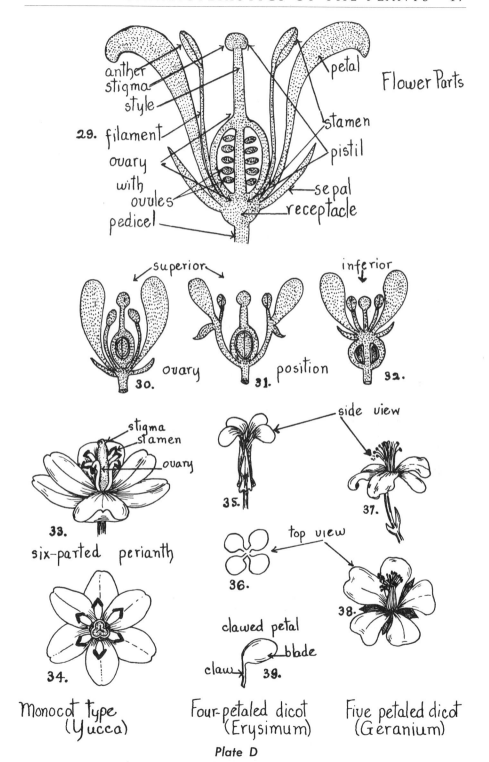

Flower Parts

anther
stigma
style
29. filament
ovary
with
ovules
pedicel

petal
stamen
pistil
sepal
receptacle

superior — superior
ovary
30.

position
31.

inferior
32.

stigma
stamen
ovary
33.

six-parted perianth

34.

Monocot type
(Yucca)

side view
35.

top view
36.

37.

38.

clawed petal
blade
claw   39.

Four-petaled dicot
(Erysimum)

Five petaled dicot
(Geranium)

Plate D

Calyx Types

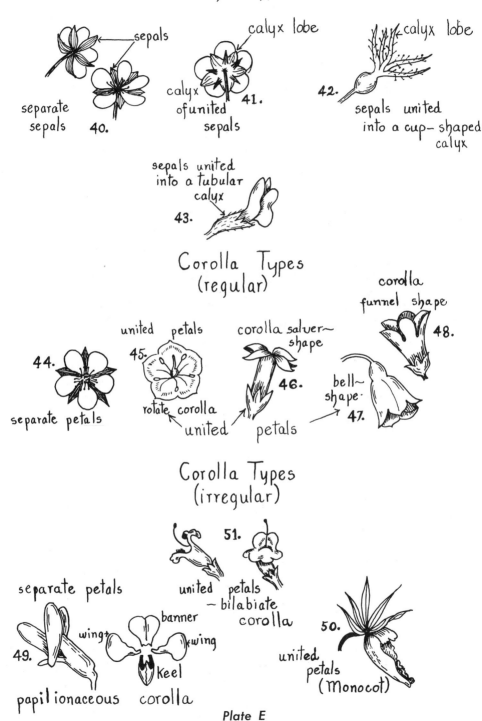

sepals

separate sepals   40.

calyx lobe

calyx of united sepals   41.

calyx lobe

42.

sepals united into a cup-shaped calyx

sepals united into a tubular calyx

43.

Corolla Types (regular)

44.

separate petals

united petals

45.

rotate corolla

united

corolla salver-shape

46.

petals

corolla funnel shape

48.

bell-shape.

47.

Corolla Types (irregular)

51.

separate petals

united petals -bilabiate corolla

banner

wing   wing

49.

Keel

papilionaceous   corolla

50.

united petals (Monocot)

Plate E

Inflorescences

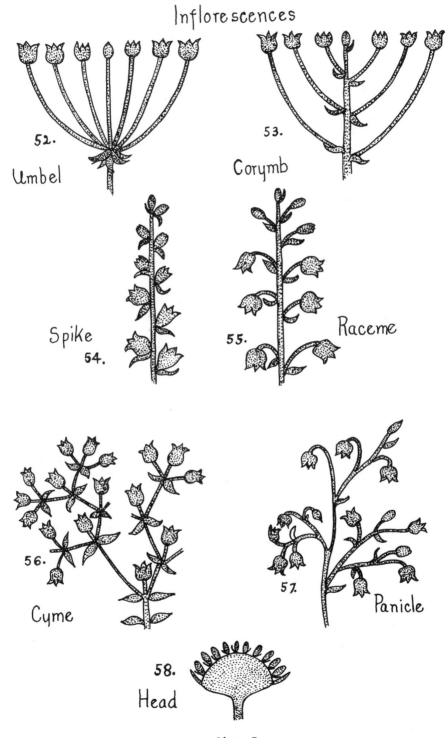

52. Umbel

53. Corymb

Spike
54.

55. Raceme

56. Cyme

57. Panicle

58. Head

Plate F

Composites

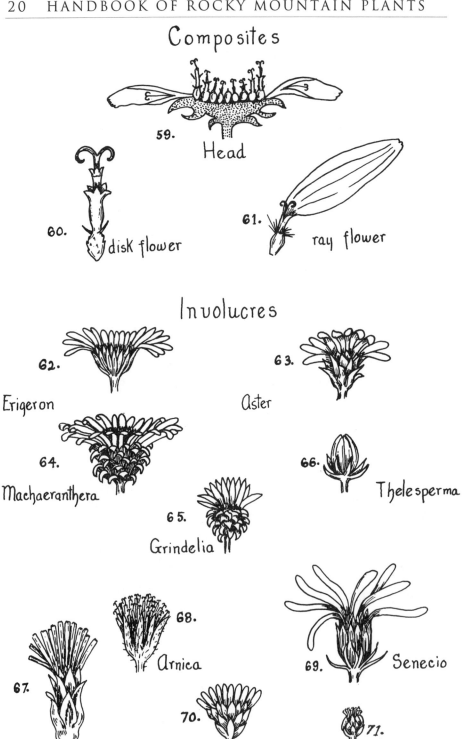

59. Head

60. disk flower

61. ray flower

Involucres

62. Erigeron

63. Aster

64. Machaeranthera

66. Thelesperma

65. Grindelia

68. Arnica

69. Senecio

67. Agoseris

70. Chrysopsis

71. Artemisia

Plate G

KEYS AND DESCRIPTIONS

The first step in identifying an unknown plant is to look down the following key to decide into which of these *major* groups it fits. It helps to read both of the coordinate statements in *any* key before making a choice; that is, the alternate choices under "1" and "1", then under "2" and "2", and so on. Then turn to the page of the group that best fits your plant and begin again with the key within that group.

KEY 1. THE MAJOR PLANT GROUPS

1. Woody plants (trees and shrubs) with needle-shaped or scale-like evergreen leaves; monoecious or dioecious cones; ovules not in a closed ovary.

Key 3. *Gymnosperms*, p. 33

1. Plants woody or herbaceous, but never with needle-shaped or scale-like evergreen leaves.
  2. Vascular plants without true flowers or seeds; reproduction by spores.

Key 2. *Pteridophytes*, p. 22

  2. Vascular plants with true flowers; reproducing by seed. Spermatophytes.
    3. Embryo with a single seed-leaf; flower parts usually in threes or sixes; leaves grass-like.

Key 7. *Monocotyledons* (Monocots), p. 383

    3. Embryo with a pair of opposite seed-leaves; flower parts mostly in fours and fives; leaves not grasslike.

Dicotyledons.

      4. Flowers without petals; calyx often wanting; flowers sometimes in aments (catkins).

Key 4. *Apetalous dicots*, p. 41

      4. Flowers with petals.
        5. Corolla of separate petals.

Key 5. *Free-petal dicots*, p. 44

        5. Flowers with petals more or less united into a regular or irregular corolla.

Key 6. *United-petal dicots*, p. 49

FERNS, HORSETAILS, AND CLUBMOSSES

These plants are included in the large group called the Pteridophytes. In addition to the true ferns, Polypodiaceae, it includes the horsetails or scouring rushes, Equisetaceae; the clubmosses or ground-pine, Lycopodiaceae; the selaginellas or little clubmosses, Selaginellaceae; and other small families not described here. The group is placed below the seed plants in the evolutionary system because reproduction is carried on by spores, which are simpler structures than seeds.

Spores, being much smaller and lighter even than seeds, can be carried by wind or water over great distances. This partly explains why many fern species are very widely distributed over the world, especially those species that are adapted to growing in moist habitats. All of our large species, and several of the smaller ones, belong in this group. In the comparatively arid areas of the American west, a group of dryland ferns has evolved, able to live under minimum moisture conditions by special morphological adaptations. All are small plants, conferring the advantage of reduced evaporating surface of the leaves. Most of them are dull green in color, indicating a tough, thick covering over the upper surface. Some have devices that cause in-rolling of the leaf margins during particularly dry periods; in other species, one or both surfaces are covered by scales or hairs. In these and other ways, evolution has made it possible for ferns to grow in desert habitats.

During the geologic period called the Carboniferous Age, ferns and their relatives attained dominance, providing the vegetative material that produced our great coal beds. But seed plants superseded them and have become the most conspicuous part of our present vegetation.

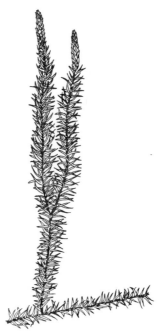

1. Mountain Club-Moss, ½ X

KEY 2. DIVISION PTERIODOPHYTA: FERNS AND FERN ALLIES

1. Stems conspicuously jointed; nodes covered by toothed sheaths; sporangia (spore-cases) on the scales of terminal dry cone-like spikes.

4. Equisetaceae, p. 24
Horsetail Family

1. Stems without conspicuously sheathed joints.
   2. Leaves (fronds) closely imbricated or very narrow; sporangia sessile, axillary.
      3. Stems short, corm-like; leaves subulate (awl-shaped) to long-linear.

3. Isoetaceae, p. 23
Quillwort Family

      3. Stems elongated, creeping or branching; leaves very short, crowded or imbricated.
         4. Sporangia of two kinds: some containing many minute spores (micro-spores), others bearing few (3–4) much larger spores (megaspores).

2. Selaginellaceae, p. 23
Little Club-moss Family

         4. Sporangia bearing uniform minute spores.

1. Lycopodiaceae, p. 23
Club-moss Family

   2. Leaves not closely imbricated; if narrow, no axillary sporangia.
      3. Leaves 4-foliate; sporocarps (receptacle inclosing the sporangia) stalked from the creeping stem.

7. Marsileaceae, p. 33
Pepperwort Family

3. Leaves not 4-foliate, simple or variously cleft; sporangia borne on the under side of ordinary leaves or on sporophylls (specialized spore-bearing leaves).
    4. Sterile segment of the frond simple; fertile segment a long-stalked simple spike.

5. Ophioglossaceae, p. 24
Adder's-Tongue Family

    4. Sterile and fertile fronds or segments more or less cleft, either essentially similar or conspicuously unlike.

6. Polypodiaceae, p. 25
Fern Family

CLUB-MOSS FAMILY: LYCOPODIACEAE

A small family of low, creeping plants having narrow, pointed leaves about 13mm long, arranged in whorls which entirely clothe the stems. MOUNTAIN CLUB-MOSS, *Lycopodium annotinum* (fig. 1), is our most common species, but rarely seen. It grows under spruce forest or in sphagnum moss at the edge of subalpine ponds. The sporangia are in cone-like structures at the tips of upright branches. Each "leaf" of the "cone" bears one round sporangium filled with tiny spores. *L. annotinum* occurs from Labrador to Alaska, and south to Pennsylvania, Michigan, Colorado, and Oregon; also in Greenland and Eurasia.

SELAGINELLA FAMILY: SELAGINELLACEAE

1. Stems very short, densely tufted.

Selaginella densa

1. Stems creeping, 5–10cm long.
    2. Leaves with a conspicuous awn

S. rupestris

    2. Leaves obtuse, without awn.

S. mutica

2. Selaginella, ½X

A small family of moss-like plants, which includes important natural ground covers on rocky or gravelly soil. They are called little club mosses because they are similar to members of Lycopodiaceae, but smaller in every way and usually more compact. Our commonest species is *Selaginella densa* (fig. 2), with stems that are much branched, densely tufted, and never more than 5cm high. The closely appressed leaves are pointed, bristle-tipped, and less than 6mm long. The fruiting branches are strongly 4-sided. In dry weather the plants are dull or grayish, but when moist they show a more lively green. This species, and similar ones, are abundant, especially on dry slopes among grasses or rocks, from the foothills to the alpine zone. From Alberta south to South Dakota, New Mexico, Utah, and Washington.

QUILLWORT FAMILY: ISOETACEAE.

Only *Isoetes*, Quillwort, is in our range.

1. Leaves acute, not long-pointed; stomata (small openings on the leaf surfaces) absent.

I. lacustris

1. Leaves with a long fine point; stomata present.

I. bolanderi

HORSETAIL FAMILY: EQUISETACEAE
> Only *Equisetum* is known in our range.
> 1. Stems bearing numerous branches in whorls at the nodes, dying down to the ground each year.
>> *E. arvense*
>> Common horsetail
>
> 1. Stems perennial, evergreen.
>> 2. Stems many grooved.
>>> 3. Stems rough with conspicuous tubercles, under 5dm tall.
>>>> *E. hyemale*
>>>> Tall scouring rush.
>>> 3. Stems nearly smooth, tubercles inconspicuous, 3–10dm tall.
>>>> *E. laevigatum*
>>>> Kansas scouring rush
>> 2. Stems with only 5–10 grooves.
>>> *E. variegatum*
>>> Scouring rush

These plants have hollow, green, striate, jointed stems, which are either unbranched or have whorls of branches. The plant tissue contains minute bits of silica, which make it unpalatable and detrimental to animals if eaten and give it an abrasive quality. The leaves are very small, like pointed teeth, and form a short crown around each joint.

Our commonest species is the COMMON HORSETAIL, *Equisetum arvense*, a plant 15–50cm tall, in which the sterile stems bear whorls of slender branches at each joint. Its fertile stems appear earlier and soon wither; they are pale brown, 5–25cm tall, and each bears one pale, brown, cone-like structure, which contains the sporangia. It is often found on moist, sandy soil, especially along railroad embankments. Two similar species with much stouter and unbranched stems 60cm to over 1m tall may be found. Both occur on moist ground, along ditches or streams, and along railroads. The KANSAS SCOURING RUSH, *E. laevigatum*, has annual aerial stems; and the TALL SCOURING RUSH, *E. hyemale*, has perennial stems, which remain green all winter. The stems of these plants, as their name implies, were used for scouring purposes.

ADDER'S-TONGUE FAMILY: OPHIOGLOSSACEAE
> Only *Botrychium* is known in our range.
> 1. Leaf above the middle of the stalk.
>> 2. Leaf near the middle of the stalk, its lobes lunate (crescent-shaped) or fan-shaped.
>>> *B. lunaria*
>>> Moonwort
>> 2. Leaf near the top of the stalk, triangular.
>>> *B. lanceolatum*
>>> Grape fern
> 1. Leaf borne near the base of the stalk.
>> 2. Sporophyll simple
>>> *B. simplex*
>>> Little grape fern
>> 2. Sporophyll multipinnate.
>>> *B. multifidum*
>>> Leathery grape fern

FERN FAMILY: POLYPODIACEAE

The leaf of a fern plant is called a *frond* and consists of a stalk, called the *stipe*, and a leafy part called the *blade*. The main extension of the stipe through the blade is the *rachis*. The blade is usually divided one or more times. The first division is a *pinna* (plural *pinnae*). The pinnae may be divided into *pinnules*, and these are sometimes further divided or variously lobed or toothed. Fronds arise from an underground stem or rootstock. If this stem is short, the fronds form a tuft or rounded clump. If the stem is creeping, the fronds may form a large, dense, irregular clump or bed; or they may occur scatteringly, as with bracken, over a large area. Some species often grow in rock crevices. This habit may result in a line of fronds following the crevice, as is seen in polypody.

The reproductive body, which contains the *sporangia* (spore cases), is called a *sorus* (plural *sori*). Sori occur on the undersides of the *fertile* fronds. They are often very numerous; but some fronds, usually the earlier ones, lack them all-together and are referred to as *sterile* fronds. The shapes and positions of sori are distinctive and useful in identification. Certain species have a very small membranous cover over the sorus called an *indusium* (plural *indusia*), which sometimes withers and disappears as the frond ages. The presence or absence of the indusium (and its shape) are also important in determining species.

1. Fronds of 2 kinds, the sterile broader than the fertile, the fertile taller.
   Cryptogramma, p. 29
   Rock brake
1. Fronds all alike, usually at least some bearing sori on the under surfaces (which appear to.be black dots).
   2. Sori without indusium.
      3. Sori dorsal, roundish.
         4. Leaves 2–3 pinnatifid.
            Gymnocarpium, p. 30
            Oak fern
         4. Leaves simply pinnate.
            Polypodium, p. 31
            Polypody
      3. Sori marginal, elongated.
         Notholaena, p. 30
         Zigzag cloak fern
   2. Sori with indusium.
      3. Indusium formed from the inflexed (in-rolled) edges of the leaf-margin.
         4. Sporangia on the underside of the reflexed portion of the leaf.
            Adiantum, p. 26
            Maidenhair
         4. Sporangia in a continuous vein-like groove or receptacle, which connects the ends of the veins.
            5. Large plants of moist copses.
               Pteridium, p. 31
               Bracken
            5. Small plants of cliffs and rock slides.
               6. Pinnae tomentose or scaly beneath.
                  Cheilanthes, p. 28
                  Lip fern
               6. Pinnae smooth beneath.
                  Pellaea, p. 31
                  Cliff-brake

3. Indusium formed from epidermal cells (not from the leaf-margin) and variously attached.
    4. Sori round.
        5. Indusium superior, kidney-shaped or shield-shaped, attached by a central stalk.
            6. Sori near the midvein of the leaf.

Dryopteris, p. 29
Wood fern

            6. Sori near the margin of the leaf.

Polystichum, p. 31
Holly fern

        5. Indusium inferior or partly so.
            6. Hood-like: attached to the inner side.

Cystopteris, p. 29
Bladder fern

            6. Cup-like: attached underneath.

Woodsia, p. 31
Woodsia

    4. Sori oblong or linear.
        5. Fronds simply pinnate.

Asplenium, p. 26
Spleenwort

        5. Fronds bipinnate.

Athyrium, p. 27
Lady-fern

(1) *Adiantum.* Maidenhair

1. Fronds bipinnate, ovate-lanceolate.

A. capillus-veneris
Venus-hair fern

1. Fronds two-forked, orbicular.

A. pedatum
Northern maidenhair

(2) *Asplenium.* Spleenwort

1. Pinnae 12–30 pairs, oval, ovate, or rhomboidal, obtuse.
    2. Rachis purple-brown.

A. trichomanes
Maidenhair spleenwort

    2. Rachis green.

A. viride
Green spleenwort

1. Pinnae only 2–5, linear cuneate (wedge-shaped).

A. septentrionale
Grass-leaved spleenwort

GRASS-LEAVED SPLEENWORT, *Asplenium septentrionale* (fig. 3), is a curious looking plant for a fern, as it hangs in tufts and fringes from crevices of dry granitic rocks. Its evergreen fronds are slender and grass-like, with one to three main divisions, usually mixed with rust-colored dry leaves of previous seasons. In the Rockies it is found in the foothill and montane zones and occurs from South Dakota and southeastern Wyoming southward through Colorado to western Oklahoma, New Mexico, and Arizona.

MAIDENHAIR SPLEENWORT, *A. trichomanes*, a small and rare fern with narrow, dainty fronds composed of roundish pinnules along a shining, dark rachis, is sometimes found in crevices of granitic rocks, especially in moist situations, in the montane zone.

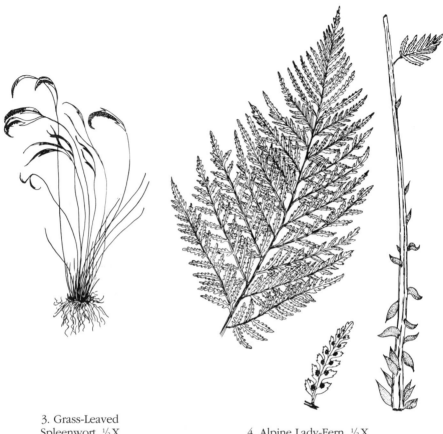

3. Grass-Leaved
Spleenwort, ½X

4. Alpine Lady-Fern, ½X

(3) *Athyrium*. Lady-fern

1. Sori with indusium.

A. filix-femina
Lady fern

1. Sori without indusium.

*A. distentifolium var.
americanum*
Alpine lady fern

The common LADY FERN, *Athyrium filix-femina*, is a fern of moist ravines, woods, and stream banks, with fronds up to 1m or more tall, growing in clumps. It may be distinguished from bracken by its unbranched main rachis, the stalk of the leaf continuing through to its apex. This plant is widely distributed throughout the world in the cooler parts of both hemispheres. In our mountains it occurs from the upper foothills to the subalpine regions.

ALPINE LADY FERN, *A. distentifolium* var. *americanum* (fig. 4), is a delicate, fragrant fern with fronds 20cm–1m tall. It grows in large clumps or beds in crevices or among rocks in the timberline region and lower part of the alpine zone in Wyoming and Colorado.

(4) Cheilanthes. Lip fern

1. Fronds tomentose beneath, but not scaly.

C. feei
Slender lip fern

1. Fronds very scaly beneath, tomentum scanty or absent.

C. fendleri
Fendler lip fern

FENDLER LIP FERN, *Cheilanthes fendleri* (fig. 5), is a small fern growing in dense clumps or beds among shaded rocks, sometimes in crevices or at the foot of dry cliffs. Its fronds are 8–25cm tall with small, roundish segments covered on their backs with thin, tapering, transparent scales. Foothills and lower montane zone from western Texas to Colorado and Arizona.

SLENDER LIP FERN, *C. feei*, is similar except that the lower surfaces of the fronds are covered with reddish-brown hairs instead of scales. It occurs on dry cliffs on the plains and foothills.

5. Fendler Lip Fern, ½X   6. Rock-Brake, ½X   7. Brittle Fern, ½X

(5) *Cryptogramma*. Rock brake

1. Stipes (leaf-stalks) tufted; fronds 3–4-pinnate.

C. acrostichoides
Rock brake or Parsley fern

1. Stipes scattered; fronds 2–3-pinnate.

C. stelleri
Stellers rock brake

ROCK BRAKE, *Cryptogramma acrostichoides* (fig. 6), is a tufted plant 15–25cm tall, conspicuous because the fertile fronds, with their slender, pod-like pinnules, over-top the sterile ones. In the mountains of northern Colorado and Wyoming, it is most commonly found at about 9,000 feet on dry, rocky slopes, often in crevices, but may occur from the foothills to the alpine zone. Its range is from Quebec across Canada, and in northern Michigan, Nebraska, the western states, and Alaska.

(6) *Cystopterias*. Bladder fern

1. Leaf blade at least twice as long as wide.

C. fragilis
Brittle bladder fern

1. Leaf blade about as long as wide.

C. montana
Mountain bladder fern

BRITTLE FERN, *cystopteris fragilis* (fig. 7), is probably the most commonly seen fern in our range and throughout the world. The fronds, 15–20cm long, are delicate, with brittle stipes that break off easily, so that one does not find many of last season's stipes still persisting on the rootstocks. It grows in most moist situations from the foothills to timberline, on stream banks, under ledges, between rocks. It appears quickly in spring and often withers early as the season advances and soil dries out, but it sometimes sends up fresh fronds during a later, moist period. It is usually not so definitely tufted as the woodsias.

NORTHERN or MOUNTAIN BRITTLE FERN, *C. montana*, which is circumboreal, extends into Colorado in a few localities. It differs from *C. fragilis* in having the two lower pinnae considerably larger than the others, giving a triangular effect to the frond.

(7) *Dryopteris*. Wood fern

1. Fronds bipinnatifid, the ultimate divisions blunt.

D. filix-mas
Male fern

1. Fronds bipinnate, the ultimate divisions pointed.

D. expansa
Mountain wood fern

MALE FERN, *Dryopteris filix-mas*, is a large, coarse fern that grows in clumps, with fronds 38–100cm tall. The stipes bear numerous, thin, brown scales, especially near the base; blades are 2-pinnately divided; the pinnae have toothed margins. Sori are round with kidney-shaped indusia attached at the sinus. This fern grows in many localities around the world in the northern

8. Oak fern, ½X

9. Zigzag Cloak Fern, ½X

hemisphere and is found in our mountains in moist situations from the upper foothills to the subalpine zone.

MOUNTAIN WOOD FERN, *D. expansa*, is a rare plant, occurring in a few moist, shaded situations in Rocky Mountain National Park and north. It is similar to, but more delicate than, the male fern, having a scaly stipe. The tips of the leaf divisions are bristle-tipped.

(8) *Gymnocarpium.* Oak fern

Only one species in our range: OAK FERN, *Gymnocarpium dryopteris* var. *disjunctum* (fig. 8), is a delicate, sweet-scented fern of moist woods or shaded seepage areas, where its slender, creeping rootstocks send up scattered, smooth fronds, 2–3-pinnate, triangular-ternate. It occurs from Newfoundland to Alaska, south into the northern United States, and in mountainous areas as far south as Virginia, New Mexico, and Arizona.

(9) Notholaena. Zigzag cloak fern

ZIGZAG CLOAK FERN, *Notholaena fendleri* (fig. 9), is the only species likely to be found in our range. It is found in rock crevices or on ledges of dry cliffs in the foothills from southeastern Wyoming to southern New Mexico. Its brittle stipe and rachis are dark brown with scattered small oval or irregularly shaped pinnules. Leaves 3–5-pinnate and covered beneath with a white or yellow powder.

(10) *Pellaea*. Cliff brake

1. Fronds herbaceous or subcoriaceous (nearly leathery).
    2. Pinnae 6–8, membranous.
        P. breweri
    2. Pinnae 2–5, subcoriaceous.
        P. occidentalis
1. Fronds coriaceous, the veins not perceptible; bipinnate, 10–20cm long.
        P. truncata

(11) *Polypodium*. Polypody

Only one species is in our range, and its name is not settled: COMMON or WESTERN POLYPODY, *Polypodium hesperium* or *P. vulgare* (fig. 10). Leaves pinnate; sori large, round, on the veins or at their ends, without indusium. Found along shaded crevices of granite rocks, usually on north exposures, in the montane zone. The young sori are well separated and pale green, becoming golden and finally orange-brown and confluent. Occurs from Newfoundland to Alaska, southward in the mountains to Mexico. Also in Eurasia.

(12) *Polystichum*. Holly fern

1. Pinnae undivided but spiny-toothed.
        P. lonchitis
        Mountain holly fern
1. Pinnae deeply cleft, with 1–3 pairs of prominent lobes.
        P. scopulinum
        Western holly fern

MOUNTAIN HOLLY FERN, *Polystichum lonchitis* (fig. 11), with dark green, tough, 1-pinnate fronds, having pinnae enlarged at base on their upper edges and sharp, bristle-tipped teeth, occurs in shaded locations on cliffs and rock slides of the subalpine zone from Newfoundland to Alaska, south to the mountains of New Mexico, Arizona, and California.

(13) Pteridium. Bracken

Only one species in our range: BRACKEN, *Pteridium aquilinum*, a coarse, stout fern with triangular-shaped fronds 1–2m in length, forming large colonies. It occurs throughout our range and is widely distributed in the world. Sori are marginal and are rarely seen. It grows in open areas on either moist or dry ground. It sometimes becomes established on burned-over areas. When killed by frost, the fronds turn a bright rusty color, which makes patches of it conspicuous in autumn.

(14) Woodsia. Woodsia

1. Fronds glandular-puberulent beneath.
        W. scopulina
        Rocky Mountain woodsia

10. Common Polypody, ½X   11. Mountain Holly Fern, ½X

1. Fronds smooth beneath.

W. oregana
Oregon woodsia

To distinguish these ferns from some of our other small species, it is helpful to have a good hand lens. The important character for identification is the indusium, which consists of small filaments attached beneath the sorus. The specimen should be in good condition, neither too young nor too old, to see this plainly. ROCKY MOUNTAIN WOODSIA, *W. scopulina*, is a small fern usually found in tufts in rather dry, rocky situations of the upper montane andsubalpine zones. The hairs beneath the fronds sometimes give a glistening appearance. The filaments of the indusia are often concealed by the spreading brown sporangia.

OREGON WOODSIA, *W. oregana* (fig. 12), is similar but lacks the hairs beneath, stands more stiffly erect, and usually has a bright brownish stipe and rachis. Grows most commonly in foothill and lower montane regions. Both occur throughout our range

PEPPERWORT FAMILY: MARSILEACEAE

Only *Marsilea* is known in our range. These are perennial, herbaceous plants rooting in mud. WATER CLOVER, *Marsilea vestita*, has leaflet broadly cuneate, entire or slightly toothed; sporocarp 4–7mm long, 3–5mm wide, densely covered with soft hair-like scales.

PINES, JUNIPERS, AND EPHEDRAS

The leaves of evergreen trees and some shrubs of this region are always very narrow. This reduction in leaf surface keeps evaporation at a minimum and permits such trees to grow under very severe climatic conditions. Their "flowers" are small and simple and are found attached to scales that form a cone. The flowers are unisexual. Junipers have small, juicy, berry-like cones; ephedras have cone-like aments (catkins).

12. Oregon Woodsia, ½ X

KEY 3. DIVISION SPERMATOPHYTA: GYMNOSPERMS

1. No true flowers: staminate and pistillate cones, the pistillate becoming dry or berrylike.
   2. Leaves linear or needle-like; cones dry and woody at maturity.
               8. Pinaceae, p. 33
               Pine Family
   2. Leaves scale-like or awl-shaped; ovulate cones small and berry-like.
               9. Cupressaceae, p. 38
               Cypress family
1. Staminate flowers in cone-like aments (catkins); pistillate flowers single or in pairs; perianth urn-shaped; low shrubs with jointed branches; leaves reduced to sheathing scales.
               10.Ephedraceae, p. 40
               Ephedra Family

PINE FAMILY: PINACEAE

1. Needles in fascicles of 2–5 (rarely 1).
   2. Needles in fascicles of 5.

3. Needles marked by white granules.

> *Pinus aristata*
> Bristlecone pine

3. Needles lacking white granules.
    4. Cones 4–7.5cm long, remaining closed.

> *P. albicaulis*
> Whitebark pine

    4. Cones 12.5–25cm long, opening at maturity.
        5. Cones with stalks 13–17mm long.

> *P. strobiformis*
> Southwestern white pine

        5. Cone stalks less than 13mm long.

> *P. flexilis*
> Limber pine

2. Needles in fascicles of 2–3 (rarely 1).
    3. Small shrub-like trees; seeds nut-like, without wings; needles 1–2.

> *P. edulis*
> Piñon pine

    3. Trees usually with single erect trunk; seeds with wings.
        4. Needles 10–17.5cm long, cones opening and falling.

> *P. ponderosa*
> Ponderosa pine

        4. Needles 2.5–7.5cm long; cones remaining closed and persistent.

> *P. contorta*
> Lodgepole pine

1. Needles attached to twig singly.
    2. Needles soft and flat.
        3. Needles attached by small stem; buds sharp-pointed; cones pendant.

> *Pseudotsuga menziesii*
> Douglas-fir

        3. Needles attached by a disk; buds blunt and pitch-covered; cones erect.
            4. Needles usually over 3cm long; ovulate cones yellow, brown, or greenish-purple.

> *Abies concolor*
> White fir

            4. Needles rarely over 3cm long; ovulate cones purplish.

> *A. lasiocarpa*
> Subalpine fir

    2. Needles rigid, 4-angled, sharp-pointed; cones pendant.
        3. Cone scales entire on the ends; needles mostly under 2.5cm long.

> *Picea glauca*
> Black Hills spruce

        3. Cone scales ragged on the ends; needles often over 2.5cm long.
            4. Cones mostly under 6cm long; younger twigs finely pubescent.

> *P. engelmannii*
> Engelmann spruce

            4. Cones mostly over 6cm long; younger twigs glabrous.

> *P. pungens*
> Colorado blue spruce

In the genus *Pinus*, the male cones are small, about 2.5cm long, and papery in texture. Sometimes reddish before they ripen, they are usually bright orange at the time pollen is being shed (often in great clouds of yellow powder). They will be found clustered around the base of the terminal shoots.

The female, or carpellate, cones are fewer in number and appear first as small purple knobs near the ends of the young shoots (or candles) just before the needles break out. At the time of fertilization, the scales of the infant cone separate so that windblown pollen can reach the ovules. After fertilization takes place, the cone scales close and will remain tightly closed over the developing ovules for about 18 months. In most pines, when these ovules have matured, at the end of the second growing season, the scales separate and release the seeds. After that, except in the cases of whitebark and lodgepole pines, the cones drop off. The length of time it takes for pine ovules to ripen into seeds explains why pine cones are such sturdy, protective structures.

13. Ponderosa Pine, $^2/_5$ X

PONDEROSA PINE, *Pinus ponderosa* (fig. 13), becomes the largest of all trees in the Rocky Mountain region. In favorable locations, the straight trunk can become more than 1m in diameter, with a few large, horizontally spreading branches, which form an open, rounded or flat-topped crown at maturity. The young trees are often branched to the ground and usually have black bark, which later becomes an orange-brown. The most widely distributed of the western pines, its eastern outposts are in the Black Hills of South Dakota, in eastern Wyoming, western Nebraska, the hills of northeastern New Mexico, and the mountains of west Texas. It is cut for lumber in Colorado, New Mexico, Arizona, eastern Utah, northern Idaho, and western Montana. It skips the Great Basin but extends through the mountains of the Pacific states into British Columbia.

LODGEPOLE PINE, *P. contorta*, is most commonly noticed when it grows in dense stands, as on the western slope in Colorado, and in Yellowstone and Glacier National Parks. Under such conditions, the trunks are slender, straight, and gray, and the lower branches die off. The lopsided cones remain closed and attached to the branches for many years. The bark on isolated trees may become partly orange-brown. This is a pioneer tree of the montane and sub-alpine zones.

Lodgepole forests grow on dry slopes. Because of the dry situations, the closeness of the stands, and the pitchiness of their wood and bark, they are very susceptible to fire. The heat of a forest fire opens the long-closed cones, and the seeds are released. For this reason, and because these seedlings thrive in full sunshine and on very poor soil, lodgepole forests often succeed themselves following repeated fires. And, because of their tolerance of sun, they also usually succeed spruce-fir forests destroyed by fire.

14. Limber Pine, ²/₅ X

LIMBER PINE, *P. flexilis* (fig. 14), got its Latin name from its flexible branches. The bark is light gray on young trunks and branches but much darker on old trunks. The needles are dark green, more bluish than those of the ponderosa pine. The cylindrical cones are often very pitchy. Each scale is rounded, with a pale border. Limber pine is found from the foothills to timberline throughout our region, most commonly on windy ridges and rock outcrops. It may be scattered among ponderosa pine and Engelmann spruce where the situation is too rugged for those species.

Limber pines occasionally attain large size. Near South Pass, Wyoming, several of these large old trees grow from an almost solid rock pavement. At timberline and other windy sites, they become gnarled and twisted. The trunk may separate into several main, upward-reaching branches. They become mature at about 300 years and some are definitely known to have lived to an age of 1300 years in Utah, Idaho, and California.

SOUTHWESTERN WHITE PINE, *P. strobiformis*, is found in the mountains of New Mexico. In exposed situations, it resembles the limber pine; but in congenial surroundings it develops a tall, straight trunk like that of the eastern white pine. Its needles are longer and more slender than those of the limber pine. On very old trees the bark becomes red-brown, resembling that of mature ponderosa pines.

BRISTLECONE PINE, *P. aristata*. The white granules on the needles are specks of pitch. The needles remain on the tree for 10 or 15 years; young branches are completely covered with them and have a brush-like appearance. Because of this characteristic, the tree is sometimes called the foxtail pine. In our area, this species is restricted to higher altitudes of the mountains from north-central Colorado southward and westward. It grows in high, windy places, and old trees are often twisted into very picturesque shapes. They commonly attain ages of 300 to 400 years but may become much older in the Rocky Mountains.

WHITEBARK PINE, *P. albicaulis*. This tree may be hard to distinguish from the limber pine except by its cones, which are short and roundish. The scales do not separate, but after several years the whole cone disintegrates and releases the large seeds. This species seems to replace the bristlecone pine in the northern part of the Rocky Mountains. It grows in rocky places and particularly near timberline, where it often sprawls on the ground in typical "wind timber" formation. It occurs in northwestern Wyoming, central and northern Idaho, and is also found in the mountains of central and northern California.

PIÑON OR NUT PINE, *P. edulis* is a bushy pine of the southern and southwestern foothills. It grows in an open formation intermingled with junipers, giving a spotted appearance to the slopes. The common form in our area normally has two needles in a bundle, but a one-needle form, which is similar in appearance, occurs throughout much of the Great Basin. The cones are short with a few thick scales, which open in August or September, releasing the large edible seeds. The shells of the seeds are soft enough to be easily cracked with the teeth. These trees are important to the Indians of Arizona and New Mexico. The wood provides fuel, posts, and material for furniture. The seeds, called piñon nuts, are a staple food for Navajo and Pueblo Indians and are available in the markets of the Southwest. Cutting of the trees for firewood in recent years may threaten the piñon groves.

The trees of the genus *Abies*, the true firs, have whorled, horizontal branches and soft, flat needles. Their cones are held upright on the topmost branches and mature at the end of the first growing season. Instead of dropping off, they disintegrate where they are, releasing the winged seeds some distance above the ground to be widely dispersed by the wind. The slender, spike-like axes to which the cones were attached persist on the branches, and one can often see these spikes stiffly erect on the upper branches of a fir tree for two or three years.

WHITE FIR, *Abies concolor*, is one of the most beautiful trees of North America when well grown. It produces a dense, silver-green, conical crown, densely clothed with branches to the ground. White firs are found in the canyons of southern Colorado and northern New Mexico, the mountains of Arizona, and westwards to California.

SUBALPINE FIR, *A. lasiocarpa*, is found in the moist shade of the subalpine forests, but also in moist forests in the upper montane in the more northern part of our range. The tree develops a slender, spire-like crown with short, horizontal branches, which often have a shelf-like appearance. At timberline, it takes various forms, frequently becoming a prostrate, matted shrub. Sometimes a slender trunk might stand erect in the midst of a cluster of low branches. The "flag trees" at timberline are frequently of this species, so called because they have a line of branches only on the lee side of the trunk, all the others having been shorn off by the severe winds. Often this species forms "hedges" or "windrows" shaped by the wind. The bark on young wood is light gray, thin, and smooth except for resin blisters; but on old trunks it becomes thick and rough. The species is found in all high parts of the Rocky Mountains.

DOUGLAS-FIR, *Pseudostuga menziesii* (fig. 15), may reach 1m in diameter. It has a pyramidal, irregularly much-branched crown. On young trees its bark is smooth and gray with resin blisters, somewhat similar to those on the true firs. On old trunks, the bark becomes very thick, patterned in dark and light brown, and deeply furrowed. This tree has almost as wide a distribution

15. Douglas-Fir, ²/₅X

in the Rockies as does the ponderosa pine and is usually found in association with it. The Douglas fir is more abundant on north-facing slopes and in ravines than the pine, and it is the dominant tree of the montane zone in the parks and valleys of Colorado's western slope, where the ponderosa pine is seldom seen. The common name of this tree commemorates David Douglas, a Scottish botanist and horticulturist who explored and collected plants in northwest America early in the 19th century.

All the spruces, *Picea*, may be distinguished from their Rocky Mountain neighbors by their 4-angled, sharp-pointed needles and by their pendant cones, which lack the fringed appearance of the Douglas fir cones.

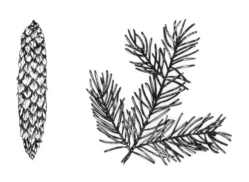

16. Colorado Blue Spruce, ²⁄₅ X

COLORADO BLUE SPRUCE, *Picea pungens* (fig. 16), is a conical tree whose trunk may reach 60cm in diameter. It may be a silver-blue color, but many are fine green specimens. The bark is gray and scaly. It is at its best in the montane canyons of central and northern Colorado, although its range extends from the Mexican border into western and central Montana. It occurs in small groves along streams and occasionally in mixed forests.

ENGLEMANN SPRUCE, *Picea engelmannii*, is the tree that forms great, unbroken forests in the upper montane and subalpine areas of the Rockies. It is an important regional lumber tree, named for Dr. George Engelmann, who practiced medicine in St. Louis and carried on botanical work. He traveled widely in the western U.S., to study particular plant groups, especially the cone-bearing trees. Although he did not collect extensively, he acquired the plant collections of others, and his personal herbarium became the basis for the great Missouri Botanical Garden Herbarium.

CYPRESS FAMILY: CUPRESSACEAE
1. Leaves awl-shaped or needle-like, 7–22mm long.

Juniperus communis
Common juniper

1. Leaves scale-like on mature branches, not over 3mm long.
    2. Creeping shrub.

J. horizontalis
Creeping juniper

    2. Small bushy trees or large upright shrubs.
        3. Trunk bark thick, broken into squares; berries usually 4-seeded.

J. deppeana
Aligator juniper

        3. Trunk bark shredding, fibrous; berries usually 1–3-seeded.
            4. Berry red-brown, seed 1.

J. osteosperma
Utah juniper

4. Berry bluish.
   5. Leaf margin smooth; seeds 1–3.

J. scopulorum
Rocky Mountain juniper or red cedar

   5. Leaf margins toothed (use lens); seed usually 1.

J. monosperma
One-seeded juniper

Except for one species in *Juniperus*, the leaves of mature individuals are short, scale-like, and closely appressed (pressed close together for the whole length). Their "berries" are really much-modified cones in which the few small, soft scales enclose one to a few seeds in a berry-like structure. If you look closely at a juniper "berry" cone, you will be able to see on its surface two or more little points, which correspond to the tips of pinecone scales. Plants of this genus show both juvenile and mature foliage.

The common juniper is the exception. It has only the juvenile type of leaf: a very sharp-pointed, awl-shaped needle. All the other species have similar awl-shaped leaves when they are young, and older plants will often show this type of leaf on young shoots near the base of the main trunk. But when seedlings of these other species are several years old, they begin to develop their adult foliage, the short, scale-like type of leaf. If you look near the base of an old juniper tree, especially if it has been injured there, you will likely find some sprouts with awl-shaped leaves. Because of this normal habit, botanists who study plant evolution have concluded that the common juniper is close to the original juniper species from which these other different, but evidently related, species have evolved.

COMMON JUNIPER, *J. communis*, is a shrub 30cm–1m tall, sometimes forming large clumps, found on north slopes and under open forest from the foothills to timberline. The sharp-pointed leaves are green with a white line on the upper surface, and the round berries are blue when ripe. This species is very widly distributed, and forms of it are found in all the northern lands of the northern hemisphere.

CREEPING JUNIPER, *J. horizontalis*, varies from a low creeper with trailing, rooting stems only a few inches high to low spreading shrubs that may be up to 30cm tall. There are both green and silver forms. When exposed to winter sun, the foliage often turns purplish. It is found on open dry slopes from northern Wyoming northward and across the northern United States. Its "berries" are dark blue with a silvery bloom and are always borne on short, curved stalks.

ALLIGATOR JUNIPER, *J. deppeana*, may reach a substantial height with a trunk 60cm in diameter, but usually is a smaller, compact tree or large spreading shrub. It takes its name from the thick, brown or grayish bark, which breaks into square plates on the trunks of old trees. Its foliage is blue-green and its berries dark red-brown. It is found in the mountains of southern New Mexico and Arizona.

UTAH JUNIPER, *J. osteosperma*, and ONE-SEED Juniper, *J. monosperma*, are the common junipers of the foothills of the southern and western parts of our region. They are not easily distinguished. Both usually have a yellowish-green foliage color; both usually have several main stems from the ground; and both usually have 1-seeded berries. (See the key.) Along the eastern side of the mountains in southern Colorado and New Mexico, only the one-seed juniper occurs. In the western third of both states, the two species are intermingled. In Wyoming, only the Utah juniper is found.

ROCKY MOUNTAIN JUNIPER, *J. scopulorum*, is sometimes called WESTERN RED CEDAR. The color is variable, from silvery or grayish to green. The fruits are light blue when mature and usually have 2 or 3 seeds. On the eastern side of the Continental Divide in Colorado, from Palmer Lake northward, it is the only upright juniper, frequently the first tree form found as one ascends the dry foothill slopes. At higher elevations on south and southwest slopes, it is mixed with ponderosa pine, though it occurs on the driest and most rocky sites. Throughout the rest of our range, it occurs mixed with piñon and oak. In general, where its geographic range overlaps that of the one-seed and Utah junipers, *J. scopulorum* will be found at a higher elevation than those species.

EPHEDRA OR JOINTFIR FAMILY: EPHEDRACEAE

1. Scales and bracts in twos, the bracts connate.

Ephedra antisyphilitica

1. Scales and bracts in threes, the bracts hardly connate.
    2. Scales short, 2–3mm; fruit scabrous.

E. torreyana

    2. Scales long, 6–12mm; fruit smooth.

E. trifurca

These low shrubs with jointed branches have axillary, cone-like catkins. The younger branches are green, resembling horsetail or scouring rush. Primarily a genus of the Great Basin, the *ephedra* may be found at lower elevations on the western side of our range.

DICOTYLEDONOUS PLANTS (DICOTS)

The plants whose seedlings have two seed leaves (*cotyledons*) belong in this group. Botanists refer to them as the Dicotyledons or Dicots. (The Monocotyledons or Monocots have only one seed leaf.) Some other characters that distinguish dicots from monocots are the number of flower parts and the type of leaf venation. Dicots usually have flowers with a basic number pattern of 4 or 5, and their leaves have branched veins, which form a network (Plate B8). Frequently the leaves are compound, that is, made up of several leaflets, and the edges of the leaves or leaflets may be serrated, notched, or more or less deeply dissected. These characters contrast with the smooth-edged, undivided monocot leaves. Even if a plant has narrow leaves and apparently parallel veins, but its flower parts are in 4s or 5s, look for it here. If its flower parts are in 3s and 6s, look for it under monocots.

For clarity in classification, the dicots are here divided into three sections. The first section includes those plants that have no petals (only sepals). The second section includes those which have free, or separate, petals. And the third section includes those with united, or fused, petals. Because beginners do not always easily distinguish between petals and sepals, it has become common to combine the first two sections into one key. We separate them here on the assumption that it is well to learn to recognize apetalous plants. On the other hand, if you are not certain whether your plant has petals or not, remember to check on it in both Key 4 and Key 5. The order of the keys reflects the belief that plants whose flowers have an indefinite number of separate parts are lower in the evolutionary scale than those whose flowers have few and somewhat united parts. Hence, the united-petal dicots appear in Key 6. Unlike the earlier editions of this book, this revised edition includes in the keys some plant families not treated in the text. They are quite uncommon, yet present. If your plant seems to fall within such a category, consult a more detailed flora for the state in which you collected the specimen.

KEY 4. DIVISION SPERMATOPHYTA: ANGIOSPERMS. APETALOUS DICOTS

1. Flowers in aments (catkins); flowers unisexual, plants either monoecious or dioecious.
  2. Staminate *or* pistillate flowers (never both) in aments.
    3. Trees; leaves alternate.
      4. Sap not milky.
        5. Fruit a samara (winged all around).
                (17) Ulmaceae, p. 74
                Elm Family
        5. Fruit a nut with bony exocarp (acorn).
                (20) Fagaceae, p. 75
                Oak Family
      4. Sap milky.
                (18) Moraceae, p. 74
                Mulberry Family
    3. Herbs or herbaceous vines; leaves generally opposite.
      4. Styles and stigmas 2.
                (18) Moraceae, p. 74
                Mulberry Family
      4. Style and stigma 1.
                (19) Urticaceae, p. 75
                Nettle Family
  2. Staminate and pistillate flowers *both* in aments.
    3. Ovary becoming a 1-celled, multi-seeded pod; seeds hairy-tufted.
                (37) Salicaceae, p. 116
                Willow Family
    3. Ovary becoming a 1-seeded nut, achene, or berry.
      4. Trees or shrubs (not parasitic).
                (21) Betulaceae, p. 75
                Birch Family
      4. Parasitic plants on trees.
                (56) Loranthaceae, p. 217
                Mistletoe Family
1. Flowers not in aments.
  2. Ovary or its cells containing 1–2 ovules (rarely 3–4).

3. Pistils more than 1 (distinct or nearly so).
    4. Stamens perigynous (inserted in a ring around the pistil); leaves with stipules.

Cercocarpus in
(48) Rosaceae, p. 166
Rose Family

    4. Stamens hypogynous (inserted beneath the ovary); calyx petal-like.

(13) Ranunculaceae, p. 52
Buttercup Family

3. Pistil 1, simple or compound.
    4. Ovary superior; calyx sometimes wanting.
        5. Stipules sheathing the stem at the nodes.

(29) Polygonaceae, p. 97
Buckwheat Family

        5. Stipules (if present) not sheathing the stem.
            6. Herbaceous plants.
                7. Aquatics, usually submerged.
                    8. Leaves whorled and dissected; style 1.

(12) Ceratophyllaceae, p. 52
Hornwort Family

                    8. Leaves opposite and entire; styles 2.

(81) Callitrichaceae, p. 283
Water-starwort Family

                7. Terrestial plants.
                    8. Style 1 (if any); stigma 1.
                        9. Flowers unisexual; ovary 1-celled.

(19) Urticaceae, p. 41
Nettle Family

                        9. Flowers bisexual; pod 2-celled and 2-seeded.

Lepidium in
(39) Cruciferae, p. 125
Mustard Family

                    8. Styles 2–3 (or branched); 1–4 carpels (cells).
                        9. Ovary and pod tricarpellate; juice milky.

(58) Euphorbiaceae, p. 219
Spurge Family

                        9. Ovary not 3-celled; juice not milky.
                            10. Flowers in an involucrate head; fruit a 3-angled achene.

Eriogonum in
(29) Polygonaceae, p. 97
Buckwheat Family

                          10. Flowers not involucrate.
                            11. Leaves (at least below) covered with stellate hairs.

(58) Euphorbiaceae, p. 219
Spurge Family

                            11. Leaves without stellate hairs.
                                12. Leaves opposite.
                                    13. Plant fleshy.

Salicornia in
(25) Chenopodiaceae, p. 82
Goosefoot Family.

                                    13. Plant not fleshy.
                                        14. Flowers in heads or spikes.

(26) Amaranthaceae, p. 84
Amaranth Family

                                      14. Flowers sessile (in forks of branching inflorescence).

Paronychia in

(28) Caryophyllaceae, p. 89
Pink Family
12. Leaves alternate.
13. Flowers and bracts scarious
(scale-like).
(26) Amaranthaceae, p. 84
Amaranth Family
13. Flowers greenish, no scarious
bracts.
(25) Chenopodiaceae, p. 82
Goosefoot Family
6. Trees or shrubs.
7. Leaves opposite.
8. Fruit tricarpellate, not winged.
(59) Rhamnaceae, p. 220
Buckthron Family.
8. Fruit winged.
9. Fruit 2-celled (double samara).
(62) Aceraceae, p. 224
Maple Family.
9. Fruit 1-celled (single samara).
(83) Oleaceae, p. 284
Olive Family.
7. Leaves alternate.
8. Ovary tricarpellate.
(59) Rhamnaceae, p. 220
Buckthorn Family
8. Ovary 1-celled.
(17) Ulmaceae, p. 74
Elm Family
4. Ovary inferior (or appearing to be so).
5. Parasites on tree branches.
(56) Loranthaceae, p. 217
Mistletoe Family.
5. Plants not parasites on trees.
6. Aquatic herbs.
(51) Haloragidaceae, p. 208, 283
Water-milfoil Family
6. Terrestial plants.
7. Herbs with corolla-like calyx.
8. Leaves opposite.
(22) Nyctaginaceae, p. 77
Four-O'clock Family
8. Leaves alternate.
(55) Santalaceae, p. 217
Sandalwood Family
7. Shrubs or trees with scurfy leaves (with minute scales or
particles).
(50) Elaeagnaceae, p. 207
Oleaster Family
2. Ovary or its cells containing many ovules.
3. Ovary (or ovaries) superior.
4. Ovaries 2 or more, separate.
(13) Ranunculaceae, p. 52
Buttercup Family
4. Ovary 1.
5. Ovary with 3–5 carpels; leaves opposite or
whorled.
(23) Aizoaceae, p. 79
Carpetweed Family

5. Ovary with 1–2 carpels.
  6. Leaves compound.

> (13) Ranunculaceae, p. 52
> Buttercup Family

  6. Leaves simple.
    7. Sepals distinct.

> (28) Caryophyllaceae, p. 89
> Pink Family

    7. Sepals more or less united.
      8. Calyx 5-toothed or cleft.

> *Glaux in*
> (43) Primulaceae, p. 149
> Primrose Family

      8. Calyx 4-toothed or cleft.

> *Synthyris in*
> (84) Scrophulaceae, p. 284
> Figwort Family

3. Ovary and pod inferior.

> *Chrysosplenium in*
> (47) Saxifragaceae, p. 158
> Saxifrage Family

KEY 5. DIVISION SPERMATOPHYTA: ANGIOSPERMS. FREE-PETAL DICOTS

1. Stamens more than 10 (more than double the sepals or calyx- lobes).
  2. Calyx separate and free from pistil (or pistils); ovary superior.
    3. Pistils several to many, either entirely distinct or united at base into a lobed or several-beaked ovary.
      4. Aquatics with peltate leaves (the petiole arises from the under surface).

> (11) Nymphaeaceae, p. 52
> Water-lily Family

      4. Terrestial plants.
        5. Filaments united into a tube; leaves alternate.

> (32) Malvaceae, p. 106
> Mallow Family.

        5. Filaments distinct, on the calyx; leaves alternate.

> (48) Rosaceae, p. 166
> Rose Family

    3. Pistils 1, styles and stigmas sometimes more numerous.
      4. Leaves punctate with pellucid (translucent) dots.

> (31) Clusiaceae, p. 105
> Garcinia Family

      4. Leaves not punctate.
        5. Ovary simple (1 placenta, 1 style, 1 stigma).
          6. Ovules 2, seed a drupe.

> *Prunus in*
> (48) Rosaceae, p. 166
> Rose Family

          6. Ovules many; leaves 2–3-ternately compound or dissected.

> (13) Ranunculaceae, p. 52
> Buttercup Family

        5. Ovary compound (2 or more placentae, styles, or stigmas).
          6. Ovary 1-celled.
            7. Placentae parietal (attached to the walls).
              8. Sepals 2 (rarely 3).

> (15) Papaveraceae, p. 72
> Poppy Family

8. Sepals 4.

(38) Capparidaceae, p. 123
Caper Family

7. Placentae central; sepals 2.

(27) Portulacaceae, p. 86
Purslane Family

6. Ovary several-celled.
7. Stamens united.

(32) Malvaceae, p. 106
Mallow Family

7. Stamens not united.

(11) Nymphaeaceae, p. 52
Water-lily Family

2. Calyx more or less adnate (united) to a compound ovary (inferior).
3. Fleshy-stemmed, without true leaves; petals numerous.

(24) Cactaceae, p. 79
Cactus Family

3. Leaves present.
4. Sepals or calyx lobes 2; ovules rising from the base of a 1-celled ovary.

(27) Portulacaceae, p. 86
Purslane Family

4. Sepals or calyx lobes more than 2.
5. Leaves opposite; no stipules.

(44) Hydrangeaceae, p. 152
Hydrangea Family
(sometimes put in
Saxifragaceae)

5. Leaves alternate.
6. Stipules present.

(48) Rosaceae, p. 166
Rose Family

6. Stipules wanting; herbage rough-pubescent.

(36) Loasaceae, p. 115
Loasa Family

1. Stamens not more than twice as many as the petals.
2. Stamens equal in number to the petals and opposite them.
3. Ovary 2-4-celled.
4. Calyx-lobes minute or vestigial; petals valvate (meeting but not
overlapping).

(60) Vitaceae, p. 223
Grape Family

4. Calyx 4-5 cleft; petals involute (rolled inward from the edges).

(59) Rhamnaceae, p. 220
Buckthorn Family

3. Ovary 1-celled.
4. Anthers opening by uplifted valves.

(14) Berberidaceae, p. 70
Barberry Family

4. Anthers not opening by uplifted valves.
5. Style 1, unbranched; stigma 1.

Lysimachia in
(43) Primulaceae, p. 149
Primrose Family

5. Styles, style branches, or stigmas more than 1; sepals or calyx lobes 2.

(27) Portulacaceae, p. 86
Purslane Family

2. Stamens not equal to the number of petals; or, if equal, then alternate with
petals.
3. Calyx free from the ovary (ovary wholly superior).

4. Ovaries 2 or more, separate or somewhat united.
  5. Stamens connate and united with large thick stigma common to the two ovaries.

Sarcostemma in
(71) Asclepiadaceae, p. 247
Milkweed Family

  5. Stamens entirely free.
    6. Stamens hypogynous (inserted at the base of the ovary).
      7. Leaves fleshy.

(46) Crassulaceae, p. 157
Stonecrop Family

      7. Leaves not fleshy.
        8. Ovary (or ovary lobes) 5, with a common style.

(66) Geraniaceae, p. 228
Geranium Family

        8. Ovaries with separate styles or sessile stigmas.

(13) Ranunculaceae, p. 52
Buttercup Family

    6. Stamens perigynous (inserted on the calyx).
      7. Plant fleshy; stamens just twice the number of pistils.

(46) Crassulaceae, p. 157
Stonecrop Family

      7. Plant not fleshy; stamens not twice as many as the pistils.
        8. Stipules present.

(48) Rosaceae, p. 166
Rose Family

        8. Stipules lacking.

(47) Saxifragaceae, p. 158
Saxifrage Family

4. Ovary 1.
  5. Ovary simple, with 1 parietal placenta.

(49) Leguminosae, p. 187
Pea Family

  5. Ovary compound.
    6. Ovary 1-celled.
      7. Corolla irregular.
        8. Petals 4, stamens 6.

(16) Fumariaceae, P. 73
Fumitory Family

        8. Petals and stamens 5.

(33) Violaceae, p. 109
Violet Family

      7. Corolla regular (or nearly so)
        8. Ovule 1; trees or shrubs.

(63) Anacardiaceae, p. 225
Sumac Family

        8. Ovules more than 1.
          9. Ovules at the center or bottom of the cell.
            10. Petals not inserted on the calyx.

(28) Caryophyllaceae, p. 89
Pink Family

            10. Petals inserted on the throat of a bell-shaped or tubular calyx.

(58) Lythraceae, p. 209
Loosestrife Family

          9. Ovules on 2 or more parietal placentae.
            10. Leaves punctate with translucent dots.

(31) Clusiaceae, p. 105
Garcinia Family

            10. Leaves not punctate.

11. Petals 4–5; stamens 6.
  12. Stamens roughly equal.
    13. Calyx persistent; fruit a capsule.
      (34) Frankeniaceae, p. 113
      Frankenia Family
    13. Calyx deciduous; fruit a stipitate pod.
      (38) Capparidaceae, p. 123
      Caper Family
  12. Stamens unequal, 2 shorter than the other 4; pod sessile.
    (39) Cruciferae, p. 125
    Mustard Family
11. Petals 4–5; stamens as many; fruit a berry.
    (45) Grossulariaceae, p. 155
    Gooseberry Family
6. Ovary 2–several-celled (flowers regular).
  7. Stamens neither just as many nor twice as many as the petals.
    8. Trees or shrubs.
      9. Woody throughout; fruit a double samara.
        (62) Aceraceae, p. 224
        Maple Family
      9. Woody at base; fruit a capsule.
        (34) Frankeniaceae, p. 113
        Frankenia Family
    8. Herbs.
      9. Petals 5.
        (31) Clusiaceae, p. 105
        Garcinia Family
      9. Petals 4.
        (39) Cruciferae, p. 125
        Mustard Family
  7. Stamens equal to, or double the number of, petals.
    8. Ovules and seeds 1–2 in each cell.
      9. Herbs
        10. Flowers imperfect.
          (58) Euphorbiaceae, p. 219
          Spurge Family
        10. Flowers perfect and symmetrical.
          11. Cells of the ovary as many as the sepals, 5.
            (66) Geraniaceae, p. 228
            Geranium Family
          11. Cells of the ovary twice as many as the sepals.
            12. Leaves abruptly pinnate.
              (64) Zygophyllaceae, p. 227
              Caltrop Family
            12. Leaves simple.
              (61) Linaceae, p. 223
              Flax Family
      9. Shrubs or trees with opposite leaves.
        10. Leaves palmately veined.
          (62) Aceraceae, p. 224
          Maple Family
        10. Leaves pinnately veined.
          (57) Celastraceae, p. 218
          Staff-tree Family.

8. Ovules (and usually seeds) several to many in each cell.
9. Leaves compound; leaflets 3, obcordate.
(65) Oxalidaceae, p. 227
Wood-sorrel Family
9. Leaves simple.
10. Style 1.
11. Stamens free from the calyx.
(41) Pyrolaceae, p. 146
Wintergreen Family
11. Stamens inserted on the calyx.
(52) Lythraceae, p. 209
Loosestrife Family
10. Styles 2–5; leaves opposite.
(28) Caryophyllaceae, p. 89
Pink Family
3. Calyx-tube adherent to the ovary, at least to its lower half (ovary inferior).
4. Tendril-bearing herbs.
Echinocystis in
(35) Cucurbitaceae, p. 113
Gourd Family
4. Not tendril-bearing.
5. Ovules and seeds more than 1 in each cell.
6. Ovary 1-celled.
7. Sepals or calyx-lobes 2; ovules at the base of the ovary.
(27) Portulacaceae, p. 86
Purslane Family
7. Sepals or calyx-lobes 4–5; placentae 2–3; fruit a berry.
(45) Grossulariaceae, p. 155
Gooseberry Family
6. Ovary 2–many-celled.
7. Stamens inserted on or about a flat disk that covers the ovary.
(57) Celastraceae, p. 218
Staff-tree Family
7. Stamens inserted in the calyx.
8. Style 1; stamens 4 or 8.
(53) Onagraceae, p. 209
Evening-primrose Family
8. Styles 2–3, stamens 5 or 10.
(47) Saxifragaceae, p. 158
Saxifrage Family
5. Ovules and seeds 1 in each cell.
6. Stamens 5 or 10.
7. Trees or shrubs.
8. Leaves simple.
Crataegus in
(48) Rosaceae, p. 166
Rose Family
8. Leaves compound or prickly.
(67) Araliaceae, p. 230
Ginseng Family
7. Herbs.
8. Fruit dry, splitting at maturity; styles 2.
(68) Umbelliferae, p. 230
Parsley Family
8. Fruit berry-like; styles 2–5, either separate or united.
(67) Araliaceae, p. 230
Ginseng Family
6. Stamens 4 or 8.
7. Style and stigma 1, fruit a drupe.

(54) Cornaceae, p. 216
Dogwood Family

7. Fruit not drupaceous.
    8. Style 1; stigma 2–4-lobed.

(53) Onagraceae, p. 209
Evening-primrose Family

    8. Styles or sessile stigmas 1 or 4.

(51) Haloragidaceae, p. 208
Water-milfoil Family

KEY 6. DIVISION SPERMATOPHYTA: ANGIOSPERMS. UNITED-PETAL DICOTS

1. Stamens more numerous than the lobes of the corolla.
    2. Stamens free from the corolla.
        3. Style 1; leaves simple.
            4. Ovary superior; fruit a capsule or berry.

(40) Ericaceae, p. 141
Heath Family

            4. Ovary inferior; fruit a berry.

Vaccinium in
(40) Ericaceae, p. 141
Heath Family

        3. Styles 5; leaves 3-foliolate; may be united at base.

(65) Oxalidaceae, p. 227
Wood-sorrel Family

    2. Stamens attached to the base or tube of the corolla.
        3. Saprophytic herbs (without green foliage).

(42) Monotropaceae, p. 148
Indian-Pipe Family

        3. Not saprophytic; foliage green.
            4. Filaments united into a tube.

(32) Malvaceae, p. 106
Mallow Family

            4. Filaments free; radical leaves ternate.

(90) Adoxaceae, p. 316
Moschatel Family

1. Stamens not more numerous than the corolla-lobes.
    2. Stamens equal in number to the corolla-lobes and opposite them.

(43) Primulaceae, p. 149
Primrose Family

    2. Stamens alternate with the corolla-lobes, or fewer.
        3. Ovary superior.
            4. Corolla regular.
                5. Stamens as many as the corolla-lobes.
                    6. Ovaries 2 (or if 1, 2-horned).
                        7. Stamens distinct.

(71) Asclepiadaceae, p. 247
Milkweed Family

                        7. Stamens united.

(70) Apocynaceae, p. 246
Dogbane Family

                  6. Ovary 1.
                    7. Ovary deeply 4-lobed.
                      8. Leaves alternate.

(77) Boraginaceae, p. 267
Borage Family

                      8. Leaves opposite.

(79) Labiatae, p. 276
Mint Family

7. Ovary not deeply lobed.
  8. Ovary 1-celled; seeds several to many.
    9. Leaves entire, opposite.
      (69) Gentianaceae, p. 241
      Gentian Family
    9. Leaves toothed, lobed, or compound.
      10. Whole upper surface of the corolla white
        bearded; leaflets 3, entire.
        (74) Menyanthaceae, p. 254
        Buckbean Family
      10. Corolla not conspicuously bearded; leaves,
        if compound, with toothed leaflets.
        (76) Hydrophyllaceae, p. 263
        Waterleaf Family
  8. Ovary 2–10-celled.
    9. Stamens free from the corolla, or nearly so.
      (40) Ericaceae, p. 141
      Heath Family
    9. Stamens on the tube of the corolla.
      10. Stamens 4.
        11. Stem with opposite leaves; corolla
          petaloid.
          (78) Verbenaceae, p. 275
          Vervain Family
        11. Stem wanting; corolla scarious, at
          least on margins.
          (82) Plantaginaceae, p. 283
          Plantain Family
      10. Stamens 5.
        11. Fruit of 2–4 nutlets.
          (77) Boraginaceae, p. 267
          Borage Family
        11. Fruit a many-seeded-capsule or berry.
          12. Styles 2.
            13. Capsule (mostly 4)-seeded.
            (73) Convolvulaceae, p. 252
            Morning-Glory Family
            13. Capsule many-seeded.
            (76) Hydrophyllaceae, p. 263
            Waterleaf Family
          12. Style 1, often branched.
            13. Branches of the style (or
              the lobes of the stigma) 3.
            (75) Polemoniaceae, p. 254
            Phlox Family
            13. Branches of the style or
              lobes of the stigma 2, or
              wholly united.
            (72) Solanaceae, p. 249
            Potato Family
5. Stamens fewer than the corolla lobes.
  6. Stamens with 4 anthers, in pairs.
    (78) Verbenaceae, p. 275
    Vervain Family
  6. Stamens with 2 anthers, rarely 3.
    7. Ovary 4-lobed.
      Lycopus in
      (79) Labiatae, p. 276
      Mint Family
    7. Ovary 2-celled, not 4-lobed.
      8. Stemless; corolla scarious.
        (82) Plantaginaceae, p. 283

Plantain Family
8. Leafy-stemmed; corolla not scarious.
Veronica in
(84) Scrophulariaceae, p. 284
Figwort Family
4. Corolla irregular.
5. Stamens with 5 anthers.

Verbascum in
(84) Scrophulariaceae, p. 284
Figwort Family
5. Stamens with 2 or 4 anthers.
6. Ovules solitary in the 1–4 cells.
7. Ovary 4-lobed; style rising from between the lobes.
(79) Labiatae, p. 276
Mint Family
7. Ovary not lobed; style from its apex.
(78) Verbenaceae, p. 275
Vervain Family
6. Ovules 2–many in each cell; ovary 1–2 celled.
7. Parasites, without green foliage.
(85) Orobanchaceae, p. 307
Broomrape Family
7. Not parasitic.
8. Ovary 1-celled; stamens 2; aquatics.
(86) Lentibulariaceae, p. 307
Bladderwort Family
8. Ovary 2-celled; placenta in the axis, usually many seeded.
(84) Scrophulariaceae, p. 284
Figwort Family
3. Ovary inferior.
4. Tendril-bearing herbs; anthers often united.
(35) Cucurbitaceae, p. 113
Gourd Family
4. Not tendril-bearing.
5. Stamens separate.
6. Stamens free from the corolla; equal in number to its lobes.
(87) Campanulaceae, p. 310
Bellflower Family.
6. Stamens inserted on the corolla.
7. Stamens 1–3, always fewer than the corolla-lobes.
(91) Valerianaceae, p. 317
Valerian Family
7. Stamens 4–5; leaves opposite or whorled.
8. Leaves opposite or perfoliate (a leaf with the stem apparently passing through it), but neither whorled nor with stipules.
(89) Caprifoliaceae, p. 312
Honeysuckle Family
8. Leaves either opposite and stipulate, or whorled and without stipules.
(88) Rubiaceae, p. 311
Madder Family
5. Stamens united by their anthers; these joined in a ring or a tube.
6. Flowers separate, not involucrate; corolla irregular.
Lobelia in
(87) Campanulaceae, p. 310
Bellflower Family
6. Inflorescence a compact head of small individual florets surrounded by an involucre.
(92) Compositae, p. 318
Composite or Aster Family

WATER-LILY FAMILY: NYMPHACEAE

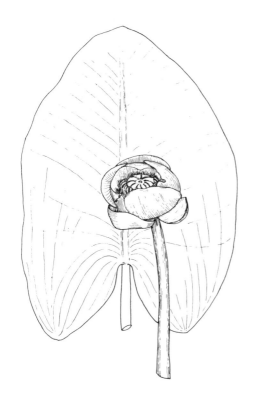

Only one genus and one species in our range. YELLOW POND LILY, N*uphar luteum* ssp. *polysepalum* (fig. 17), grows in the cold ponds and small lakes of the montane and subalpine zones. Its large shining leaves are pushed up from the root, which is embedded in mud, to the surface of the water, where they uncurl and lie flat on the surface. The yellow, cup-like flowers on strong stalks lie at the surface or just below it. The leaves of seedling pondlilies are very thin and look like bunches of lettuce growing at the bottom under several feet of water. This subspecies is found in Rocky Mountain, Yellowstone, and Glacier National Parks and in occasional ponds elsewhere in the Rocky Mountains. Other subspecies of *Nuphar* are widely distributed in the United States. Their seeds, called *wokas*, were an important food for the Indians of our Northwest, and moose are said to feed on the large leaves.

17. Yellow Pondlily, ²⁄₅ X

HORNWORT FAMILY: CERATOPHYLLACEAE

Only one genus and one species in our range: HORNWORT, *Ceratophyllum demersum*. Hornwort will be found submerged in ponds or slow-flowing streams. The leaves are sessile, cut into narrow linear divisions, and whorled.

BUTTERCUP FAMILY: RANUNCULACEAE

A large family including many plants of differing appearances, all of which have two characteristics in common: all flower parts are distinct and separate from each other, and the stamens are of an indefinite number. The pistils are from 3 to many, and the fruits may be achenes, follicles, or berries. The leaves are sometimes simple and entire, but more often they are lobed or divided, and sometimes compound. Petals are often lacking, but in several genera the sepals are petal-like. In general, plants of this family grow in cool, moist situations.

1. Flowers irregular (bilaterally symmetrical).
   2. Upper sepal spurred; petals 4.

          Delphinium, p. 61
          Larkspur

   2. Upper sepal hooded; petals 2.

          Aconitum, p. 54
          Monkshood

1. Flowers regular (radially symmetrical).
   2. Petals spurred at base.
      3. Tiny annual plants; fruit are achenes.

          Myosurus, p. 63
          Mousetail

      3. Showy perennial plants; fruit are follicles.

          Aquilegia, p. 57
          Columbine

   2. Petals not spurred.
      3. Carpel 1; fruit a berry.

          Actaea, p. 54
          Baneberry

      3. Carpel more than 1; fruit an achene or follicle.
         4. Petals present (very reduced in *Trollius*).
            5. Sepals over 1cm long, whitish; fruit a follicle.
                    Trollius, p. 70
                    Globeflower
            5. Sepals mostly less than 1cm long, yellow or green; fruit an achene.
                    Ranunculus, p. 64
                    Buttercup
         4. Petals absent (sepals often petaloid).
            5. Cauline leaves opposite or whorled.
               6. Leaves opposite; sepals usually 4.
                    Clematis, p. 61
                    Clematis
               6. Leaves whorled; sepals usually 5 or more.
                  7. Sepals over 2cm long; styles becoming elongated (to
                      3cm) and plumose.
                      Pulsatilla, p. 63
                      Pasque flower
                  7. Sepals never over 2cm long; styles short, not elongating
                      in fruit, not plumose.
                      Anemone, p. 54
                      Anemone
            5. Leaves cauline and alternate, or leaves basal.
               6. Leaves simple, either lobed or not lobed.
                  7. Leaves not lobed; flowers white, showy.
                      Caltha, p. 60
                      Marsh marigold
                  7. Leaves palmately-lobed or parted; anthers oval or
                      ovate, about 1mm long; flowers whitish or greenish,
                      inconspicuous.
                      Trautvettaria, p. 70
                      False bugbane
               6. Leaves decompound (more than once compound); anthers
                linear, well over 1mm long; flowers whitish, greenish, or
                purple inconspicuous.
                  Thalictrum, p. 68
                  Meadow rue

(1) *Aconitum*. Monkshood.

1. Flowers blue-purple.

Aconitum columbianum var. columbianum

1. Flowers greenish-white.

A. columbianum var. ochroleucum

MONKSHOOD (fig. 18), is found in wet meadows and in thickets of the montane and subalpine zones throughout the Rocky Mountains. Its stems are from 30cm to 13dm tall, with hooded flowers widely spaced along them. Usually these are dark blue, but the white-flowered plants are occasionally seen.

(2) *Actea*. Baneberry

Only one species in our range. BANEBERRY, *Actea rubra* (fig. 19), is a tall plant with thrice compound leaves, the primary divisions long-stalked with pinnately arranged, toothed or lobed leaflets. The small white flowers are on an erect raceme. They mature into shiny, short-stalked, poisonous berries, which may be either cherry-red or white. Found in moist, shaded areas in the montane and subalpine zones.

18. Monkshood, ½ X

(3) *Anemone*. Windflower

Plants of this genus have no petals, but their sepals are petal-like in appearance. There are many stamens and pistils. Each pistil ripens into a separate achene, and the achenes are smooth, pubescent, or woolly. In this genus, there are always leafy collars called involucres, which are folded around the flower bud, and which later encircle the stalk below the flower. Most of the species have their leaves dissected into narrow segments. The pasque flower, often called an anemone, develops a different fruit (long, plumose tails) and is now generally put in the genus *Pulsatilla*.

fruits red or
white

19. Baneberry, ½X

1. Achenes (1-seeded fruit) densely woolly.
    2. Stem low, simple, 1-flowered.
        3. From slender rootstocks.

                                          A. parviflora
                                          Northern anemone

        3. From a short erect caudex.

                                          A. lithophila
    2. Stems generally branching above, 1–3dm high; flowers mostly more than 1.
        3. Head of carpels globose.

                                          A. multifida var. multifida
                                          Pacific anemone

        3. Head of carpels cylindrical.

                                          A. cylindrica
                                        Candle anemone

1. Achenes pubescent only or glabrate.
    2. Leaves of the involucre sessile.
        3. Achenes appressed-pubescent.

                                          A. canadensis
                                          Meadow anemone

        3. Achenes wholly glabrous.

                                          A. narcissiflora ssp. zephyra
                                          Alpine anemone
    2. Leaves of the involucre petiolate; stems from an erect caudex.
                                          A. multifida var. tetonensis

NORTHERN ANEMONE, *A. parviflora*, is a small, dainty plant with 3-parted leaves and a single white or purple-tinged flower. It grows in the arctic and in our range it is found in cold, moist woods in the high mountains, sometimes above timberline.

ALPINE ANEMONE, *A. narcissiflora* ssp. *zephrya* (fig. 20), has usually 3 or more white flowers on stalks 5–10cm long, all of which rise from the involucre. All the leaves are finely dissected and more or less hairy. The ripened achenes are flattened and black. This plant is usually found above timberline in the tundra but may occasionally be seen in subalpine meadows. It grows from Alaska south through the Rockies to central Colorado, and also in the Alps in Europe.

MEADOW ANEMONE, *A. canadensis* (fig. 21), is a plant of ditch banks in the foothills and of montane meadows throughout the eastern part of our range. Its white flowers are held on long stalks above the leaves, which are parted into sharply pointed and toothed divisions. The leaves of the involucre are sessile.

20. Alpine Anemone, ½X

PACIFIC ANEMONE, *A. multifida* var. *multifida*, usually has pink or red sepals, but sometimes they are yellowish. The ripening seed head is roundish, and the achenes are densely hairy. This grows from the montane to the alpine zone throughout our range.

CANDLE ANEMONE, *A. cylindrica* (fig. 22), has whitish sepals and an elongated receptacle covered by densely woolly achenes. Its involucral leaves are petioled. It grows throughout our range in the foothill and montane zones.

(4) *Aquilegia.* Columbine.

This genus is easy to recognize when in flower because the 5 petals are elongated backwards into knob-tipped spurs. The nectar in the spurs is sought by long-tongued insects and humming-birds. Sometimes other insects steal it by biting holes in the knobs. The petals alternate with 5 petal-like sepals, many stamens, and 5 or more pistils, which ripen into many-seeded pods called follicles.

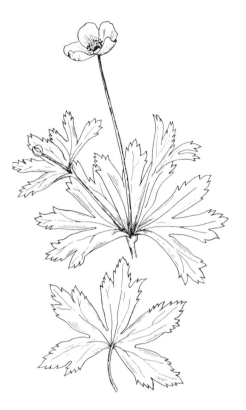

21. Meadow Anemone, $^2/_5$ X

1. Spur longer than the calyx.
   2. Flowers red in whole or in part.

                                           A. elegantula
                                           Western red columbine

   2. Flowers never red.
      3. Flowers blue or white.
         4. Flowers blue.
            5. Plants mostly over 2dm tall; leaflets mostly over 15mm wide.

                                         A. caerulea var. caerulea
                                         Colorado columbine
            5. Plants mostly under 2dm tall; leaflets mostly 15mm or less wide.

                                         A. scopulorum
                                         Rock columbine

         4. Flowers white.

                                         A. caerulea var. ochroleuca

      3. Flowers yellow or yellowish.
         4. Spurs long (5–7cm), yellow.

                                         A. chrysantha
                                       Golden columbine

         4. Spurs shorter (1.5–3cm), yellowish.

                                         A. micrantha var. micrantha

1. Spur shorter than the calyx.
    2. Spurs reduced to short sac-like structures or lacking entirely.
        3. Sepals 3–4cm long, blue.

                A. caerulea var. *dailyae*

        3. Sepals 1–1.5cm long, white to pinkish.

                A. micrantha var.
                mancosana

    2. Spurs present as slender structures.
        3. Stem leafy or not wholly scapose (naked); flowers several.
            4. Flowers yellow or yellowish.
                5. Plant over 2dm tall.

                        A. flavescens
                        Northern yellow
                        columbine

                5. Plant less than 2dm tall.

                        A. laramiensis
                        Laramie columbine

            4. Flowers blue.

                        A. saximontana
                        Alpine columbine

        3. Stem scapose; one-flowered.

                        A. jonesii
                        Jones columbine

COLORADO COLUMBINE, *A. caerulea* var. *caerulea* (fig. 23), is a handsome plant with its showy blue or lavender and white blossoms, 5–10cm broad, and is the state flower of Colorado. It grows tall, with large blue blossoms, in the shade of moist aspen groves; but shorter and bushier, with paler flowers, among rocks of the subalpine and timberline areas. In some areas, especially in Cedar Breaks National Monument, the white variety is found. The species is found in moist foothill canyons up to over 12,000 ft. through the Rocky Mountains. Aven Nelson, Rocky Mountain botanist, called it the queen of columbines.

In the mountains of southwestern Wyoming, Utah, and Nevada, a species named ROCK COLUMBINE, *A. scopulorum*, may be found, which has similar but somewhat smaller flowers with pale blue sepals and yellow or

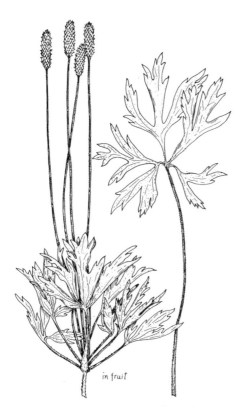

22. Candle anemone, in fruit, ²⁄₅X

white petals. It is very densely tufted with short-
stalked leaves.

ALPINE COLUMBINE, *A.
saximontana* (fig. 24), is
a charming miniature
species found among
rocks at high altitudes on
the peaks of central and
north-central Colorado. Its
drooping flowers have
very short spurs.

JONES COLUMBINE, *A.
jonesii*, is another rare,
small, alpine species,
which occurs in northern
Wyoming and Montana.
Its tiny bluish-gray leaves
are densely tufted, and the
erect, large blue flowers
are held above on leafless
stalks 2.5–8cm tall.

GOLDEN COLUMBINE, *A.
chrysantha,* is a handsome,
much-branched plant
1–1.5m tall, with large,
long-spurred, clear yellow
flowers. It is found in
moist ravines of the foothill
and montane zones (and
sometimes up to 11,000
feet) from southern
Colorado through New
Mexico and Arizona.

THE NORTHERN YELLOW
COLUMBINE, *A. flavescens*,
grows in montane and
subalpine woods (some-
times above timberline)
from Wyoming north-
ward and westward, the
common species in
Yellowstone and Glacier
National Parks. The flow-
ers are pale yellow to
salmon pink.

23. Colorado Columbine, ½X

24. Alpine Columbine, ½X

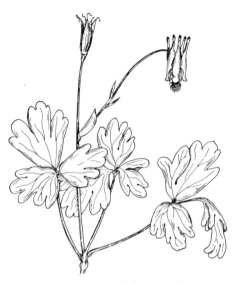

25. Western Red Columbine, ²⁄₅ X

WESTERN RED COLUMBINE, *A. elegantula* (fig. 25), a medium-sized plant rarely more than 3dm tall, with nodding red and yellow flowers, is one of the two species that go by this common name. It occurs in southwestern Colorado, northern New Mexico, and Utah. Its flowers are about 2.5cm long, with straight spurs, which are abruptly narrowed near the tip. A larger species, which also has red and yellow nodding flowers, is *A. formosa*, which barely comes into our range from the northwest.

✔(5) *Caltha*. Marsh-marigold.

One species in our range: WHITE MARSH-MARIGOLD, *Caltha leptosepala* var. *leptosepala* (fig. 26), is without petals. The flowers comprise 5–15 oblong or oval sepals, white or greenish or bluish-tinged, around numerous yellow-anthered stamens and several separate pistils, which develop into a cluster of green seedpods. The leaves are all basal and are roughly oval or roundish, with heart-shaped bases and dentate margins. One of the commonest plants of the subalpine and lower alpine regions, where it often grows in large beds in wet situations and around melting snowbanks.

26. White Marsh-Marigold

(6) *Clematis.* Clematis

Clematis is one of the genera that has no petals, but its sepals are petal-like. Our species are usually bluish, purplish, or white. There are usually 4 sepals and an indefinite number of pistils, which develop into plume-tailed achenes. The leaves are compound.

1. Plants with woody stems, climbing or trailing.
  2. Flowers white or yellow.
    3. Flowers numerous, in clusters, white.

> *C. ligusticifolia*
> Western virgin's bower

    3. Flowers mostly solitary, yellow tinged with green.

> *C. orientalis*
> Oriental clematis

  2. Flowers blue or lavender.
    3. Leaves with 3 leaflets, not cleft.

> *C. occidentalis var.*
> *grosseserrata*
> Blue clematis

    3. Leaves biternately compound (twice 3 times compound).

> *C. columbiana*
> Rocky Mountain clematis

1. Plants appearing bushy but stems herbaceous.
  2. Stems erect; leaves with linear leaflets.

> *C. hirsutissima var. hirsutis-*
> *sima*
> Sugarbowls

  2. Stems sprawling; leaves with ovate leaflets.

> *C. hirsutissima var. scottii*
> Scott clematis

(7) *Delphinium.* Larkspur

Wild larkspurs are easily recognized by anyone familiar with the garden varieties, for the distinctive structure of the flower is similar. The irregular flowers have 5 showy sepals and 4 smaller petals. The upper sepal is spurred. The 2 upper petals extend back into the spur. The 4 petals are usually paler than the sepals and, together, make a knot in the center of the flower, which in the cultivated forms is called the *bee*. The leaves are palmately divided or lobed. Some of the species are poisonous to stock.

1. Flowers whitish or pale.

> *D. virescens ssp. penardi*
> Plains larkspur

1. Flowers bright or dark blue, or purplish.
  2. Plants 1–3dm tall.
    3. Flowers drab, brownish on back; low caespitose plants, alpine.
> *D. alpestre*
    3. Flowers deep bluish-purple; stems mostly single; montane into subalpine.
      4. Flowers entirely blue, slightly villous.

> *D. nelsonii*
> Nelsons larkspur

      4. Petals yellowish and heavily blue-veined.

> *D. bicolor*

  2. Plants 3dm–2m tall.
    3. Plants 3–6dm tall.
      4. Stems pubescent to base.
        5. Leaves notably dimorphic; upper portion of stem leafless.

> *D. burkei*

        5. Leaves not dimorphic.
            6. Radical leaves numerous, tufted; canescently tomentulose.
                                                    D. geyeri
                                                    Geyer larkspur
            6. Radical leaves few, open; stems puberulent.
                                                    D. carolinianum
        4. Stems glabrous, at least below; leaves all basal.
                                                    D. scaposum
    3. Plants 4dm–2m tall.
        4. Plants not all viscid (sticky).
            5. Stems grayish strigose (pointed, appressed hairs).
                6. Leaves divided into cuneate, merely cleft, segments.
                                                    D. geranifolium
                6. Leaves repeatedly divided into linear lobes.
                                                    D. robustum
            5. Stems glabrous or glaucous, at least below.
                6. Inflorescence dense; sepals bicolored.
                                            D. occidentale ssp. cucullatum
                                            Tall mountain larkspur
                6. Inflorescence loose and slender; sepals not bicolored.
                                                    ✓*D. ramosum*
                                                    Branched larkspur
        4. Plants more or less viscid, notably the pedicels; sepals not bicolored.
            5. Flowers narrow, the upper sepal and its spur forming a nearly
                straight line.
                                            D. occidentale ssp. occiden-tale.
                                            Tall mountain larkspur
            5. Flowers not especially narrow, the upper sepal flaring.
                                                    D. barbeyi
                                                    Subalpine larkspur

NELSON LARKSPUR, *D. nelsonii* (fig. 27), is one of the spring and early summer flowers of foothill and montane zones throughout our range. Its few leaves are usually dark-colored and palmately deeply-parted into narrow segments. The stem is 15–30cm tall, sometimes taller, and bears a raceme of 4–10 bright purplish-blue flowers. It is found abundantly on open, sunny hillsides, in meadows, and under ponderosa pines. This species was named in honor of Aven Nelson, Professor of Botany at the University of Wyoming for over fifty years. *D. bicolor* and *D. scaposum* are similar and closely related species.

GEYER LARKSPUR, *D. geyeri*. The leaves of this plant, which are finely divided into narrow segments, appear in tufts in early spring. At this time, on poor pasture where there is little other forage, they are a serious stock poison. The flowers, which come in early summer, are a brilliant and beautiful blue. It grows on the high plains, mesas, and foothill slopes of northern Colorado and Wyoming.

PLAINS LARKSPUR, *D. virescens* ssp. *penardi*, is a plant with tall, unbranched stems and long racemes of very pale blue or whitish flowers. It is found on the plains and lower foothills along the east base of the southern Rocky Mountains.

SUBALPINE LARKSPUR, *D. barbeyi*, is a stout plant with stems 1–2m tall topped by short, dense racemes of very dark purple flowers. It grows in large clumps or beds in wet meadows and bogs of the subalpine zone in Colorado, Wyoming, and Utah.

One or the other of the 2 subspecies of TALL MOUNTAIN LARKSPUR, *D. occidentale*, is found throughout our area. There are several other tall species, similar in appearance and difficult to distinguish, as they all grow in aspen groves and meadows of the montane zone and are 1–2m tall. The racemes are usually long, not crowded, and sometimes branched. In western Colorado and north-westward, *D. ramosum* is common. *D. geranifolium*, with dense flower spikes, occurs in northern Arizona.

(8) *Myosurus*. Mousetail

1. Sepals 1-nerved; spikes less than 1cm long.
   M. aristatus
1. Sepals 3–5-nerved (may be faint); longest spikes over 1cm long.
   M. minimus ssp. montanus
   Mousetail

MOUSETAIL, *M. minimus* ssp. *montanus*, is a tiny plant with a tuft of linear leaves and white-petaled flowers. It has numerous pistils on a much elongated receptacle, which is so long and narrow as to suggest the tail of a mouse. It is occasionally found on muddy banks and shores of ponds in the foothill and montane zones of our region.

(9) *Pulsatilla*. Pasque flower.

1. Flowers purple or violet; involucre leaves sessile.
   P. patens ssp. multifida
   Pasque flower
1. Flowers white or tinged with purple; involucre leaves petioled.
   P. occidentalis

27. Nelson Larkspur, ⅖ X

PASQUE FLOWER, *Pulsatilla patens* ssp. *multifida* (fig. 28), sometimes found within *Anemone*, is one of the earliest flowers of spring, pushing its furry, lavender-blue, golden-centered cups up into the sun even before the snow is entirely gone. Its stems and finely dissected foliage are covered with silky hairs. When they first appear above ground, the buds are protected by the involucral leaves; but as the flower develops, its stalk lengthens until, by the time the silvery head of long-tailed achenes is mature, it is several inches above the involucre. This is the state flower of South Dakota and is found most abundantly on open slopes of foothills and montane zones, some-

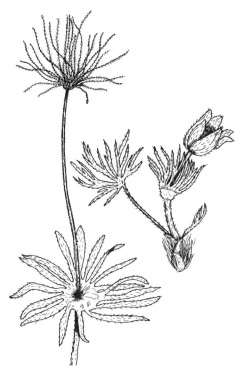

28. Pasque Flower, ²/₅ X

times up to timberline, but only on the eastern side of the Continental Divide.

(10) *Ranunculus.* Buttercup.

This is a large genus in our range. Buttercups have 5 separate sepals and usually 5 petals, which are always separate. The sepals are easily detached and frequently drop off before the petals fall. Most species have bright yellow flowers. The shiny, varnished look on the inner side of the petals is one of the best characters distinguishing buttercups from other yellow-petaled flowers. They have many stamens and pistils, and each pistil develops into an achene. Most species grow in moist locations.

1. Plants aquatic (submerged).
  2. Leaves in filiform divisions; petals white.

            R. aquatilis var. capillaceus
            White water buttercup

  2. Leaves reniform, 3-lobed; petals yellow.

            R. natans var. intertextus
1. Plants terrestial (often growing in wet places); petals yellow.
  2. Plants with stolons, or rooting at the nodes, growing in mud or water.
    3. Leaves simple, entire.
      4. Leaves linear to lanceolate.

            R. flammula var. ovalis
            Creeping buttercup

      4. Leaves cordate, reniform, or ovate.

            R. cymbalaria var. saximon-
              tanus
            Trailing buttercup

    3. Leaves finely dissected.

            R. gmelinii var. hookeri
            Gmelin buttercup

  2. Plants without stolons, not rooting at the nodes.
    3. Leaves compound.
      4. Stems glabrous, at least in maturity.
        5. Plants small, 4–10cm tall; basal leaves usually lacking; sepals glabrous; petals barely exceed broader sepals; leaf divided into 3-5 narrowly oblanceolate and entire lobes.
           R. jovis

5. Plants taller, 7–35cm tall; basal leaves on long petioles; petals
  exceed sepals in length.
  6. Petals twice the length of sepals; sepals subvillous; leaves
    2–3-ternately parted into linear lobes; 1-flowered; plant
    10–20cm tall; found at 11,000-13,000 feet.
                                            R. adoneus ✓
                                          Snow buttercup
  6. Petals 1/3 longer than sepals; sepals petaloid; leaves
    biternate, deeply parted into oblong or linear lobes;
    flowers few; plant 7–35cm tall; found at 5,000-7,200 feet.
                                            R. ranunculinus
                                          Nuttall buttercup.
4. Stems variously pubescent.
  5. Stems appressed pubescent; sepals reflexed and shorter than the
    petals.
                                            R. acriformis var. acriformis
                                          Sharp or Wyoming
                                          buttercup
  5. Stems hirsute or hispid, spreading hairs.
    6. Sepals reflexed and longer than the petals.
                                            R. pensylvanicus
    6. Sepals spreading, somewhat shorter than the petals.
                                            R. macounii
                                          Macoun buttercup
3. Leaves simple, entire to deeply cleft.
  4. Leaves all simple, entire to merely serrate or toothed, not deeply
    parted or divided; petals longer than the sepals.
    5. Leaves entire to somewhat serrate; sepals yellowish-green;
      usually 10 petals.
                                            R. alismaefolius var. mon-
                                            tanus
                                          Caltha-flowered buttercup
    5. Leaves 3–10-toothed at apex; sepals dark-villous; petals 5.
                                            R. macauleyi
                                          Macauley buttercup
  4. Leaves (at least some of them) deeply parted or divided.
    5. Some leaves entire and some 3–5-lobed.
      6. Basal leaves elliptic to oblanceolate, entire; cauline leaves
        entire to 3-lobed; stems not fistulose (not hollow).
                                            R. glaberrimus var. ellipticus
                                          Sage buttercup
      6. Basal leaves reniform-oval to cordate, deeply crenate; cauline
        leaves deeply divided into 3–5 linear or lanceolate segments.
                                            R. abortivus
                                          Small-flowered buttercup
    5. None of the leaves entire.
      6. Stems 2–7cm tall.
                                            R. pygmaeus
                                          Dwarf buttercup
      6. Stems more than 7cm tall.
        7. Leaves (at least some of the basal leaves) deeply 3-parted.
          8. Apex of ultimate leaf segments obtuse.
                                            R. sceleratus var. multi-
                                            fidus
                                          Blister buttercup
          8. Apex of ultimate leaf segments acute.
            9. Stems stout, hispid, the hairs reddish-brown,
              fisulose.
                                            R. uncinatus var. parviflorus
                                          Little buttercup

9. Stems glabrous; sepals pilose on outside; not fisulose.
>
> *R. eschscholtzii*
> Subalpine buttercup

7. Leaves often deeply and many-parted, but not deeply 3-parted.

8. Petals 2–8mm long; leaves obovate to ovate.
>
> *R. inamoenus*
> Ugly buttercup

8. Petals 8–15mm long; leaves cordate.

9. Basal leaves petioled, irregularly parted into 5–7-lobed main segments; upper cauline leaves sessile, divided into 5–7 linear lobes.
>
> *R. pedatifidus*
> Bird-foot buttercup

9. Basal leaves petioled, deeply crenate to shallowly lobed or toothed; upper cauline leaves sessile, deeply 5–7-parted linear bracts.
>
> *R. cardiophyllus*
> Heart-leaved buttercup

TRAILING or SHORE BUTTERCUP, *R. cymbalaria*, has small flowers, usually less than 12mm across, and its roundish leaves have heart-shaped bases. It's most common around ponds or on wet ground at middle or lower altitudes. CREEPING BUTTERCUP or SPEARWORT, *R. flammula*, is found on muddy shores and stream banks in the montane and subalpine regions. Its flowers are under 12mm broad.

CALTHA-FLOWERED BUTTERCUP, *R. alismaefolius* (fig. 29), is easily distinguished from all other species by the combination of undivided leaves and 10-petaled flowers. It sometimes covers large areas in wet meadows, blooming in early summer. Found in the subalpine zone from southern Wyoming through Colorado.

SAGE BUTTERCUP, *R. glaberrimus* var. *ellipticus* (fig. 30), is a very early-blooming species of the foothill and montane zones, found throughout our range. Its basal leaves are elliptic or roundish, but some of the stem leaves are 3-lobed with the middle lobe the largest. Its sepals are tinged with lavender.

HEART-LEAF BUTTERCUP, *R. cardiophyllus*, is a hairy plant of montane and subalpine meadows with at least some of its basal leaves heart-shaped. The flowers are comparatively

29. Caltha-Flowered Buttercup, ²/₅X

30. Sagebrush Buttercup, ²/₅X

large, 2.5cm or more across. The sepals are pubescent with long hairs, and sometimes there are hairs on the rounded petals. UGLY OR HOMELY BUTTERCUP, *R. inamoenus*, is similar, but has smaller flowers with narrower petals. WYOMING BUTTERCUP, *R. acriformis*, is a plant 30–50cm tall with appressed-pubescent leaves twice 2 or 3 times divided into lanceolate-linear lobes. Its flowers are nearly 2.5cm across, and its achenes are smooth and flattened with short, hooked beaks. This grows in wet montane meadows in Montana, Wyoming, and adjacent Idaho.

SUBALPINE BUTTERCUP, *R. eschscholtzii*, has deeply 3-lobed leaves; the middle lobe may be 3-lobed or undivided, and the lateral lobes are 3–7-parted. The plant is very smooth. Its flowers vary in size from 12 to 25mm in diameter. The large-flowered form has been called var. *eximius*, and the small one var. *alpeophilus*. The species occurs throughout our range at altitudes from 9,000 to 12,000 feet.

There are several species that could be called alpine buttercups, but the commonest and most conspicuous in the central part of our region, and the one that is most strictly alpine, is SNOW BUT-TERCUP, *R. adoneus* (fig. 31). It is found in areas where snow remains late into the summer, and in many such places it is the first plant out of the ground as the snow melts. Its poppy-like bright yellow blooms open before the finely-cut leaves have a chance to unfold. It may be seen on Pikes Peak, Mount Evans, and on Trail Ridge in Colorado.

31. Snow Buttercup, ½X

MACAULEY BUTTERCUP, *R. macauleyi*, another large-flowered, showy species, with 3-toothed leaves and black, hairy sepals, is found in the high mountains of southern Colorado and New Mexico. In northern Wyoming and Montana (especially in Yellowstone and Glacier National Parks and the Beartooth Range), handsome large-flowered forms of *R. eschscholtzii* var. *eximius*, described above, are found in the alpine zone.

BIRD-FOOT BUTTERCUP, *R. pedatifidus*, is a pubescent plant with erect stems and small flowers, which grows in meadows of the montane and subalpine zones, but occasionally in the tundra above timberline. It may be distinguished from *R. eschscholtzii* by pubescent achenes.

NUTTALL BUTTERCUP, *R. ranunculinus* (fig. 32). The basal leaves of this erect, completely smooth plant are 3-times-divided, and each division is again divided. It has several to many flowers in a branching inflorescence. Both petals and sepals are yellow and shiny on the inner surfaces, and both fall, or wither, early, leaving heads of plump, green achenes, each with its style forming a little hook at the tip. Frequently found among rocks in the foothills and ponderosa pine zone east of Rocky Mountain National Park, and occurring occasionally from Wyoming through mountainous regions into northern New Mexico.

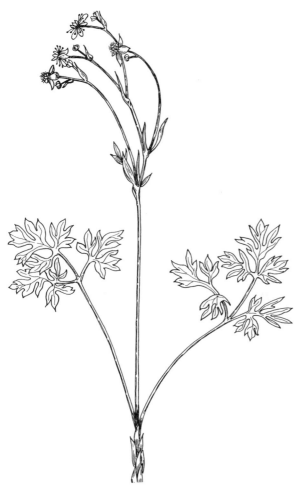

32. Nuttall Buttercup, ½X

WHITE WATER BUTTERCUP or WATER CROWFOOT, *R. aquatilis*, often forms submersed, brownish masses of branched stems and finely-divided leaves, which send up numerous dainty, white flowers. It is found in ponds and slow streams throughout the Rockies, extending to an altitude of 10,000 feet in Colorado.

FLOATING BUTTERCUP, *R. natans*, with shiny yellow petals and floating 3-lobed leaves, is found in standing or slow-moving water from the plains to the sub-alpine zone, from the Yukon through the Rockies into Idaho and into southern Colorado. The lobes of the leaves are more or less notched.

33. Alpine Meadowrue, ½X

(11) *Thalictrum*. Meadow rue

Most of these plants have ternately compound leaves with small leaflets much like those of columbine; but their apetalous, small, green or purplish flowers, which grow in clusters, are very different. Some species have perfect flowers (bisexual), but in others they are imperfect (unisexual). The staminate flowers are composed of tassels of yellowish-green stamens. The fruit is a cluster of achenes.

1. Flowers perfect.
    2. Stem scapose; sepals grayish-purple.

    T. alpinum
    Alpine meadow rue

    2. Stem leafy; sepals whitish or greenish.

    *T. sparsiflorum var.
    saximontanum*
    Few-flowered meadow rue
1. Flowers imperfect, plants either dioecious (with staminate and pistillate flowers on separate plants) or polygamo-dioecious (mostly dioecious, but with a few perfect flowers)
    2. Achenes (seeds) flattened, 2-edged.
        3. Leaves ovate or ovate-oblong, gibbous.

        T. fendleri
        Fendler's meadow rue
        3. Leaves lanceolate, acuminate (taper-pointed), the 2 edges nearly alike.
        *T. occidentale var.
        palousense*
        Western meadow rue
    2. Achenes terete (cylindrical) or very slightly flattened.
        3. Plants dioecious; leaves glabrous and glaucous.

        T. venulosum
        Veiny meadow rue
        3. Plants polygamo-dioecious; leaves obscurely glandular or waxy.
        T. dasycarpum
        Purple meadow rue

ALPINE MEADOW RUE, *T. alpinum* (fig. 33), is a tiny plant when it grows in the alpine tundra; but in subalpine bogs, it may be 15–30cm tall. Its leaves are mostly basal and are divided into firm, roundish, 3–7-lobed leaflets. The few perfect flowers are arranged singly along an upright stalk. Found from arctic regions through the Rockies as far south as New Mexico.

FENDLER'S MEADOW RUE, *T. fendleri* (fig. 34), has unisexual flowers, with staminate and pistillate usually on separate plants. It is found in the montane and subalpine zones, often in aspen groves and meadows. Its foliage is bluish-green, and its achenes are compressed and strongly nerved. This species is common from northern Wyoming through the southern part of our range.

The FEW-FLOWERED MEADOW RUE, *T. sparsiflorum* var. *saximontanum*, has perfect flowers. Its foliage tends to be a more yellowish-green than *T. fendleri*, and its achenes are not strongly nerved. In Colorado, the two species are often found in the same locations, but *T. sparsiflorum* extends farther north and west, occurring in Yellowstone Park.

34. Fendler Meadowrue, ½X

WESTERN MEADOW RUE, *T. occidentale*, is another species of the northern and western part of our region. It has slender achenes 7–20 mm long and grows to about 1m tall. *T. venulosum* is a widespread species with strongly-veined, bluish-green leaflets and oblong, turgid achenes, not over 7mm long.

PURPLE MEADOW RUE, *T. dasycarpum*, is sometimes seen along ditch banks and in wet meadows of the lower montane valleys. It is tall, with large, purple, pyramid-shaped flower clusters.

(12) *Trautvettaria*. False bugbane.

FALSE BUGBANE, *Trautvettaria carolinensis* var. *occidentalis*, is the only species of this genus in our range. A stout plant with stems 30–100cm tall, it is apetalous with 4–5 sepals, forming showy panicles of small white flowers. A pompom of white stamens is the comspicuous part of each flower. The seeds are achenes less than 6mm long. It grows on wet ground in partial shade in the subalpine zone of southern Colorado, northern New Mexico, Utah, and Idaho.

(13) *Trollius*. Globeflower

GLOBEFLOWER, *Trollius laxus* (fig. 35), is the only species in our range. The plant usually forms clumps of smooth, long-petioled leaves with several 1-flowered stalks. Cauline leaves are short-petiolate below, sessile above. The sepals are a pale yellow or cream when the flowers open, but may become dingy white in fading. There are some very small nectar-bearing petals between the ring of many golden stamens and the sepals. Leaf blades are roundish in outline but are palmately divided into several lobes, which are themselves cut or deeply toothed.

BARBERRY FAMILY: BERBERIDACEAE

Our plants of this family are shrubs with racemes of yellow flowers: 6 separate sepals,

35. Globeflower, ½X

6 separate petals, 10 stamens, and 1 pistil, which ripens into a few-seeded berry. The roots produce a fine yellow dye. Some species are alternate hosts for stem rust, a fungal disease that attacks wheat.

1. Stems spine-bearing; leaves simple.

Berberis fendleri
Fendler barberry

1. Stems unarmed; leaves pinnately compound.
    2. Upright shrubs; leaflets not over 3cm long, fewer than 5 teeth on each side, but occasionally up to 9.

Mahonia fremontii
Fremont mahonia

    2. Trailing shrubs; leaflets over 3cm long, over 5 teeth on each side.

M. repens
Creeping hollygrape

FENDLER BARBERRY, *Berberis fendleri*, with spiny stems and simple, deciduous leaves, is found in the upper foothill and lower montane areas of southern Colorado and northern New Mexico.

CREEPING HOLLYGRAPE, *Mahonia repens* (fig. 36), is a subshrub that spreads by underground stems. It is rarely over 30cm tall and has evergreen, compound leaves of 5 or 7 holly-like leaflets. Its juicy berries, which make good jelly, are dark blue. On dry, open hillsides exposed to winter sun, its leaves turn red or maroon; but in shaded, moist locations, where it grows more luxuriantly, they remain green. It is a useful low-growing evergreen for home planting.

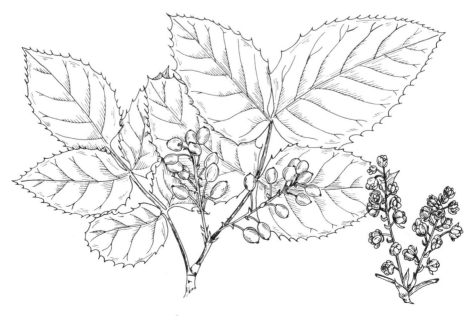

36. Creeping Hollygrape, ½X

FREMONT MAHONIA, *M. fremontii*, is a wide-spreading, erect shrub often 1–1.5m tall with compound, leathery evergreen leaves. The leaflets are light blue-green, and the abundant yellow flowers attract bees and humming birds. Navajo Indians called this species *yellow wood* and used the bark and roots to dye their wool. It grows in the foothill zone, often among piñon pines and junipers, in southwestern Colorado and adjacent Utah and Arizona.

POPPY FAMILY: PAPAVERACEAE

The flowers of this family have 2–3 sepals which usually fall off when the bud opens, 4–6 separate petals, which are packed so tightly in the bud that they are always crinkled when they open, many stamens, and 1 large pistil with a disk-like, sessile stigma. They have a yellowish or milky juice, which dries into a gummy substance. This may form a seal when the flowers are picked, making it impossible for the stems to absorb water, the reason cut poppies often wilt even when placed in water. If the cut end of the stem is put in boiling water or held over a flame for an instant, the substance will be dissolved, and the flowers will remain fresh when put in water.

1. Herbage, sepals, and capsules prickly; leaves sinuate-dentate or sinuate-pinnatifid (like a thistle); sepals with hornlike appendages; flowers usually white.

Argemone, p. 72
Prickly poppy

1. Herbage, sepals, and capsules not prickly; leaves compound or dissected (not thistle-like); sepals without hornlike appendages.
　　2. Leaves dissected into linear divisions; sepals coherent; capsule linear-elongate.

Eschscholzia, p. 72
California poppy

　　2. Leaves pinnately compound; sepals distinct; capsule obovoid.

Papaver, p. 72
Poppy

(1) *Argemone.* Prickly poppy

1. Stems and leaves with few or no hairs between the prickles.

A. polyanthemos
Prickly poppy

1. Stems and leaves hispidulous-pubescent (with bristle-like hairs) between the prickles.

A. hispida
Prickly poppy

(2) *Eschscholzia.* California poppy

One species, *E. californica*, has escaped from cultivation in our range. Leaves ternately dissected, smooth and glaucus; petals orange or yellow.

✓(3) *Papaver.* Poppy

1. Alpine tundra plant; pale yellow to near white; caespitose; stems blackish-hirsute; leaves basal, deeply lobed or divided. Endangered.

P. kluanense
Alpine poppy

1. Escaped cultivar; bright orange to scarlet; foliage stiff-hairy but not spiny.
P. orientale
Oriental poppy

PRICKLY POPPY, *Argemone polyan-themos* (fig. 37), is a gray-leaved, prickly plant with large, white, golden-centered blossoms, frequently seen along roadsides and on stony slopes of the high plains and foothills from New Mexico to Wyoming. In the middle of the central tuft of bright yellow stamens is the large, black, 4-lobed stigma. The buds become about 2.5cm long before the 3 separate horned sepals drop off and the wrinkled white petals unfold. *A. hispida* is a similar plant that has lots of bristles in addition to the spines on its sepals, stems, and leaves.

ALPINE POPPY, *Papaver kluanense*, is a small plant, very rare in our range, with pubescent, deeply-cut foliage and yellow flowers 2.5–5cm across. The sepals are thickly set with black hairs, a character easily seen on the buds. It occurs in alpine regions on the highest peaks from Glacier Park southward into northern New Mexico.

37. Prickly Poppy, ½X

FUMITORY FAMILY: FUMARIACEAE

The plants of this family have smooth, finely dissected, often glaucous foliage and irregular flowers with 2 scale-like sepals and 4 petals. The outer petals spread, and one or both are spurred; the inner petals are united at their tips and enclose the 6 stamens. The cultivated bleeding heart and the eastern woodland wild flowers known as dutchmans breeches and squirrel corn are members of this family. *Corydalis* is the only genus in our range.

1. Corolla white, rose, or purplish; stout stalk, usually over 50cm tall.
C. caseana
1. Corolla yellow; slender stalk, less than 50cm tall.
    2. Spur 4–5mm long, 1/3 the length of the body of the flower; capsule pendulous.
C. aurea var. aurea
Golden smoke
      2. Spur 5–9mm long, nearly as long as the body of the flower; capsule erect or ascending.
C. aurea var. occidentalis

38. Golden Smoke, ½X   39. Wild Hop-Vine, ½X

GOLDEN SMOKE, *Corydalis aurea* (fig. 38), is a plant with bluish, fern-like foliage and racemes of yellow flowers. The stems are spreading, so the plants are in low clumps. Frequently seen on the disturbed soil of banks and in moist ravines of the montane zone throughout our range. C. caseana is a tall, stout plant, with flowers similar in form but pink in color, which grows in wet situations of the subalpine zone in Utah and southwestern Colorado.

ELM FAMILY; ULMACEAE

Only one native genus, *Celtis,* is found in our range. HACKBERRY, *C. occidentalis,* has leaves that are pointed and petiolate. Its flowers, greenish and axillary, perfect or imperfect, solitary or in pairs, and appear with the leaves. Calyx 5–6-parted; stamens 5–6; ovary 1-celled; fruit a globular drupe.

MULBERRY FAMILY: MORACEAE

WILD HOP-VINE, *Humulus lupulus* var. *neomexicanus* (fig. 39), is the only species of this family found growing wild in our range. Leaves mostly opposite, palmately 5–9 parted; the ovary 1-celled, becoming a glandular achene. The vine, with its papery clusters of fruits, is sometimes found clambering over shrubs in the canyons.

NETTLE FAMILY: URTICACEAE
1. Leaves alternate, entire; no stinging hairs.

Parietaria pensylvanica
Pellitory
1. Leaves opposite, serrate; with stinging hairs; plants rhizomatous.
    2. Leaf blades narrowly to broadly lanceolate.

Urtica dioica ssp. holosericea
    2. Leaf blades ovate-lanceolate to ovate.

U. dioica ssp. gracilis
Common nettle

COMMON NETTLE, *Urtica dioica*, may be found along streams or on waste ground in the foothill and montane zones. It is a stinging, bristly plant, usually in clumps, with sharply angled stems, and small clusters of inconspicuous greenish flowers at the nodes.

OAK FAMILY: FAGACEAE
    Only one genus is found in our range. Mature acorns are necessary for identification.
1. Upper scales of the cup (involucre) prolonged.

Quercus macrocarpa

1. Upper scales of the cup not prolonged.

Q. gambelii
Gambel oak

    In the foothills from central Colorado and southwestern Wyoming southward, large areas are covered by oak brush. There is variation in the shape of the leaves and the height of the trunks. Some individuals grow into small trees. GAMBEL OAK, *Quercus gambelii* (fig. 40), is the more common species throughout our range.

BIRCH FAMILY:
BETULACEAE
    A family of trees and shrubs. Its flowers are of two kinds, but both kinds are on the same plant. Except for the pistillate flowers of the hazelnut, the flowers are in catkins. The staminate catkins of all species release great quantities of pale yellow pollen in early spring, after which they drop off. Most of the members of this family grow only on moist soil, especially along streams.

40. Gambel Oak, ½X

1. Pistillate flowers with a calyx; nut enclosed in a large, thick involucre.

Corylus, p. 76
Hazelnut

1. Pistillate flowers without a calyx (catkins); the nut naked.
    2. Stamens 2; filaments bifurcate; nutlets winged.

Betula, p. 76
Birch

    2. Stamens 4; filaments not forked; nutlets not winged.

Alnus, p. 76
Alder

(1) *Alnus.* Alder

Only one species in our range: THINLEAF ALDER, *A. incana* ssp. *tenuifolia.* Leaves ovate with prominent veins, sharply double-toothed.

(2) *Betula.* Birch

1. Low bushy shrubs, less than 2m tall; leaves orbicular to oval or obovate, the teeth usually rounded; nutlets orbicular-winged.

B. glandulosa
Bog birch

1. Taller and slender, more than 2m tall; leaves subcordate to ovate, the teeth usually pointed; nutlets with broad lateral wings.

B. occidentalis
Western red or river birch

(3) *Corylus.* Hazelnut or filbert

41. Thinleaf Alder, ²/₅ X

Only one species in our range. BEAKED HAZELNUT, *C. Cornuta.* A shrub 6–16dm tall; leaves ovate or ovate-oblong, somewhat heart-shaped, abruptly pointed.

THINLEAF ALDER, *Alnus incana* ssp. *tenuifolia* (fig. 41), a shrub or small tree, is found in moist situations from Alaska and the Yukon south into New Mexico and Utah. In our range, it is restricted to streamsides from the upper foothill to the lower subalpine zones and is often associated with the red birch, usually having more than 1 trunk. It is stouter than the birch and has gray, smooth bark. When mature, the pistillate catkins are oval, about 2cm long, dark brown, woody, and persistent.

WESTERN RED or RIVER BIRCH, *Betula occidentalis*, a shrub or small tree, is common along most mountain streams from the lower foothills through the montane zone. It grows in graceful clumps, and its several stems are clothed in glossy, reddish-brown bark marked by horizontal lenticels. The smallest twigs are roughly glandular. Leaves are thin, ovate, and toothed or lobed, tapering to a slender tip, usually turning a clear, light yellow in autumn. The cylindrical staminate catkins are 5–6.5cm long at flowering time. The pistillate ones are oval and ripen into light-brown papery cones about 2.5cm long, which hang on the twigs after the flat, winged seeds are dispersed.

BOG BIRCH, *Betula glandulosa* (fig. 42), is one of the shrubs found in all the arctic and high mountain areas of North America. It varies in

42. Big Birch, ½ X

size from a low spreading shrub in very exposed locations to a height of about 2m along streams or in subalpine bogs. Its leaves are thick, roundish, and finely toothed. Its bark is dark and rough, not glossy. It has brilliant autumn coloring in shades from orange to maroon.

BEAKED HAZELNUT, *Corylus cornuta*, is a shrub with smooth, light-brownish bark and ovate or obovate, pointed leaves, 5–13cm long. It is found across our northern states to the foothills of the Rockies. It is common in the Black Hills and occurs along streams in some of the deep canyons of the east slope of the Continental Divide in Colorado. The husk of the nut has a long snout. Squirrels evidently harvest the nuts, as they seem to disappear before they ripen.

FOUR-O'CLOCK FAMILY: NYCTAGINACEAE

Our plants of this family have opposite, simple, entire leaves; stems with enlarged joints; and small, tubular flowers subtended by involucres of separate or united, sometimes petal-like, bracts.

1. Involucral bracts more or less united; flowers mostly pedicellate (with a stalk).
   Mirabilis, p. 78
1. Involucral bracts distinct; flowers sessile in heads.
   Abronia, p. 77

(1) *Abronia.* Sand-verbena

1. Perennials.
   2. Flowers white, 2–3cm long.
      A. fragrans
      Sand-verbena

   2. Flowers greenish, 2cm or less long.
      3. Involucral bracts 4–5mm long.
         A. ammophila
         Yellowstone sand-verbena

      3. Involucral bracts 7–15mm long.
         A. elliptica
         Red-stem sand-verbena

1. Annuals.
    2. Perianth rose-color; wings of fruit double.

                                A. carletonii
                                Carleton sand-verbena
    2. Flowers inconspicuous, reddish-green; wings of fruit single.
                                A. micrantha
                                Sandpuffs

(2) *Mirabilis.* Four-o'clock

1. Fruits smooth or slightly 5-ribbed (not 5-angled), glabrous or nearly so.
    2. Perianth 3–6cm long; stamens 5.

                                M. multiflora
                                Wild four-o'clock
    2. Perianth less than 1cm long; stamens 3.

                                M. oxybaphoides
1. Fruits strongly 5-angled longitudinally, pubescent.
    2. Leaves linear or linear-lanceolate, less than 1cm wide, sessile or short-petioled.
                                M. linearis
                                Fringe-cup four-o'clock
    2. Leaves lanceolate or wider, over 1cm broad; stems densely short-pilose or hirsute
       below; fruit pubescent; leaves sessile or short-petioled.
                                M. hirsuta
                                Hairy four-o'clock

43. Sand Verbena, ²⁄₅ X

SANDPUFFS, *Abronia micrantha*, has clusters of flowers 2.5cm long, which ripen into balls of 3-winged seeds. It is common on sandy plains of the west and may be found in Colorado up to 8,000 feet. YELLOWSTONE SAND-VERBENA, *A. ammophila*, with pale yellow or greenish flowers, is found on the shores of Yellowstone Lake.

SAND-VERBENA, *A. fragrans* (fig. 43), is a sprawling plant with whitish, sticky stems and heads of white or pinkish, fragrant flowers. The slender tubular perianths are 2.5–4cm long, and shorter, papery bracts form a collar beneath the head. The scent is especially noticeable at night. A common plant of the western plains extending into the foothills on sandy soil.

WILD or SHOWY FOUR-O'CLOCK, *Mirabilis multiflora*, is a rounded, compact, leafy plant with large, magenta-colored, funnel-shaped flowers resembling morning-glory blossoms, 6–8 in each involucre. Common in the piñon-junipers of Colorado and Utah and south, especially in Grand Canyon and Zion National Parks.

FRINGE-CUP FOUR-O'CLOCK, *Mirabilis linearis*, is a sprawling, sticky, plant with whitish stems, linear leaves, and a branched inflorescence of scattered pink to purple flowers, which open in late afternoon. Found on dry ground throughout our range below 9,000 feet. HAIRY or LANCE-LEAVED WILD FOUR-O'CLOCK, *Mirabilis hirsuta* (fig. 44), similar but with broader leaves, is sometimes seen.

CARPETWEED FAMILY: AIZOACEAE

Only two species of this family are likely to be found in the southern part of our range.

1. Fleshy plants; capsules circumscissile; leaves opposite, linear to spatulate, entire; flowers axillary, sessile; calyx-tube turbinate.

Sesuvium verrucosum
Sea purslane

1. Plants not fleshy; capsules valvate; leaves opposite, spatulate to linear-oblanceolate, entire; flowers axillary on long pedicels.

Mollugo verticillata
Carpetweed

CACTUS FAMILY:
CACTACEAE

The plants of this family are succulents. They store water in their thick stems. Not all succulents are cacti. A true cactus must have spines on its stems arranged in little groups called *areoles*, which are arranged in definite patterns. Cactus flowers have many sepals, petals, and stamens, and a large pistil. Their petals usually have a satiny sheen. Some species never have leaves; others have small leaves that appear when moisture is plentiful, but soon shrivel and drop off. The usual functions of leaves are carried on by the green-colored bark of the stems. Only a few species of cacti occur naturally in our mountains.

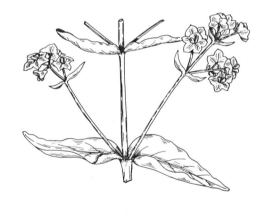

44. Lance-Leaved Wild Four-O'Clock, ²/₅X

1. Stems globose, oval, or ovoid, not jointed; leaves wanting; spines not barbed.
    2. Flowers from between the tubercles.

Coryphantha, p. 80
Nipple cactus

    2. Flowers from the tubercles or the ribs.
        3. Plants with parallel longitudinal ridges or ribs.

Echinocereus, p. 80
Hedgehog cactus

3. Plants with nipple-like tubercles not united into longitudinal ribs.

Pediocactus, p. 80
Pincushion or Ball cactus

1. Stems flat or cylindrical, conspicuously jointed; bristles or spines barbed; leaves early deciduous.

Opuntia, p. 80
Prickly pear

(1) *Coryphantha.* Nipple cactus

1. Flowers pink-purple; mature fruit green.

C. vivipara var. vivipara
Ball nipple cactus

1. Flowers yellowish-green; mature fruit scarlet.

C. missouriensis var. missouriensis
Mesa nipple cactus

(2) *Echinocereus.* Hedgehog cactus

1. Flowers yellowish-green; fruit greenish.

E. viridiflorus var. veridiflorus
Green-flowered hedgehog cactus

1. Flowers scarlet or orange-red; fruit red when ripe; spine definitely angled.

E. triglochidiatus var. melanocanthus
Kings crown cactus

(3) *Opuntia.* Prickly pear

1. Stem joints flattened.

O. polyacantha
Hunger cactus

1. Stem joints nearly cylindrical.

O. fragilis var. fragilis
Brittle cactus

(4) *Pediocactus.* Pincushion or Ball cactus

Only one species likely to be found in our range. The flowers are usually bright pink, but sometimes yellowish or whitish.

P. simpsonii var. simpsonii
Mountain ball cactus

KINGS CROWN CACTUS, *Echinocereus triglochidiatus* var. *melanacanthus,* belongs to the group referred to as hedgehog cacti. It forms mounds, sometimes 2 or 3 feet in diameter, and is very handsome in bloom. The flowers are a rich, brilliant red with a knot of green stigmas in the center. Its spines are angled. Grows on warm, rocky hillsides of foothill canyons from southern Colorado to Texas and New Mexico.

GREEN-FLOWERED HEDGEHOG CACTUS, *E. viridiflorus* (fig. 45), grows singly or in small clusters. The stems are 7–8cm tall, longitudinally ribbed, with the areoles along the ridges, which are sometimes more or less spiral. The chartreuse-colored flowers come out from the ribs, often near the base of the plant. It grows on grassy mesas and hillsides of the foothill zone from Wyoming south to New Mexico, mainly on the eastern side of the Continental Divide.

PINCUSHION or MOUNTAIN BALL CACTUS, *Pediocactus simpsonii* (fig. 46). The fragrant, pink satiny flowers of this prickly ball grow from its top like a crown. Instead of ribs, its surface is covered with tubercles arranged in rows, which are somewhat spiral. Its stem is a slightly flattened ball 5–15cm in diameter. This species reaches the highest altitude of any of our cacti and is frequently seen on sunny, rocky situtations in the montane zone, even as high as 10,000 feet. It blooms from mid-April to late May.

BALL NIPPLE CACTUS,

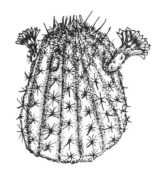

45. Green-Flowered Hedgehog, ½X

46. Mountain Ball Cactus, ½X

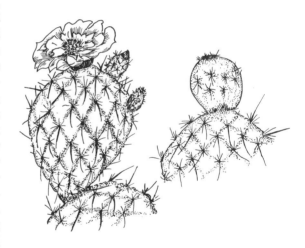

47. Hunger Cactus, ½X

Coryphantha vivipara, is similar, but the globular plant body is not depressed, and its tubercles are more conspicuous. The flowers are pink or rose-purple. Grows 4–6.5 cm tall and is found on mesas and dry slopes of the foothill and lower montane zones. Its relative, MESA NIPPLE CACTUS, *C. missouriensis*, has inconspicuous yellow or greenish flowers and occurs on mesas and foothill slopes from Montana south to Texas.

PRICKLY PEAR, *Opuntia*, is a large genus. Two species extend into the mountains. HUNGER CACTUS, *O. polyacantha* (fig. 47), has flat joints 7–15cm long and yellow flowers, that appear on the edges of the joints or pads. They turn orange as they fade. The fruits are spiny. BRITTLE CACTUS, *O. fragilis*, has shorter, nearly cylindrical joints and yellow or pinkish flowers. Both species grow on sunny, rocky slopes in the foothills.

GOOSEFOOT FAMILY: CHENOPODIACEAE

This is a large family of plants that lack showy petals. In some cases, though, the fruits become conspicuous. Many of the species are classed as weeds, growing on waste ground, and some grow only on wet, alkali soil. The pollen of several species is believed to cause hay fever. Strictly speaking, only a few properly belong to the mountain flora, but some occur in the mountain valleys where there is alkaline soil. Given these circumstances, a key is provided only for the numerous genera, with commentary on only a few of the species most likely to be found. The family includes shrubs and herbs; annuals and perennials; leaves alternate or opposite; the small, green, or greenish flowers perfect or imperfect. Usually 5 petals and stamens, but rarely fewer; the ovary superior, 1-celled with 1–3 stigmas; the fruit 1-seeded.

1. Stem jointed; leaves scalelike.

> *Salicornia*
> Glasswort

1. Stem not jointed; leaves variable but not scalelike.
   2. Leaves or branches spiny at tips in maturity.
      3. Plants annual.

> *Salsola*
> Russian thistle

      3. Plants perennial; a shrub.

> *Sarobatus*
> Greasewood

   2. Leaves or branches not spiny; spines, if any, are on stems.
      3. Flowers perfect.
         4. Without bracts.
            5. Perianth wingless, persistent.
               6. Perianth 2–5 parted.

> *Chenopodium*
> Goosefoot

               6. One sepal only.
                  7. Herbaceous and bract-like.

> *Monolepis*
> Poverty weed

                  7. Hyaline (nearly transparent) and scale-like.

> *Corispermum*
> Bugseed

            5. Perianth horizontally winged.
               6. Plant annual.

> *Cycloloma*
> Winged pigweed

               6. Plant perennial.

> *Kochia*
> Summer cypress

         4. Flowers with bracts

> *Suaeda*
> Seepweed

      3. Flowers imperfect.
         4. Fruiting bracts compressed (flattened on two opposite sides.)

> *Atriplex*
> Saltbush

         4. Fruiting bracts obcompressed (flattened the opposite of the usual way).
            5. Pericarp (wall of the ovary or fruit) naked.
               6. Orbicular, wing-margined.

> *Grayia*
> Hop-sage

6. Nearly hastate, 2-toothed at apex.

Suckleya
Poison suckleya

5. Pericarp silky-pubescent.

Ceratoides
Winterfat

FOURWING SALTBUSH OR CHAMISO, *Atriplex canescens* (fig. 48), a shrub of sandy and saline soil, widely distributed at altitudes up to 8,000 feet from South Dakota and Oregon to northern Mexico. (New Mexico's "chamisa" is a different plant.) The foliage is covered with minute silvery scales, giving the bush a gray appearance. The 4-winged fruits occur in conspicuous, often rose-tinged clusters and are very nutritious. It is a valuable food plant for browsing animals, especially sheep. In Arizona's caliche deserts, it is an indicator of tillable soil. SPINY SALTBUSH OR SHADSCALE, *A. confertifolia*, is similar in appearance and distribution, but its twigs become spines, and its fruits are flat and wingless. GARDEN ORACHE, *A. hortensis*, an annual with round, flat fruits about 12mm in diameter, escapes from gardens in some areas.

48. Fourwing Saltbush, ½X

SUMMER CYPRESS, *Kochia scoparia*, is an introduced annual weed found along most roadsides. It is finely branched with narrow leaves, forming an erect, bushy plant, which often turns red in late summer. If your *Kochia* is a perennial, found in alkaline soil, it is likely *K. americana.*

RUSSIAN THISTLE, *Salsola kali*, is one of the commonest "tumbleweeds". It is not a true thistle. An annual plant, as all tumbleweeds are, much branched into a rounded form and covered with very slender, stiff, spine-tipped leaves. The stem and main branches are often reddish. It is common in Eurasia from where it has been introduced into our area. It is now common along roadsides and on waste ground.

GREASEWOOD, *Sarcobatus vermiculatus*, is a spiny shrub, 6dm–1.5m tall, with very slender, fleshy leaves, well-described by the specific name *vermiculatus*, meaning worm-like. Its staminate flowers are in small greenish or reddish catkin-like spikes at the ends of the branches. The winged fruits, in the axils of the leaves, are sometimes tinged with red. This is a bright green, white-barked shrub growing in large colonies in very drab situtations. In general it belongs at altitudes below our range, but it occurs occasionally in Colorado and Wyoming up to 8,500 feet, always on flat, alkali land. In the West, many different plants are called greasewood, one example being the creosote bush of low Southwestern deserts.

GLASSWORT OR SAMPHIRE, *Salicornia europaea* ssp. *rubra*, is a spreading, apparently leafless plant with a fleshy, jointed, oppositely-branched stem under 30cm tall. Its leaves are reduced to triangular scales at the joints. The flowers are partially embedded in fleshy stems. The fruiting spikes, 2.5–5cm long, turn ruby red at maturity and are are so numerous that the plants, growing on the borders of alkali ponds, sometimes color the whole shoreline, or make patches of brilliant red where a pond has dried up.

SEEPWEED, *Suaeda calceoliformis*, is a decumbent, branched perennial with cylindrical leaves, growing around alkali lakes where its rusty fall color shows up conspicuously against the white soil.

POVERTY-WEED, *Monolepis nuttalliana*, is a low annual occurring in rosettes on mud or sand along streams and as a weed in gardens. Its leaves are spear-shaped, the green flowers in axillary clusters. *M. pusilla* has entire leaves with flowers on slender pedicels.

49. Strawberry Blite, ½ X

GOOSEFOOT, *Chenopodium*, is a genus containing many weeds, but also some edible wild plants such as lambs quarters, and several cultivated vegetables, including spinach and beets. The tolerance of alkalinity, a family characteristic, is a favorable factor for the cultivation of sugar beets in the West, where thousands of tons are grown annually on irrigated land. Three widespread species all have leaves lanceolate or broader. FREMONT GOOSEFOOT, *C. fremontii*, is an erect plant with thin, triangular leaves somewhat lobed at base and small clusters of green flowers in flexible, sometimes branched spikes. These plants grow in colonies, often under pine trees, and in mid- or late summer turn pinkish or orange. Found throughout our range up to 9,000 feet, furnishing feed for cattle in autumn. STRAWBERRY BLITE, *C. capitatum* (fig. 49), has bright green, smooth, triangular and hastately-lobed leaves, and deep-red, fleshy, berry-like clusters of fruits. It grows in rich, moist soil, often around old barnyards, but also among rocks in the high mountains up to 11,000 feet. The leaves of the similar *C. overi* are more lanceolate-ovate to ovate.

AMARANTH FAMILY: AMARANTHACEAE

This family includes several weedy species besides the horticultural forms of amaranth and cockscomb, *Celosia*. The flowers are minute but closely compacted in variously formed clusters, which in cultivated species are conspicuous. The commonest native species, often called ragweeds, are garden and roadside weeds, the pollen of which bothers many allergic people.

1. Anthers 2-celled, leaves alternate; flowers imperfect; plants annuals.
Amaranthus, p. 85
Amaranth

1. Anthers 1-celled; leaves mostly opposite; flowers perfect; plants annuals.
    2. Leaves sessile; flowers spicate.

    Froelichia, p. 85
    Snakecotton

    2. Leaves petiolate; flowers axillary.

    Tidestromia, p. 85

(1) *Amaranthus.* Amaranth

1. Utricle (thin-walled 1-seeded fruit) circumscissile.
    2. Flowers in dense spikes, mostly terminal; stamens 5.
        3. Plants pubescent, at least above.
            4. Spikes thick, erect.

            A. retroflexus
            Redroot pigweed

            4. Spikes slender, panicled.

            A. hybridus

        3. Plants glabrous, or nearly so.
            4. Spine-bearing in the leaf-axils.

            A. spinosus

            4. Not spine-bearing.

            A. powellii

    2. Flowers in small axillary clusters or spikes; stamens 3.
        3. Plant erect but bushy-branched.

        A. albus

        3. Plant prostrate matted.

        A. blitoides
        Flat pigweed

1. Utricle indehiscent (not splitting open).
    2. Bracts not exceeding the sepals.

    A. torreyi

    2. Bracts twice as long as the sepals.

    A. palmeri

(2) *Froelichia.* Snakecotton

1. Spikes all opposite.

    F. floridana var. campestris

1. Spikes alternate, at least in part.

    F. gracilis
    Slender snakecotton

(3) *Tidestromia.*

One species in our range, an annual. Perianth of 5 erect, equal, oblong, rigid-scarious petals; flowers 3-bracted; leaves round-obovate to rhomboidal.

T. lanuginosa

REDROOT PIGWEED, *Amaranthus retroflexus*, is an erect, stout plant 3dm–1.5m tall with dense, irregular, bristly spikes of green flowers. Its stem and the midribs of the pointed, ovate leaves are often reddish and rough with bristly hairs.

FLAT PIGWEED, *A. blitoides*, is a much smoother plant, with many prostrate, sometimes reddish branches. Its leaves are oval or spatula-shaped and often white-margined, larger toward the ends of the branches. It produces many shiny, black seeds.

PURSLANE FAMILY: PORTULACACEAE

Our plants of this family are somewhat succulent with very delicate white, pink, or rose-red petals. The flowers open only in sunshine or bright light. Leaves simple and entire, opposite or alternate; the flowers perfect.

1. Ovary superior; mostly perennial plants.
   2. Sepals herbaceous; 3 stigmas.
      3. Calyx deciduous, the petals ephemeral.

> *Talinum*, p. 87
> Fame flower

      3. Calyx persistent, the petals often marcescent (withering without falling off).
         4. Roots cormose or fleshy.

> *Claytonia*, p. 86
> Spring beauty

         4. Roots fibrous or rhizomatous.

> *Montia*, p. 87
> Indian or Miner's lettuce

   2. Sepals somewhat scarious; 2 stigmas.
      3. Stamens not more than 3.

> *Calyptridium*, p. 86
> Calyptridium

      3. Stamens 5–40.

> *Lewisia*, p. 87
> Bitterroot

1. Ovary partly inferior (adnate to calyx below the middle); plants annual.

> *Portulaca*, p. 87
> Purslane

(1) *Calyptridium*. Calyptridium

1. Annuals; style short or lacking; taprooted.

> *C. roseum*
> Rosy calyptridium

1. Perennials; style very long, flowers in dense capitate clusters.

> *C. umbellatum*
> Pussy-paws

(2) *Claytonia*. Spring beauty

1. Radical leaves (coming from the roots) and the stems from a tuber-like corm.
   2. Flowers white or pink, with pink or purple veins.
      3. Stem leaves narrowly lanceolate or lance-linear; usually with a basal leaf.

> *C. rosea*
> Spring beauty

      3. Stem leaves narrowly lanceolate to broadly lanceolate; plants without a basal leaf.
         4. Stem leaves eliptic-lanceolate to ovate-lanceolate.

> *C. lanceolata var. lanceolata*
> Spring beauty

         4. Stem leaves narrowly lanceolate, rarely over 1cm broad.

> *C. lanceolata var. multiscapa*
> Spring beauty

   2. Flowers deep yellowish-orange; leaves lanceolate or narrowly oblong.

> *C. lanceolata var. flava*
> Spring beauty

1. Radical leaves and stems from a thick caudex; leaves broadly spatulate, obtuse, tufted; alpine.

C. megarhiza var. megarhiza
Alpine spring beauty

(3) *Lewisia*. Bitterroot

1. Sepals large, 6–8.

L. rediviva var. rediviva
Bitterroot

1. Sepals smaller, only 2.
    2. Roots vertical, some spindle-shaped.

L. pygmaea
Tiny or Least lewisia

    2. Root a small corm.

L. triphylla
Three-leaf lewisia

(4) *Montia*. Indian or Miner's lettuce

1. Perennial by stolons, also rooting at the joints.

M. chamissoi
Indian lettuce

1. Annuals.
    2. Leaves all linear.

M. linearis

    2. Cauline leaves connate, forming a disk-like involucre.

M. perfoliata
Miner's lettuce

(5) *Portulaca*. Purslane

One species in our range. COMMON PURSLANE, *P. oleracea*, is a much-branched, matted, and prostrate plant; leaves fleshy and spatulate; flowers small, yellow.

(6) *Talinum*. Fame flower

One species in our range. FAME FLOWER, *T. parviflorum,* has a short stem from thick branching roots; leaves terete (cylindrical) or nearly so; peduncles 5–20cm high; flowers pink in loose cymes.

BITTERROOT, *Lewisia rediviva* (fig. 50), an amazing plant brought back from Montana by the Lewis and Clark expedition in 1806, revived after many months in a dried condition and, when planted, bloomed. It was given the name rediviva, meaning to live again. The new genus was named in honor of Meriwether Lewis, leader of the expedition. The little

50. Bitterroot, ½X

green tufts of slender leaves come up in winter on bare ground or under snow, but they have withered almost completely away by blooming time. The showy, many-petaled pink blossoms dot gravelly terraces and stony places that receive abundant moisture in early spring. The thick roots were a staple food for the Indians of the northwest, who boiled them until the bitterness disappeared. The plant, the state flower of Montana, is known to grow in only a few places in Colorado and southern Wyoming, but it is abundant farther north and west, and in the Grand Canyon area.

TINY LEWISIA, *L. pygmaea*, is a small rosette plant with narrow leaves about 2.5–5cm long. The white, pink, or rose-red flowers are less than 13mm broad and have 2 sepals whose upper edges are fringed with tiny, glandular teeth. Some of the plants found in the western part of our range do not have the glandular teeth. The species is found on moist slopes of the subalpine zone and occasionally in cold, wet situations of the upper montane zone such as the Kaibab Plateau. THREE-LEAF LEWISIA, *L. triphylla*, with similar flowers and 2–4 leaves in a whorl on the stalk, occurs in the subalpine zone from northern Colorado northwestward.

51. Lance-Leaved Springbeauty, ⅖ X

ALPINE SPRING BEAUTY, *Claytonia megarhiza*, a fleshy plant of stony ground and rock crevices of the alpine zone, has several flowers to each stem, arranged in a one-sided raceme. Each flower is 13–26 mm across and has 2 greenish sepals, 5 fragile white petals veined with red, 5 stamens each attached to a petal, and 3 thread-like styles. The succulent leaves are in a basal tuft, and they vary from spatula-shaped to almost round. A very long, thick, purple or reddish-colored root is tucked away between rocks. Blooms soon after the snow is gone, late June into July.

SPRING BEAUTY, *C. lanceolata* (fig. 51), has a pair of lanceolate leaves below the terminal raceme of delicate flowers. (See the key for the 3 varieties.) It is an early spring flower found throughout our range in moist ravines of the foothills, on open montane fields, and in subalpine meadows. *C. rosea* is similar, but its stem leaves are narrower, and there often are one or more basal leaves.

INDIAN LETTUCE or WATER SPRING BEAUTY, *Montia chamissoi* (fig. 52), is a plant seen along slow moving streams and in wet meadows of the foothills and montane zones. It spreads by long stems, which root at the nodes, and has paired, oblanceolate leaves. Flowers white or pink, and pink-veined.

MINER'S LETTUCE, *M. perfoliata*, is a clump plant with long-petioled, roundish basal leaves and a pair of united leaves forming a collar below the small, white flower. Used by Indians and pioneers as a pot herb and salad plant. Commonly found in Arizona along brooks and around springs up to 6,000 feet, but less commonly in other parts of the Rockies.

FAME FLOWER, *Talinum parviflorum*, has an erect stalk holding an open cluster of pink flowers, which are delicate and short-lived. It grows on dry soil in ponderosa pine forests.

PUSSY-PAWS, *Calyptridium umbellatum*. The stalks of this plant lie flat, radiating from a rosette of dark leathery leaves. Each tiny, pink flower has 2 papery white or pinkish, persistent sepals, and many are gathered together into tightly-coiled, ball-like spikes. Several of these balls are clustered on each stalk, giving the effect of a cat's foot. It grows on geyser formations and volcanic gravel from northwestern Wyoming into the high mountains of Idaho and westward. In the montane zone, its stems may be 13–15cm long; on very dry or alpine situations, they may be only 2.5cm long.

52. Water Springbeauty, ⅖ X

PINK FAMILY: CARYOPHYLLACEAE

Plants of this family have opposite, simple, entire leaves; their stems are swollen at the nodes; 5 (4) separate or united sepals; 5 (4) separate petals, but in some species, petals are minute or even absent; the flowers mostly perfect; the ovary superior and 1-celled. Cultivated pinks and carnations belong to this family.

1. Sepals united, forming a tubular or ovoid calyx; petals clawed, usually with appendages.
    2. Ribs or nerves of calyx 10 or more.
        3. Calyx-teeth foliaceous.

Agrostemma, p. 90
Corn cockle

        3. Calyx-teeth short.
            4. Styles 3.

Silene, p. 95
Catchfly or Campion

            4. Styles 5.

Lychnis, p. 92
Cockle

    2. Ribs or nerves of calyx 5.

Saponaria, p. 94
Wheat cockle or Soapwort

1. Sepals distinct, or united only at the very base; petals (usually present) not definitely clawed or appendaged.
    2. Petals deeply emarginate or bifid (two-cleft).
        3. Styles mostly 3.

Stellaria, p. 96
Chickweed or Starwort

3. Styles mostly 5.

Cerastium, p. 91
Chickweed

2. Petals entire or barely emarginate.
   3. Stipules not present.
      4. Styles 5.

Sagina, p. 94
Pearlwort

      4. Styles 3.
         5. Leaves lanceolate to ovate, 5 pairs or more per stem.
            Moehringia, p. 93
            Sandwort
         5. Leaves linear, mostly basal.
            6. Leaves filiform or grasslike, over 3cm long.
               Arenaria, p. 90
               Sandwort
            6. Leaves linear, but short and thickish, less than 1cm long.
               Minuartia, p. 93
               Sandwort
   3. Stipules present, usually scarious and conspicuous.
      4. Petals absent or very minute; fruit a 1-seeded achene or utricle.
         Paronychia, p. 94
         Nailwort
      4. Petals present and definite; fruit a several-to many-seeded capsule.
         5. Styles and capsule valves 5.

Spergula, p. 96
Spurry

         5. Styles and capsule valves 3.

Spergularia, p. 96
Sandspurry

(1) *Agrostemma.* Corn Cockle

One species in our range, *A. gigatho.* Densely silky-pubescent, 3–5dm high, somewhat branched; leaves 5-10cm long; flowers few, large, long-peduncled.

(2) *Arenaria.* Sandwort

1. Plants annual; leaves ovate.

A. serpyllifolia

1. Plants perennial; leaves usually much narrower; valves of the capsule 2-cleft at summit.
   2. Leaves oblong or elliptic, not pungent.
      3. Leaves less than 1cm long.

A. lanuginosa ssp. saxosa

      3. Leaves more than 1cm long.

A. lanuginosa ssp. lanuginosa

   2. Leaves linear, more or less rigid, and pungent.
      3. Petals distinctly longer than the sepals.
         4. Cyme congested, usually capitate.

A. congesta
Ballhead sandwort

         4. Cyme open.
            5. Densely glandular-pubescent above.

A. fendleri var. tweedyi
Fendler sandwort

            5. Nearly glabrous throughout.

A. kingii ssp. uintahensis

3. Petals shorter than or equaling the sepals.
    4. Plants low, broadly and densely caespitose; cyme very dense.
                      A. hookeri
                      Hooker sandwort
    4. Plants tufted, usually 1–2dm high; cyme open.
                      A. fendleri var. fendleri
                      Fendler sandwort

Fendler Sandwort, *A. fendleri* (fig. 53), was named for August Fendler (see the reference to him within the Saxifragaceae). A plant of dry, open hillsides from the foothills to the alpine zone, it has tufts of slender, sharp leaves, and erect stems 5–30cm high, depending upon altitude. The inflorescence is open and bears scattered flowers under 8mm broad of 5 white petals, against which the 10 red anthers show as dark spots. *A. kingii* is a closely related species, with shorter petals, found in western Wyoming, Utah, and Idaho.

Ballhead Sandwort, *A. congesta*, has a tuft of sharp-pointed, grass-like leaves at the base, and erect stems with only 2–4 pairs of shorter leaves. The white flowers are in dense heads. Found from Montana to western Colorado and westward, in the foothills to alpine zones. Hooker Sandwort, *A. hookeri*, has similar dense flower heads and very sharp leaves, which are densely tufted on woody stems, often forming large mats. Its flower stalks are 5–15cm tall. A dryland plant extending from Montana through Wyoming into eastern Colorado.

53. Fendler Sandwort, ½X

(3) *Cerastium*. Chickweed
1. Annuals.
    2. Leaves less than 20mm long; pedicels not longer than the capsules.
                      C. nutans var.
                      brachypodum
                      Nodding mouse-ear
    2. Leaves more than 22mm long; pedicels much longer than the capsule; capsule long exserted.
                      C. nutans var. nutans
1. Biennials or perennials.
    2. Petals about equal to the sepals in length.
                      C. fontanum
                      Common mouse-ear
    2. Petals nearly twice as long as the sepals.
        3. Leaves narrow, linear or linear-lanceolate, usually acute at the apex.
                      C. arvense
                      Mouse-ear chickweed
        3. Leaves wider, oblong, ovate, or oval, usually obtuse at the apex.
                      C. berringianum
                      Alpine mouse-ear

54. Mouse-Ear Chickweed, ½X

MOUSE-EAR CHICKWEED, *Cerastium arvense* (fig. 54), takes its common name from its small, velvety leaves. It has stems less than 30cm tall, which tend to spread or lean, and white flowers about 8mm broad. The petals are deeply cleft. Abundant and widely distributed throughout our range, it begins blooming on the mesas at the base of the foothills in April and proceeds up the slopes to the alpine zone as the season progresses. It is often the most abundant white flower on open fields and dry meadows in spring and early summer. The ALPINE MOUSE-EAR, *C. berringianum*, with shorter stems, broader leaves, and shorter petals, is restricted to high altitudes.

(4) *Lychnis.* Cockle

1. Dwarf and caespitose plants, usually under 20cm high; stems usually 1-flowered; high altitudes.
    2. Open flower erect; petals usually exserted.

    L. kingii
    Kings cockle or Campion

    2. Open flower nodding; petals usually included.

    L. apetala var. montana
    Alpine lanterns
1. Tall erect plants with simple stems, over 25cm tall; usually more than 1 flower; rarely found higher than 10,000 feet.
    2. Calyx 12–18mm long; leaves ovate-lanceolate to broader.

    L. alba
    White campion

    2. Calyx 10–12mm long; leaves linear to oblanceolate.

    L. drummondii
    Drummonds cockle

Plants of this genus are similar in general appearance to those of *Silene* and sometimes hard to distinguish. Usually they have 5 styles and a 10-ribbed calyx. ALPINE LANTERNS, *L. apetala* var. *montana*, is a small, compact alpine plant with nodding flowers on stalks usually not over 13cm tall. The petals are very short, but the enlarged, 10-striped calyxes (fig. 55), suggest Chinese lanterns. *L. kingii* is a similar species, with erect, larger flowers, but its calyx does not become inflated. Both are rare plants of the alpine zone.

55. Alpine Lanterns, ½X

DRUMMONDS COCKLE, *L. drummondii*. Stems more or less sticky, 3–6dm tall, basal leaves oblanceolate, stem leaves linear; calyx cylindric, about 12mm long; petals included or slightly protruding. May be distinguished from similar species of *Silene* by examination of the young ovary. In this species, it is enlarged near the tip. A plant of dry hillsides and meadows, foothills to alpine.

(5) *Minuartia*. Sandwort

1. Glandular-pubescent plant.
   2. Plants annuals; petals short, inconspicuous.

   M. stricta
   2. Plants perennials.
      3. Stout taproot with a much-branched caudex; forms mat up to 3dm broad.

      M. nuttallii
      3. Small taproot and a branched crown; forms cushions up to 1dm broad.

      M. rubella
1. Glabrous plants, or nearly so.
   2. Sepals obtuse.

   M. obtusiloba
   Alpine sandwort

   2. Sepals acute.
      3. Petals 2mm longer than sepals; leaves 1-nerved.

      M. macrantha
      3. Petals lacking or shorter than the sepals; leaves 1–3-nerved, triquetrous (with 3 sharp angles).

      M. rossii

Some botanists put this genus in the similar *Arenaria*. ALPINE SANDWORT, *Minuartia obtusiloba* (fig. 56), is a cushion or mat-forming plant. It makes carpets up to 1m in diameter of crowded, light green, very tiny leaves, starred with numerous white flowers. The petals are oblanceolate and do not have distinct claws, but the rounded blade tapers to its base. This is an arctic species, which extends south along the high mountains.

56. Alpine Sandwort, ½X

(6) *Moehringia*. Sandwort

1. Leaves oblong to oval; sepals obtuse; petals twice as long as the sepals.

   M. lateriflora
   Blunt-leaved sandwort
1. Leaves lanceolate to oblanceolate; sepals acute to acuminate; petals variable, but usually

shorter than the sepals.

M. macrophylla

This is another genus that is sometimes put in *Arenaria*. Blunt-leaved Sandwort, *Moehringia lateriflora*, is a plant with thin, oblong leaves and white petals about twice as long as its sepals. It looks more like a starwort than a sandwort. It occurs in moist, shaded situations of the foothill and montane zones.

(7) *Paronychia*. Nailwort

1. Leaves eliptic or oblong; only at the highest altitudes, about 10,000 feet.
   P. pulvinata
   Alpine nailwort
1. Leaves linear or linear-awl-shaped; 3,500–10,000 feet.
   2. Flowers 1–2; leaves 4–6mm long, about the length of the bracts.
      P. sessiliflora
      Whitlow-wort
   2. Flowers clustered; leaves 6–20mm long, much longer than the bracts.
      3. Stems prostrate, forming mats, up to 6cm high.
         P. depressa
      3. Stems ascending to erect, mostly more than 10cm high.
         P. jamesii
         James nailwort

Paronychia, *Spergula*, and *Spergularia* are genera distinguished from other members of the Pink Family by the presence of thin, silvery stipules. Nailworts have very woody stems and form mats or cushions. They have inconspicuous flowers and a single seed. Most of them are plants of dry regions, and, in these areas, they sometimes extend to high altitudes. Alpine Nailwort, *P. pulvinata*, is a low, densely caespitose plant of dry tundra with elliptic or oblong leaves less than 6mm long. The similar Whitlow-Wort, *P. sessiliflora*, growing at lower altitudes, is slightly larger and has spine-tipped leaves. James Nailwort, *P. jamesii*, has stems up to 30cm tall. The flowers are in forking clusters, each cluster having a central, stalkless flower. It occurs from Nebraska and Colorado to Arizona, on dry plains and hills of the foothill and lower montane zones.

(8) *Sagina*. Pearlwort

Only one species in our range: Arctic Pearlwort, *Sagina saginoides*, is a glabrous tufted perennial of wet, cool situations, growing among rocks or in mud on the borders of ponds. Its minute white flowers are on hooked, thread-like stalks, 2–10cm tall; flowers 3–5mm broad; petals, sepals, and styles 5; stamens 10.

(9) *Saponaria*. Wheat cockle or Soapwort

Bouncing Bet, *Saponaria officialis*, is the only species likely to be found in our range. Stems 30–80cm tall, stout; inflorescence a dense terminal cyme; petals pale pink or white; and forming patches.

(10) *Silene*. Campion or Catchfly

Some plants of this genus are night-blooming and are visted by night-flying insects. Their leaves and stems are often more or less sticky; the calyx is united, sometimes inflated, and often 10-nerved; the petals are clawed.

1. Annuals.
    2. Viscidly hirsute or pubescent throughout.

> *S. noctiflora*
> Night-flowering catchfly

    2. Nearly glabrous; glutinous on one or more of the upper internodes.

> *S. antirrhina*
> Sleepy catchfly

1. Perennials.
    2. Acaulescent-caespitose.

> *S. acualis ssp.*
> *subacaulescens*
> Moss campion or moss
> pink

    2. Caulescent, stems 1–7dm high.
        3. Flowers in a paniculate, leafy cyme.

> *S. menziesii*
> Menzies catchfly

        3. Flowers in a narrow, mostly leafless cyme or thyrse.
            4. Petals cleft into 4 or more segments.
                5. Petals white.

> *S. oregana*

                5. Petals crimson.

> *S. laciniata*
> Mexican campion

            4. Petals 2-cleft.
                5. Plants low (1–3dm); flowers few, mostly in 5-flowered long-pedunculate cymes.
                    6. Calyx short, less than 1cm long.

> *S. parryi*
> Parry silene

                    6. Calyx 10–15mm long.

> *S. douglasii*

                5. Plants taller (3–7dm); flowers usually many, in a spike-like thyrse; petals purple.

> *S. scouleri ssp. hallii*
> Halls catchfly

MOSS CAMPION, *S. acaulis* ssp. *subacaulescens* (fig. 57), is a cushion plant, starred with pink or rose almost stemless flowers. Found on exposed, rocky, or gravelly fields above timberline. The blades of its petals are oblong and slightly notched, and they usually show a little space between them (compare alpine sandwort, which has white, rounded

57. Moss Campion, ½X

petals and shows no space between them). Narrow, pointed leaves 1–2cm long. This is a circumpolar species, found at sea level in the arctic and at progressively higher elevations along the mountains southward into New Mexico.

It is one of the first colonizers of barren gravel areas in the alpine zone. Each mat of stems and leaves, accumulating over many years, eventually becomes a nursery for other alpine plants. In the slow progress of the vegetation pattern toward a climax community, moss campion, which helped to make the beginning possible, is eliminated and is not found in the final stage of the closed sedgegrass community.

SLEEPY CATCHFLY, *S. antirrhina*, is a slender annual plant 3–6dm tall, green and smooth except for dark bands of sticky substance between the upper nodes, which prevent insects such as ants from climbing up the stem to steal nectar from the flowers. Insects obtaining the nectar in this way do not pay for their food by effecting pollination as flying insects do. The petals are white or pink-tipped, usually short, sometimes lacking.

HALLS CATCHFLY, *S. scouleri* ssp. *hallii*, is a tall, branching, sticky, hairy plant with nodding, pinkish flowers, which open in late afternoon. Its calyxes are cylindrical in bud but become inflated at flowering time. After flowering they become erect. In aspen groves and on fields of the montane and subalpine zones.

MENZIES CATCHFLY, *S. menziesii*, has repeatedly a 2-forked, sticky-hairy stem and calyxes less than 6cm long. Occurs throughout our range, often among bushes.

MEXICAN CAMPION or INDIAN PINK, *S. laciniata*, of the pine forests of New Mexico and Arizona, is the showiest species of the genus within our range. It has brilliant scarlet flowers over 2.5cm across, and each petal is slashed into about 4 divisions.

S. douglasii is a catchfly with several slender, sticky stems up to 6dm tall, and white or pale red flowers in clusters of three. Found in the northwestern part of our range, especially in Yellowstone and Glacier National Parks.

(11) *Spergula*. Spurry

Only one species in our range. FIELD SPURRY, *S. arvensis*, has stems 15–60cm tall, branching near the base; leaves 2–5cm long, linear-filiform; cyme loose; flowers on slender pedicels.

(12) *Spergularia*. Sandspurry

1. Stamens 7–10; leaves mostly fascicled in the axils.
    2. Seeds minutely papillose (bearing nipple-shaped projections), not winged.
                        S. rubra
                        Red sandspurry
    2. Seeds virtually smooth, usually winged.
                        S. media
1. Stamens 2–5; leaves not fascicled, some nodes with 1–2 axillary leaves.
                        S. marina

(13) *Stellaria*. Chickweed or Starwort

1. Annuals; leaves ovate, petioled at least on lower ones.
                        S. media

1. Perennials: leaves all sessile, ovate to linear.
   2. Floral bractlets small and scarious.
      3. Petals absent; pedicels subumbellate.

      S. umbellata
      3. Petals present, equaling or surpassing the calyx.
         4. Flowers in a diffuse cyme, on filiform, widely spreading or deflexed pedicels.

         S. longifolia
         Long-leaved starwort
         4. Flowers solitary or few, on erect or ascending pedicels.
         S. longipes
         Longstalked starwort
   2. Floral bractlets foliar, or the pedicels axillary to true leaves.
      3. Petals shorter than the sepals (or absent).
         4. Stems erect; leaves lanceolate.

         S. calycantha
         Northern starwort

         4. Stems prostrate; leaves ovate.

         S. obtusa
      3. Petals equaling or exceeding the sepals.
         4. Glabrous.

         S. crassifolia
         Thick-leaved starwort

         4. Viscid-pubescent.

         S. jamesiana
         James starwort

This genus includes several small plants of confusing similarity. Their white petals are deeply 2-cleft. LONG-LEAVED STARWORT, *S. longifolia*, has linear or lance-linear leaves 2.5cm or more in length, and many flowers in a spreading cluster. The LONG-STALKED STARWORT, *S. longipes* (fig. 58), has shorter, shining leaves and few flowers on erect stalks. Both are found throughout our range, but only the latter extends into the alpine zone. JAMES STARWORT, *S. jamesiana*, a glandular-pubescent plant with notched petals, is found in ravines of the foothill and montane zones.

58. Long-Stalked Starwort, ½X

BUCKWHEAT FAMILY: POLYGONACEAE

This is one of the few families among the dicots that may have flower parts in 3s or 6s. Perianth 2–6-parted; stamens 2–9; stigmas 2–3. The fruit is an achene, usually 3-sided but sometimes flattened. The individual flowers are always small. The family includes many weedy species. In all of ours (except *Eriogonum*), a papery sheath surrounds the stem at the nodes.

1. Papery sheaths present above the swollen joints of the stems; flowers without involucres or umbels.

2. Plants alpine, reddish, under 2.5cm tall.

Koenigia, p. 101
Konigia

2. Plants various but more than 2.5cm tall.
   3. Flowers green or reddish, in terminal panicles.
      4. Leaves reniform (kidney-shaped).

Oxyria, p. 101
Alpine sorrel

      4. Leaves longer than wide; plants stout.

Rumex, p. 104
Dock

   3. Flowers white or pinkish, in axils, terminal spikes, or heads.

Polygonum, p. 102
Knotweed

1. No papery sheaths present; flowers often in umbels; involucres subtend flowers.

Eriogonum, p. 98
Eriogonum

(1) *Eriogonum.* Eriogonum.

Group 1: umbrella form.
1. Bracts foliaceous; umbellate at summit of naked, scape-like peduncle.
   2. Perianth glabrous externally; leaves glabrous to cobwebby above, densely pubescent below.
      3. Flowers bright yellow.

E. umbellatum var. aureum
Sulphur flower

      3. Flowers white to pale yellow.

E. umbellatum var. majus
Subalpine buckwheat

   2. Perianth pubescent externally; leaves pubescent on both surfaces.
      3. Inflorescence compound, 2 or more sets of foliaceous bracts.
         4. Flowers bright yellow.

E. jamesii var. flavescens
James wild buckwheat

         4. Flowers cream-colored.

E. jamesii var. jamesii

      3. Inflorscence not compound, a single set of foliaceous bracts.
         4. Leaves grayish above, white beneath.

E. flavum ssp. flavum

         4. Leaves greenish above.

E. flavum ssp. piperi
Yellow wild buckwheat

1. Bracts not foliaceous.
   2. Leaves linear, about 2.5mm wide, usually glabrous above.

E. exilifolium

   2. Leaves elliptic to oval or ovate, more than 2.5mm wide, usually pubescent above.
      3. Plants alpine or subalpine.

E. ovalifolium var. depressum

      3. Plants from the plains into the montane.

E. ovalifolium var. ovalifolium
Cushion wild buckwheat

Most of these eriogonums form loose or compact mats of woody stems. The undersides, and sometimes the upper sides, of the leaves are cotton-felted. Their flower colors are yellow, cream, white, or pinkish to begin with; but the yellow forms turn orange to reddish or rust as they age, and the cream and white ones turn pink to rusty red. The perianths and the involucres are similar in texture and color, and as the seeds ripen, they lengthen somewhat and become dry, resembling tissue paper.

SULPHUR FLOWER, *E. umbellatum* var. *aureum* (fig. 59), has several stalks, 1–3dm tall, erect, leafless; umbels usually simple; involucres hairy but perianths smooth; flowers bright yellow; leaves usually with white tomentum on their undersides. *E. flavum* is similar but usually has shorter stalks and leaves tomentose on both sides. Its perianths are hairy. Both are found throughout our range, with *E. flavum* the common one above timberline. *E.f.* var. *piperi* is common in Yellowstone and Glacier National Parks.

SUBALPINE BUCKWHEAT, *E. umbellatum* var. *majus*, has erect stalks, usually simple umbels with cream-colored flowers that turn rose, smooth perianths, and leaves green on upper surfaces. It is abundant, forming large patches in the montane and subalpine zones from southern Colorado northward.

CUSHION ERIOGONUM, E. ovalifolium, with its two varieties, is a compact plant with stalks up to 3dm tall or much shorter, depending upon the

59. Sulphur Flower. ½X

altitude. Mat-forming and caespitose; yellow, cream, or pink flowers; smooth perianths; petiolate leaves usually less than 2.5cm long with roundish or elliptic blades. It is common on dry, rocky ground throughout our range. *E.o.* var. *depressum* was called "silver plant" by prospectors and miners who believed that it indicated the presence of silver ore. *E.o.* var. *ovalifolium* may be confused with *E. acaule*, a low mat-forming eriogonum of the high plains with stalks so short that the flowers are sometimes bedded in the leaves.

JAMES BUCKWHEAT, *E. jamesii*, with its two varieties, has cream, pale greenish, or yellow flowers with hairy perianths, stalks 2–3-times branched, and conspicuous leaf-like bracts below each branch. Common in the foothills and montane zones of Colorado and occuring from Wyoming southward through New Mexico and Arizona.

Group 2: solitary stalks 30cm–1m tall.
1. Annuals; stems leafy, white-tomentose; leaves 3–5cm long, alternate; inflorescence an open and much-branched cyme.

> *E. annuum*
> Annual eriogonum

1. Perennials; woody below.
    2. Stems not woody above ground, usually 1 to a plant; inflorescence a branched panicle; leaves mostly basal; plants conspicuously hairy.

> *E. alatum*
> Winged eriogonum

    2. Stems solitary or few; inflorescence spike-like or branched, the involucres racemose; leaves tomentose below, loose tufts of woolly hair above.

> *E. racemosum*
> Redroot eriogonum

60. Winged Erigonum, ½X

WINGED ERIOGONUM, *E. alatum* (fig. 60), is a green plant with elongated basal leaves and tiny yellowish flowers arranged in a large open panicle. When ripe, the 3-winged achenes hang from rust-brown involucres. These tall, sparse plants stand erect, scattered singly over the foothill and montane regions of Colorado and southward.

ANNUAL ERIOGONUM, *E. annuum* (fig. 61), is a whitish plant with ragged patches of tomentum on the erect stem; leaves narrow and often twisted. The inflorescence is an uneven corymb of slightly arching branches with numerous white or pinkish flowers usually along their upper sides. It grows abundantly on sandy areas of the plains and into the foothills from eastern Montana to Mexico.

REDROOT ERIOGONUM, *E. racemosum*. Stalks up to 7.5dm tall, whitish at least on the lower portions which rise from short woody stems. Its leaves, white at least beneath, are mostly basal on petioles as long or longer than the oval, ovate, or oblong blades. Pinkish flowers in narrow, elongated racemes. Found from southwestern Colorado through New Mexico, Arizona, and Utah, especially in ponderosa pine areas.

Group 3: bushy, wiry stems, rarely over 30cm tall.
1. Plants annual; no woody caudex.

> *E. cernuum*
> Nodding wild buckwheat

1. Plants perennial; woody caudex.
    2. Leaves all near the base, some of the petioles 1–3cm long.

> *E. brevicaule*
> Shortstemmed wild
> buckwheat

    2. Leaves on the stems, sessile to short-petiolate.
        3. Leaves linear to narrowly oblong; inflorescence white to rose.

> *E. effusum*
> Bushy eriogonum

        3. Leaves oblanceolate to oblong-lanceolate; inflorescence white to yellow.

> *E. microthecum*
> Small-fruited eriogonum

(2) *Koenigia*. Konigia

The only species in our range, KONIGIA, *Koenigia islandica*, is a tiny arctic annual occurring in a few arctic-alpine situations in Colorado and Wyoming. A rarely seen plant with threadlike stems, stamens 3, styles 2. One of only several annual species found in the alpine zone. Grows at the edges of small pools or in gravel along streams, forming rosettes of reddish leaves with few minute white flowers.

(3) *Oxyria*. Alpine sorrel

Only one species in our range. ALPINE SORREL, *Oxyria digyna* (fig. 62), is a smooth plant 10–30cm tall. Erect stems from thick taproots; leaves mostly basal, long-petiolate, reniform to cordate, entire; elongated panicle of red or greenish-red flowers or fruits; stamens 6; styles 2-parted; achenes with membranous

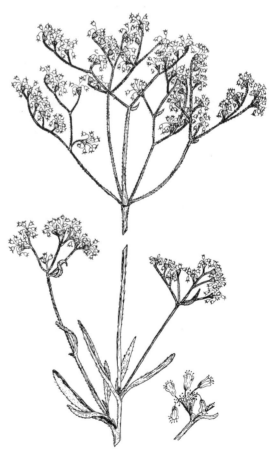

61. Annual eriogonum, ½X

winged fruit

62. Alpine Sorrel, ½X

wings. A plant of cold, wet situations, growing in gravel or around rocks near timberline. Found throughout the arctic and alpine regions of the northern hemisphere. The leaves have a pleasant acid flavor.

(4) *Polygonum.* Knotweed

This is a large and varied genus with many species in the western United States, usually recognized by the enlarged nodes with papery sheaths surrounding the stem. Because of these large nodes, some of the species are called knotweeds. The achenes are lenticular or triangular, often black and shining. This group includes the "smart-weeds," the "sidewalk weeds," and many other small or inconspicuous weedy species.

1. Flowers in terminal spikes, with scarious bracts.
  2. Roots tuberous, or fleshy rhizomes.
    3. Rootstock elongated; spike floriferous throughout.

                             P. bistortoides
                             American bistort
    3. Rootstock cormlike; spike bearing bulblets.

                             P. viviparum
                             Alpine bistort
  2. Roots fibrous; spike usually solitary, short and thick.

                             P. amphibium
                             Water buckwheat
1. Flowers axillary, or racemose with foliar bracts; mostly small annuals.
  2. Plants prostrate-spreading.

                             P. aviculare
                             Doorweed
  2. Plants erect.
    3. Branched throughout, the branches spreading.
      4. Leaves from oblong to oval.

                             P. erectum
                             Erect knotweed
      4. Leaves from oblong to linear.

                             P. ramosissimum
                             Bushy knotweed
    3. Branched only from the base.
      4. Pedicels erect.

                             P. sawatchense
                             Sawatch knotweed
      4. Pedicels reflexed.
        5. Leaves linear to lanceolate or broader.
          6. Achenes 2–2.5mm long; perianth 1.5–2.5mm long.
                             P. douglasii var. austinae
          6. Achenes 2.5–3.5mm long; perianth 3–3.5mm long.

7. Leaves linear to narrowly lanceolate.
P. douglasii var. douglasii
Douglas knotweed

7. Leaves oblong to elliptic.

*P. douglasii var.
latifolium*
Mountain knotweed

5. Leaves mostly linear; achenes 1.5–2.5mm long.
P. engelmannii
Englemann knotweed

AMERICAN BISTORT, *P. bistortoides* (fig. 63), is one of the most common plants in meadows from the montane to the alpine zones. Its compact, 5cm-long head of small white or pink flowers is held on an almost leafless stem 3dm or more tall. On tundra, these conspicuous white heads occupy a layer above most of the other alpine flowers and are usually being swayed by the wind. In lower meadows, they mingle with other lush vegetation.

ALPINE BISTORT, *P. viviparum* (fig. 64), is a smaller, less conspicuous plant of circumpolar distribution. It has erect stems 10–25cm tall with very narrow, spike-like racemes 2.5–10 cm long. On the upper part of the raceme, there are small white or pinkish

63. American Bistort, ½X   64. Alpine Bistort, ½X

flowers, which on the lower part are replaced by bulblets. Found in the alpine zone and occasionally at lower elevations in very cold situations.

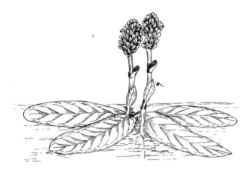

65. Water buckwheat, ⅖X

WATER BUCKWHEAT, *P. amphibium* (fig. 65), has oval, floating leaves, which resemble those of some species of pondweeds, and erect spikes of bright pink flowers 12–25mm long. It grows in ponds, sometimes on their muddy banks, from the plains to the montane zone in the Rockies, and is widely distributed in other parts of the United States.

(5) *Rumex.* Dock

This genus includes several large, coarse species, most conspicuous when in fruit. Their fruiting panicles are usually reddish at maturity, turning rust or dark brown in autumn. Usually the individual achenes are triangular, and each has 2–3 membranous wings.

1. Flowers mostly imperfect (dioecious); inflorescence slender and leafless.
  2. Inner sepals becoming winged; achene granular; leaves hastate; plant with rhizomes.

R. acetosella
Sheep sorrel

  2. Inner sepals not winged; achene smooth; leaves tapered at both ends; thick taproot.

R. paucifolius
Mountain sorrel

1. Flowers mostly perfect; inflorescence leafy and stout.
  2. Valves not granular.
    3. Perennials with horizontal rootstocks or rhizomes.
      4. Axillary shoots usually present; in dry or sandy areas.

R. venosus
Begonia dock

      4. Axillary shoots lacking; in moist meadows and on stream banks.

R. densiflorus

    3. Perennials without rhizomes; stems without axillary shoots.
      4. Single vertical taproot.

R. occidentalis var. occidentalis
Western dock

      4. Clusters of tuberous roots.

R. hymenosepalus
Canaigre

  2. Some of the valves granular.
    3. Low spreading annuals; valves finely toothed.

R. maritimus var. fueginus

    3. Perennials; valves entire or slightly toothed.
      4. Leaf margins crisped; plants without axillary shoots.

R. crispus
Curly dock

      4. Leaf margins not crisped; plants with axillary shoots.

R. salicifolius ssp. triangulivalvis
Willow dock

WESTERN DOCK, *Rumex occidentalis*, is a stout plant that grows from a strong tap-root, with a stem 6dm–2m tall. Its lower leaves are oblong or lance-shaped, sometimes more than 3dm long. The long, narrow panicle is reddish or rust-colored. Frequently seen in wet meadows and on stream banks of the upper montane and subalpine regions.

BEGONIA-DOCK, *R. venosus* (fig. 66), a plant of sandy ground on the high plains and mesas, often forms patches on roadsides. It is very showy in fruit because of its bright rose-colored clusters of papery-winged achenes. Each fruiting calyx may be 2.5–5cm broad; stems 15–40cm tall.

66. Begonia-Dock, ½X

PLUMBAGO FAMILY: PLUMBAGINACEAE

Only one species of this family grows in our range: THRIFT, *Armeria maritima* ssp. *labradorica*. Flowers 5-parted; calyx dry and scarious, funnelform; petals long-clawed, united only at base, usually pinkish; styles 5, distinct, glandular-hairy at the base; scapose perennials; dense tufts of linear basal leaves.

This plant is found in Labrador, Quebec, Newfoundland, and near Hudson's Bay. Yet, it has been collected in central Colorado over 12,000 feet.

GARCINIA FAMILY: CLUSIACEAE.

Only one genus grows in our range. Herbs with opposite, entire leaves, punctate with translucent or dark-colored glandular dots; flowers perfect, 5-parted; sepals imbricate (over-lapping); petals convolute (rolled up lengthwise) and glandular-punctate; styles 2–5; stamens numerous, in three bundles.

1. Plants heavily branched; leaves narrowly oblong; flowers in a flat-topped cluster.
Hypericum perforatum
Klamath weed

1. Plants sparingly branched, slender; leaves broadly elliptic; flowers in axillary cymes.
Hypericum formosum
Western St. Johnswort

WESTERN ST. JOHNSWORT, *Hypericum formosum* (fig. 67), is a smooth plant with 5-petaled, bright yellow flowers about 2.5cm broad with many stamens. The buds are often tinged with red. Stems are 2–6cm tall. Grows in meadows and on moist banks throughout our range. A form with shorter stems and larger flowers is found in alpine meadows of Glacier National Park, where it sometimes forms mats or clumps covered with golden flowers.

67. Western St. Johnswort, ½ X

MALLOW FAMILY: MALVACEAE

This family is represented in our gardens by the familiar hollyhocks and hibiscus. Cotton and okra also belong here. Flowers regular and perfect. Sepals 5, united at base and sometimes a calyx-like involucre outside of them; 5 petals united to each other and to the united stamens, which form a column. The superior ovary, which is inside the stamen column, is made up of several carpels with the lower part of their styles united. The fruit is dry, and the carpels usually separate from each other at maturity.

1. Style branches filiform, longitudinally stigmal, not at all capitate at ends.
    2. Petals cuneate (wedge-shaped), 2–3cm long, crimson when fresh; bractlets of involucel 8–10mm long.
Callirrhoe, p. 107
Poppy mallow
    2. Petals obcordate (heart-shaped, but with the broad, notched end at the apex), less than 2cm long, white, yellow, or rose-purple; bractlets absent or very small.
        3. Leaves merely crenate; annuals.
Malva, p. 107
Mallow
        3. Some or all of the leaves lobed or parted; perennials.
Sidalcea, p. 108
Checkermallow
1. Style branches with capitate stigmas.
    2. Carpels of differentiated parts: basal part more or less reticulated (net-like), upper part smooth, the 2 parts separated by a ventral notch; petals salmon-red or brick-red.
Sphaeralcea, p. 109
Globe mallow
    2. Carpels either completely smooth or completely reticulated; petals white, yellow pink, rose-purple, or red, but never brick-red.
        3. Ovules and seeds 1 per carpel; carpels more or less reticulated; petals white to ochroleucous; plants densely whitish-stellate.
Sida, p. 108
Sida

3. Ovules and seeds 2 or more per carpel; carpels not reticulated; petals never whitish, but variously colored; plants sparingly whitish-stellate or not at all.
    4. Petals much less than 15mm long; involucel none.

Abutilon, p. 107
Indian mallow

    4. Petals large, 22mm long or more, involucel of 3 persistent bractlets.

Iliamna, p. 107
Wild hollyhock

(1) *Abutilon.* Indian mallow

1. Stems stout, erect, branched.

A. theophrastii
Velvetleaf

1. Stems slender, spreading or trailing.

A. parvulum

(2) *Callirrhoe.* Poppymallow

One species in our range. Hirsute; stem branching, trailing on ground; leaves 3–5-parted, covered with stellate hairs; petals scarlet or purplish-red.

C. involucrata
Purple poppymallow

(3) *Iliamna.* Wild hollyhock

MOUNTAIN HOLLYHOCK, *Iliamna rivularis* (fig. 68), is a large plant with clumps of stems 6dm–2m tall. Its leaves are heart-shaped or roundish and deeply 5–7-lobed. The flowers are 2.5–5cm across, white to pale purple. Grows on banks and in meadows on the western slope in Colorado, and from New Mexico, Arizona, and Nevada to Canada. Montane and subalpine. Often seen along roadsides in Grand Teton, Yellowstone, and Glacier National Parks.

(4) *Malva.* Mallow

1. Upper cauline leaves dissected into linear segments; petals 2–3cm long.

M. moschata
Musk mallow

1. Upper leaves shallowly lobed; petals less than 2cm long.
    2. Petals 1.5–2cm long; small calyx bracts ovate or oblong.

M. sylvestris
Common mallow

    2. Petals mostly less than 1.5cm long; small calyx bracts linear.
        3. Carpels rounded and smooth (with some puberulence).

M. neglecta
Dwarf mallow

        3. Carpels flattened and wrinkled on back.

M. parviflora
Alkali mallow

(5) *Sida*. Sida

One species in our range. Leaves reniform or orbicular, wider than long, rounded at the apex; petals about 1cm long, whitish; stems decumbent or prostrate, white-stellate.

S. hederacea
Alkali mallow

(6) *Sidalcea*. Checkermallow

1. Flowers white.

S. candida
White checkermallow

1. Flowers rose-color.
   2. Stems glabrous below.

S. oregana var. oregana
Oregon checkermallow

   2. Stems hirsute-pubescent.

S. neomexicana
Purple checkermallow

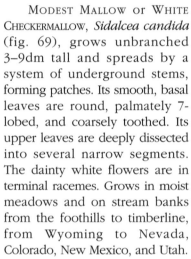

MODEST MALLOW or WHITE CHECKERMALLOW, *Sidalcea candida* (fig. 69), grows unbranched 3–9dm tall and spreads by a system of underground stems, forming patches. Its smooth, basal leaves are round, palmately 7-lobed, and coarsely toothed. Its upper leaves are deeply dissected into several narrow segments. The dainty white flowers are in terminal racemes. Grows in moist meadows and on stream banks from the foothills to timberline, from Wyoming to Nevada, Colorado, New Mexico, and Utah.

PURPLE CHECKERMALLOW, *S. neomexicana*, grows from a woody root with one or more erect stems, which are often branched. It does not spread freely, as *S. candida* does, by underground stems. Its foliage is hairy. The lower leaves are round and scalloped or shallowly lobed, the upper ones deeply dissected. Its range is similar to the last except that it does not extend much above the montane zone.

68. Mountain Hollyhock, ²⁄₅X

(7) *Sphaeralcea*. Globe mallow

1. Leaves greenish, with cuneate or oblong segments; petals copper-scarlet or brick-red.
S. coccinea
Copper mallow
1. Leaves gray-greenish, crenate, ovate-cordate; petals usually scarlet.
S. munroana
Munroe globemallow

69. Modest Mallow, ½X   70. Copper Mallow, ½X

COPPER MALLOW, *Sphaeralcea coccinea* (fig. 70), is a low plant of the roadsides and fields with gray-greenish foliage and brick-red blossoms. Its deeply dissected leaves are covered with small, star-shaped scales or hairs, which give the grayish color. These can be seen with the aid of a hand lens. A taller, gray-greenish species with leaves less dissected, MUNROE GLOBE-MALLOW, *S. munroana*, grows in Montana, Wyoming, and Utah.

VIOLET FAMILY: VIOLACEAE
    This family has simple leaves with petioles and stipules, and irregular flowers with 5 separate petals. Garden pansies and violas belong here. The

plants may be either leafy-stemmed (caulescent) or acaulescent, that is, with all leaves and flower stalks rising from a thick, short stem at ground level. Most are spring-blooming plants, but those found at the higher altitudes may be seen in bloom during the summer months.

1. Sepals auriculate (with ear-like appendages) at base; lower petal spurred.
    Viola, p. 110
    Violet
1. Sepals not auriculate at base; lower petal merely enlarged or concave at base.
    Hybanthus, p. 110
    Green violet

(1) *Hybanthus.* Green violet

One species in our range. Leaves sessile, simple, entire, linear-lanceolate.
    H. verticillatus

(2) *Viola.* Violet

1. Flowers primarily yellow; plants perennial.
  2. Leaves compound and deeply dissected.
    V. sheltonii
    Shelton's violet

  2. Leaves entire, either lobed, crenate, or dentate.
    3. Leaves cordate-reniform to ovate-cordate.
      4. Leaves rounded at apex.
      V. biflora
      Northern yellow violet

      4. Leaves abruptly acute at apex.
      V. glabella
      Stream violet
    3. Leaves not cordate or reniform.
      4. Leaf blades coarsely-lobed or few-toothed, not serrate or dentate; coarsely veined.
      V. purpurea
      Goosefoot violet
      4. Leaf blades entire to finely serrate or dentate; not coarsely-veined.
        5. Leaves narrowly lanceolate to elliptic-lanceolate, mostly under 15mm wide, the base cuneate with long petiole; sandy sites.
        V. nuttallii
        Yellow prairie violet or Nuttall violet
        5. Leaves broadly lanceolate, elliptic, deltoid, or ovate, often more than 15mm wide, the base cuneate to truncate; moist or grassy sites.
          6. Leaves often more than 5cm long, cuneate at base; capsules hairy.
          V. praemorsa
          Upland yellow violet
          6. Leaves usually under 5cm long, truncate to subcordate at base capsules glabrous.
          V. vallicola
          Valley yellow violet
1. Flowers not primarily yellow; either blue, violet, or white.
  2. Flowers blue or violet.
    3. Plants annual; petals sometimes purple-spotted.
      V. rafinesquii
      Field pansy

3. Plants perennial.
 4. Leaves deeply divided or dissected; acaulescent; not stoloniferous.

       V. pedatifida
       Birdfoot violet

 4. Leaves entire to serrate or crenate.
  5. Plants caulescent; tardily stoloniferous; head of style bearded.
   6. Leaves usually pubescent; plants usually more than 5cm tall; mostly below 10,000 feet.

       V. adunca var. adunca
       Early blue violet

   6. Leaves glabrous; plants under 5cm tall; mostly above 10,000 feet.

       V. adunca var. bellidifolia
       Mountain blue violet

  5. Plants acaulescent; not stoloniferous; head of style without beard.
   6. Upper leaf surface minutely hirsute.

       V. selkirkii
       Great-spurred violet

   6. Upper leaf surface glabrous.

       V. nephrophylla
       Northern bog violet

2. Flowers white.
 3. Plants caulescent, flowers axillary.
  4. Leaves usually wider than long, ciliate-margined; stoloniferous.

       V. canadensis var. rugulosa
       Western Canada violet

  4. Leaves longer than wide, not ciliate-margined; stolons lacking.

       V. canadensis var. canadensis
       Eastern Canada violet

 3. Plants acaulescent.
  4. Plants stoloniferous; leaves mostly glabrous.
   5. Flowers 10–13mm long; leaves rounded at apex, 2.5–3.5cm broad, margins crenulate.

       V. palustris
       Marsh violet

   5. Flowers 5–10mm long; leaves slightly pointed at apex, mostly more than 2.5cm broad, deeply crenate.

       V. macloskeyi var. pallens
       White marsh or small white violet

  4. Plants without stolons; leaves more than 3cm broad, pilose beneath.

       V. renifolia
       Kidney-leaved violet

GOOSEFOOT VIOLET, *V. purpurea*. A small plant with its stems partly underground, sometimes only 2.5–7.5cm tall, but may be up to 20cm above ground level. The leaves are variable, from entire to rather deeply lobed, and on their lower surfaces the veins are purplish. Petals yellow, the backs of the upper ones tinged purplish and the lower ones veined with purple. Found on open slopes in pine forests and sometimes above timberline from Montana south to New Mexico and Arizona.

SHELTON'S VIOLET, *V. sheltonii*. The palmately, deeply-lobed leaves of this plant are mostly in a basal tuft and on petioles 2.5–13cm long. Its bright

71. Nuttal Violet, ½X

72. Northern Yellow Violet, ½X

yellow flowers are held slightly above the leaves. Found on meadows and hillsides in rich soil, sometimes in partial shade, from Montana to south-western Colorado and westward in the montane zone.

NUTTALL VIOLET, *V. nuttallii* (fig. 71), appears acaulescent in spring when first coming into bloom, but it later develops several leafy stems 13–50cm tall. Its leaves are 5–10cm long and vary from narrowly lanceolate to elliptic-lanceolate in shape. The yellow petals may be veined with purple or the upper ones tinged with red on the back. The UPLAND YELLOW VIOLET, *V. praemorsa*, is similar. *V. nuttallii* grows on banks and open hillsides throughout our range, especially in sandy soil.

Thomas Nuttall, a naturalist in the early 19th century, was interested in both birds and plants. He was professor of botany at Harvard and collected in Wyoming and other parts of the West.

NORTHERN YELLOW VIOLET, *V. biflora* (fig. 72), is a delicate, leafy-stemmed plant with ascending stems and rounded or kidney-shaped leaves. The peduncles rise from the axils and hold yellow flowers, usually 2, above the leaves. A plant of arctic regions, occurring occasionally in cold, wet situations of the high Colorado mountains. SMOOTH or STREAM YELLOW VIOLET, *V. glabella*, which is similar but larger, occurs in moist mountain woods from Alaska to Montana, northern Idaho, and California. It is found in Glacier National Park, where it blooms during the

73. Canada Violet, ⅖X

summer and often forms large beds near snow banks at timberline.

✔ CANADA VIOLET, *V. canadensis* var. *canadensis* (fig. 73), is a leafy-stemmed species, 15–30cm tall, with broadly heart-shaped leaves that have a slender tip. The flower stalks rise from leaf axils. The petals are white on their faces but more or less purplish-red on the reverse side. A plant of moist, shaded ravines and slopes of the foothill and montane zones. *V. canadensis* var. *rugulosa* has larger flowers with more purple on the backs of the petals.

WHITE MARSH VIOLET, *V. macloskeyi* var. *pallens,* is a smooth plant spreading by slender stems, which creep over wet, mossy banks. The leaves are rather light green, 2.5–5cm long, and usually broader, heart-shaped, round, or kidney-shaped. The petals are white, sometimes finely veined with purple or lightly purple-tinged. Found in wet meadows or forest bogs across the northern part of North America and in the sub-alpine zone of the mountains.

NORTHERN BOG VIOLET, *V. nephrophylla* (fig. 74), is a smooth, acaulescent plant with flower stalks 8–25cm tall. Its leaves are heart-shaped or kidney-shaped, often purplish beneath, with finely toothed margins and sometimes short-pointed tips. The flowers are nearly 2.5cm broad with thick spurs and violet petals, which are paler and purple-veined at the base. A common spring flower of meadows and mountain valleys throughout our range. BIRDFOOT VIOLET, *V. pedatifida*, is a dark purple-flowered, acaulescent species with much dissected leaves, which occurs on the high plains and reaches the lower slopes of the foothills in Colorado and New Mexico.

74. Northern Bog Violet, ½X

SUBALPINE or MOUNTAIN BLUE VIOLET, *V. adunca* var. *bellidifolia* (fig. 75), is a quite compact plant with short stems, small roundish-ovate leaves, and comparatively large bluish-violet flowers. Frequently found along streams, at edges of subalpine meadows, and on moist slopes above timberline, where it is seen

75. Subalpine Blue Violet, ½X

in bloom during the summer. *V. adunca* var. *adunca,* grows taller, with a definitely leafy stem, and occurs throughout the Rockies from moist foothill ravines to timberline.

FRANKENIA FAMILY: FRANKENIACEAE

Only one species occurs in our range. *Frankenia jamesii* is a shrub 20–60cm tall, scabro-puberulent (rough and minutely pubescent); leaves linear, crowded and fascicled in the axils; flowers sessile in the axils and in terminal leafy buds; sepals 5, united in a tube; petals 5, clawed, bearing a crown, white; stamens 6; ovary superior. A salt-tolerant family.

GOURD FAMILY: CUCURBITACEAE

All the squashes, melons, pumpkins, gourds, and cucumbers belong to this family. In general they are plants of warm climates.

1. Flowers solitary, large, yellow; corolla campanulate; fruit smooth; perennial.
>*Cucurbita foetidissima*
>Wild gourd
1. Flowers racemose, small, greenish-white; fruit prickly; annual.
>*Echinocystis lobata*
>Mock cucumber

WILD GOURD, *Cucurbita foetidissima* (fig. 76), is a coarse, trailing plant with large, triangular, grayish leaves, and yellow, bell-shaped flowers from 2.5 to over 5cm long. It grows on roadsides, dry banks, and along railroad embankments. The fruits are mottled dark and light green but turn yellowish after the first frost, which, by killing the leaves, often reveals several of the baseball-like gourds lying on the ground, attached to a withered stem. A plains plant that approaches the eastern mesas and foothills from Colorado south to Mexico.

MOCK CUCUMBER, *Echinocystis lobata*, is an annual plant that climbs by tendrils over shrubs and low trees. Its leaves are thin and palmately-lobed into 3–7 divisions. The light greenish-yellow staminate flowers are small but numerous in clusters from the leaf axils. Pistillate flowers are solitary or few from the same leaf axils. Egg-shaped, pendant fruits are about 5cm long, covered with soft spines, and, when ripe, contain several large, flat, black seeds. Commonly found in the eastern United States and seen along the high plains and foothills from Montana to Texas and Arizona.

76. Wild Gourd, ½X

LOASA FAMILY: LOASACEAE.

The members of this family have leaves covered with stiff, barbed hairs, which cause them to stick tightly to clothing. The leaves are usually pinnately-lobed or dissected, and the stems are often white and shining. The flowers are regular, with 5 calyx lobes, 5 or 10 petals, an inferior ovary, the stamens numbering 10 to many. In some species, some of the filaments become petal-like. Frequently found in sandy soil. Petals and filaments are yellow or whitish. In some species, the flowers open only late in the afternoon. In our range, all belong to the genus *Mentzelia.*

1. Petals 5, 2–6mm long, small; filaments all filiform; seeds not winged.
   2. Leaves petiolate.
                                    M. oligosperma
   2. Leaves sessile.
      3. Seeds tuberculate.
         4. Leaves narrow, entire to sinuate-pinnatified.
                                    M. albicaulis
                                    Small-flowered stickleaf
         4. Leaves broad, merely toothed.
                                    M. dispersa var. latifolia
      3. Seeds smooth or striate.
         4. Stems slender and sparsely branched.
                                    M. dispersa var. dispersa
         4. Stems low, numerously and compactly branched.
                                    M. dispersa var. compacta
1. Petals 5 or 10, large, 8–80mm long; filaments often dilated-petaloid.
   2. Flowers white or merely yellowish.
      3. Seeds margined but not winged.
                                    M. decapetala
                                    Giant eveningstar
      3. Seeds conspicuously winged.
                                      M. nuda
                                    White eveningstar
   2. Flowers yellow.
      3. Seeds margined but not winged.
         4. Stems smooth.
                                    M. chrysantha
         4. Stems puberulent (fine, short down).
                                  M. multiflora
                                  Many-flowered
                                eveningstar
      3. Seeds evidently winged.
         4. Petals 10, large; filaments filiform.
            5. Seeds tuberculate.
                                  M. laevicaulis
             5. Seeds smooth.
                                  M. speciosa
         4. Petals 5, smaller; some of the filaments dilated and antherifrous
            (anther-bearing).
                                  M. pumila
                                  Yellow eveningstar

SMALL-FLOWERED STICKLEAF, *Mentzelia albicaulis*, has small yellow flowers, which bloom in the daytime, and white stems. Most of its leaves are pinnately-lobed. Grows on dry banks of the foothills from Wyoming south to New Mexico. *M. dispersa* is similar and has about the same range. Most of its

77. Many-Flowered Evening-Star, ½X   78. White Evening-Star, ½X

leaves are entire or slightly toothed. The seeds of both species were used for food by the Indians.

MANY-FLOWERED EVENINGSTAR, *M. multiflora* (fig. 77), is a branching plant 30–60cm tall, with bright yellow flowers 4–5cm across, which usually open in the afternoon and close before the sun is high the next forenoon. Grows on dry banks and roadsides of the foothill and montane zones from Wyoming south through our range. YELLOW EVENINGSTAR, *M. pumila*, is similar, but not so tall.

WHITE EVENINGSTAR, *M. nuda* (fig. 78), is a rough-looking, much-branched plant 3–8dm tall, which, in late afternoon, covers itself with cream-colored blossoms 5–10cm broad. It grows on dry, sandy banks of the foothills and mesas, often along highways. If one passes during the middle of the day and sees only sparse, weedy vegetation, one may be amazed on returning at dusk to find great beds of these luminous blossoms. GIANT EVENINGSTAR, *M. decapetala*, is similar but with much larger blossoms. It is less commonly seen, as it grows only on a certain type of shale soil. It can be seen along Woodman Valley Road north of the Garden of the Gods near Colorado Springs.

WILLOW FAMILY: SALICACEAE

This is a family of trees and shrubs that belongs to a group of wind-pollinated plants. Its petalless flowers are in catkins, which botanists call *aments*, with pistillate and staminate catkins usually on different plants. The seeds bear tufts of soft hair. The family contains only two genera, and, in both of them, it is common for leaves on young sprouts to be much larger than those on older branches. Sizes given in the descriptions are for average, normal leaves on second year growth.

1. Winter buds covered by several scales; bracts of the aments (catkins) lacerate; catkins pendulous; flowers with a broad or cup-shaped disk; stamens numerous; trees.

Populus, p. 117
Poplar

1. Winter buds with a single scale; bracts of the aments entire or merely dentate; catkins upright; flowers with small glands, but no disks; stamens few; mostly shrubs, a few species becoming trees.

Salix, p. 117
Willow

(1) *Populus.* Poplar or Cottonwood

1. Petioles flattened laterally.
   2. Leaves suborbicular.

P. tremuloides
Aspen

   2. Leaves broad, more or less deltoid.
      3. Leaves abruptly acuminate, crenately serrate.

*P. deltoides ssp.
monilifera*
Plains cottonwood

      3. Leaves gradually acuminate, deeply sinuate-dentate (wavy and toothed).

P. deltoides ssp. wislizenii
Valley or Rio Grande cottonwood

1. Petioles round or furrowed.
   2. Leaves pale beneath.

P. balsamifera
Balsam poplar

   2. Leaves green, scarcely lighter beneath.
      3. Leaves oblong-lanceolate.

P. angustifolia
Narrowleaf cottonwood

      3. Leaves ovate, abruptly long-acuminate.

P. acuminata
Lanceleaf cottonwood

(2) *Salix.* Willow

Group I:

Stamens 3–8; shrubs or small trees; leaves more than 3 times as long as wide, finely serrulate; scales yellowish; filaments hairy at base; flowers develop with the leaves.

1. Petiole with several coarse glands on upper side, near base of blade; leaves green on both sides.

S. lasiandra var. caudata
Whiplash willow

1. Petiole without obvious glands; leaves pale or glaucous beneath.

S. amygdaloides
Peach-leaf willow

Group II: Stamens 2 (or 1); plants depressed, 2-8cm tall, forming mats near or above timberline.

1. Flowers develop after the leaves; scales pale; leaves reticulate-veined, the apex obtuse or slightly notched.

> *S. reticulata ssp. nivalis*
> Net-veined willow

1. Flowers develop with the leaves; scales dark; leaves pinnate-veined, the apex acute.
   2. Pistillate aments 0.7–2cm long; leaves 0.5–2cm long, green both sides.

> *S. cascadensis*
> Cascade willow

   2. Pistillate aments 2–5cm long; leaves 1–4cm long, glaucous beneath.

> *S. arctica*
> Arctic or Rock willow

Group III: Stamens 2 (or 1); plants ascending to erect; scales generally yellowish; filaments hairy at base; leaves 5–15 times as long as wide; the 2 stigmas lobed and nearly sessile on the ovary.

> *S. exigua*
> Sandbar willow

Group IV: Stamens 2 (or 1); plants ascending to erect; scales brown to blackish; filaments usually glabrous at base; ovaries and capsules glabrous, sometimes sparsely villous toward the tip.

1. Leaves lanceolate to oblanceolate, or oblong-elliptic, apex short-acuminate to acute; capsules glabrous.
   2. Branchlets with a glaucous bloom; pistillate aments 2.5–4cm long, capsules 3–4mm long.

> *S. irrorata*
> Bluestem willow

   2. Branchlets bright chestnut to dark brown, lustrous and glabrous; pistillate aments 1–5cm long; capsule 5–7mm long.

> *S. boothii*
> Booth willow

1. Leaves not lanceolate or elliptic.
   2. Leaves about the same shade of green on both surfaces, silky pubescent to near glabrous on both surfaces; young twigs yellow to orange.

> *S. wolfii var. wolfii*
> Wolf willow

   2. Leaves a lighter shade beneath, glaucous; young twigs yellow to yellowish-green.

> *S. monticola*
> Mountain willow

Group V. Stamens 2 (or 1); plants ascending to erect; scales brown to blackish; filaments mostly glabrous at base; ovaries and capsules densely short-hairy, more sparsely so in age.

1. Twigs of the previous year pruinose (with a coarse waxy covering), especially at the nodes.
   2. Leaves densely white-hairy below, sparsely pubescent above; twigs yellowish; catkins more than twice as long as wide.

> *S. drummondii*
> Blue willow

   2. Leaves white-hairy on both surfaces, becoming glabrous; twigs reddish; catkins less than 2 times longer than wide.

> *S. geyeriana*
> Geyer willow

1. Twigs of previous year not pruinose.
   2. Leaves oblanceolate to narrowly oblong, densely white-tomentose beneath, dark
      green and usually glabrate to thinly tomentose above; new twigs densely
      tomentose, older twigs pubescent.

      S. candida
      Hoary willow
   2. Leaf blades usually broader, usually not densely white-tomentose beneath; twigs
      glabrous to pubescent but not tomentose.
      3. Leaves usually obovate to oblanceolate, widest above the middle; occurs in
         drier areas.

         S. scouleriana
         Scouler willow
      3. Leaves usually elliptic or oblong, widest at or below the middle; occurs in
         moist to wet areas near water.
         4. Twigs appressed hairy, red-purple the first year, later white-streaked in
            appearance; mature buds with depressed margins.

            S. bebbiana
            Beaked willow
         4. Twigs not appressed hairy or red-purple, nor white-streaked later;
            mature bud not with depressed margins; leaves glaucous or definitely
            lighter beneath than above; not pubescent on both surfaces.
            5. Most leaves toothed, dark green and shiny above.

               S. planifolia
               Planeleaf willow
            5. Most leaves entire, evidently pubescent at maturity.
               6. Petioles more than 3mm long; leaves sparsely to moderately
                  pubescent.

                  S. glauca
                  Glaucous willow
               6. Petioles less than 3mm long; leaves densely pubescent.
                  S. brachycarpa
                  Short-fruited or Barren-
                  ground willow

Populus includes most of the deciduous trees of our area. Because
poplars, often called cottonwoods, were the only broad-leaved trees found
growing here by the pioneers, they were the first trees planted in the towns.
Along streams, and where planted and watered, they grow to very large size.
They gave welcome shade to the early settlers, but the female trees are now
considered undesirable because of the abundance of "cotton" that fills the air
as their seeds are dispersed. On a hot day on the plains or in the foothills, a
cottonwood grove is still a welcome sight.

ASPEN, *P. tremuloides* (fig. 79), is a small or medium-sized tree, which
forms groves and thickets along streams and on moist soil. It is the only
deciduous tree commonly seen in the montane and subalpine zones, and the
white or light-greenish bark and small roundish leaves make it easy to
recognize. Because the petiole is flattened oppositely to the leaf blade, the
slightest breeze causes the leaves to quiver, accounting for the name "Quaking
Aspen". Its smooth bark is a favorite food of the beaver. Beavers also use the
small logs and branches in building their dams. Along streams and ponds one
often sees beaver-cut aspen stumps. Elk browse on the young shoots in winter
and damage the bark by biting and rubbing it with their antlers. It is found
from 6,000 to 7,000 feet up to timberline and along the higher mountains from

Mexico to Alaska. It spreads by root suckers and often covers burned areas, where there is sufficient moisture.

NARROWLEAF COTTONWOOD, *P. angustifolia* (fig. 80), is a tree of the canyons and streamsides, extending from the plains up to 10,000 feet throughout the region. The leaves are lance-shaped, tapering to the tips, 5–10cm long. The brown terminal bud usually has 5 scales, and it is very resinous and somewhat fragrant. The bark is smooth and cream-colored except near the base, where it becomes broken by dark furrows into broad ridges. It is a small tree in Colorado and northward, but it grows somewhat larger in New Mexico.

79. Aspen, ½X

LANCELEAF COTTONWOOD, *P. acuminata*, is a small tree of the foothill canyons. It often grows with the narrowleaf cottonwood, and it is sometimes difficult to distinguish the two. Leaves are rhombic-lance shaped to ovate, 5–10cm long, with an abruptly acuminate tip. The terminal bud in winter has 6–7 scales, and it is somewhat sticky but not fragrant.

BALSAM POPLAR, *P. balsamifera*, is a rare tree in the central Rockies but becomes more abundant northwards, extending into Alaska and across Canada and the northeastern United States. In Colorado and southern Wyoming it occurs in small groves on valley bottoms of the montane zone. The leaves are usually ovate with sharp-pointed tips, 7.5–15cm long, dark green above and lighter colored beneath. The young leaves and winter buds are fragrant. It can be recognized when not in leaf by the pointed, sticky, terminal bud, 2.5cm long.

PLAINS COTTONWOOD, *P. deltoides* ssp. *monilifera*, and VALLEY or RIO GRANDE COTTONWOOD, *P. deltoides* ssp. *wizlizenii*, are both large trees, sometimes more than 1m in diameter. They grow along river bottoms and approach the mountains where the canyons open onto the plains. They have thick, dark, furrowed bark except on the young growth, which is light yellowish or pale brown. The leaves are leathery, somewhat triangular in shape, with toothed or wavy margins. *P.d.* ssp. *monilifera* extends along the east side of the mountains from Canada south to central New Mexico. Its leaves usually have more than 10 teeth on each side. *P.d.* ssp. *wizlizenii* occurs from central Colorado southward through New

80. Narrowleaf Cottonwood, ½X

Mexico. The leaves usually have fewer than 10 teeth on each side.

Salix is the largest and most widely distributed genus of woody plants in our region. Thirty-one species are known from Wyoming, thirty from Colorado. They range in size from a few centimeters to small trees, but most are medium-sized shrubs. Some species are "pussy willows," that is, their silvery catkins appear before the leaves unfold. Others put out their catkins either with the young leaves or after the leaves have developed. The male and female catkins are on different plants. Branching is alternate, and the leaves are always simple. Members of this genus may be easily recognized by one character: all willows have only one bud scale. The bud covering is in the form of a little cap, which is pushed off in one piece when the bud starts to enlarge. Buds of other shrubs have two or more scales, which separate when growth starts.

Learning to recognize the different kinds of willows is much more difficult. Many of them are quite similar in general appearance, and identification depends on technical characters. Nearly all of them grow only on moist ground. They border streams from the plains to high mountains and make thickets in wet meadows at all altitudes, especially above timberline. Deer, elk, mountain sheep, and domestic animals feed on their twigs and buds and find shelter in their thickets. Only the most widely distributed species are discussed here, those occurring throughout the Rocky Mountain region.

PEACH-LEAF WILLOW, *S. amygdaloides*, is the only willow native to our region that regularly becomes a tree. Its thin leaves are lanceolate to ovate-lanceolate with long pointed tips, 5–13cm long, and pale bluish on their undersides. The twigs are shining dark orange or red-brown, becoming light orange-brown. Bark on the trunk and large branches is brown, often tinged with red. It grows on moist soil along stream banks of the upper foothills and lower ponderosa pine zones throughout the Rocky Mountains.

BEBB WILLOW, *S. bebbiana*, sometimes called BEAKED WILLOW, is a shrub or small tree, inclined to have a single trunk and bushy top, even when quite small. Its leaves are elliptical to oblanceolate, 2.5–8cm long, hairy when young but smooth and strongly veined when old. The catkins appear with the leaves. It is often found along streams, but it is one of the few species that is also found on dry slopes. It occurs throughout our range at altitudes of 5,000–10,000 feet, and in Arizona up to 11,000 feet.

SCOULER WILLOW, *S. scouleriana*, is a common shrub or small tree. Its blooming precedes the leaves, and the staminate shrubs are often quite showy when covered with the fluffy, pale-yellow catkins. When the yellow-anthered stamens are extended, the catkins appear oval, about 4cm long and 2.5cm wide. They offer a feast to hungry, early-season bees. The leaves are oblanceolate with a blunt apex, green above but pale beneath. Grows throughout our range from 8,000 to 10,000 feet, mainly along streams but also on slopes that dry out in late summer.

BLUESTEM WILLOW, *S. irrorata*, a shrub with dark, straight stems covered with a bluish coating and thickly set with black buds, which push out in early spring into silvery catkins. The leaves are dark green, smooth above, usually paler beneath, about 2.5–8cm long and tapering at both ends (they may be much larger on sprouts in late summer). Grows along foothill and montane streams. Common in Colorado, especially on the eastern slope, from 6,000 to 8,000 feet. It also occurs in Arizona. This is our best native pussy willow.

Another species with bloom on the twigs is GEYER WILLOW, *S. geyeriana*, which has smaller leaves and catkins, and bears its catkins on short leafy stems. In Colorado it occurs at altitudes of 8,000–10,000 feet, extending northward through Montana.

81. Nelson Willow, ½X

82. Barrenground Willow, ½X

YELLOW-TWIGGED or MOUNTAIN WILLOW, *S. monticola*, is the common yellow-colored, very intricately branched shrub of streamsides in meadows and canyons of the montane zone. In early spring it has many small, silvery-white pussies along its crooked yellow twigs. Its leaves are 2.5–5cm long, oblong-lanceolate or oblong-oblanceolate with pointed tips and finely toothed margins. In summer the twigs are yellowish green.

PLANELEAF WILLOW, *S. planifolia*, has a very wide range in altitude and size. From the upper foothill canyons to timberline it will be a shrub 1.5–3m tall; but above timberline it forms large, low patches 3–6dm high or even lower. Its young branches are glossy, dark, reddish brown; its elliptical or obovate leaves are about 2.5–5cm long and 1.5cm wide, green and smooth above, whitish beneath. At high altitudes the catkins precede the leaves; at lower altitudes they emerge together. The former is *S. p.* var. *planifolia*, Nelson Willow (fig. 81), frequently found along the Front Range in Wyoming and Colorado, especially in Rocky Mountain National Park. The latter is *S. p.* var. *monica*, widely distributed throughout the Rockies.

SHORTFRUITED or BARRENGROUND WILLOW, *S. brachycarpa* (fig. 82), grows with the two planeleaf willows, having much the same habit and range throughout our area. But whereas the planeleaf willow has leaves that are smooth and green on the upper surfaces, the barrenground willow has leaves that are gray-hairy above. Its thickets can be recognized by their overall gray-green appearance. Very similar, and difficult to distinguish from it, is GLAUCOUS WILLOW, *S. glauca*. The two often occur together.

The tiny alpine willows found on the tundra are curiosities in this genus. They are closely related to arctic species. Their small stems are woody, but instead of growing upright they creep, forming intricately branched mats. Their twigs are usually 2.5–5cm tall and bear leaves and catkins. NETVEINED or ROCKY MOUNTAIN SNOW WILLOW, *S. reticulata* var. *nivalis*, has very short catkins, 7mm or less, with few flowers. Its leaves are oval or roundish, dark green above, whitish beneath, and very strongly net-veined. It occurs on the highest peaks of northern New Mexico and northward in alpine situations to the arctic.

83. Rock Willow, ²⁄₅ X

ROCK WILLOW, *S. arctica* (fig. 83), has many-flowered catkins 2.5–5cm long; and strongly veined, but not netted, leaves, which are paler beneath, usually somewhat hairy along the margins, and have yellowish petioles and mid-veins. At timberline this is a small shrub, up to 20cm tall, with relatively heavy underground stems. Higher in the tundra, it becomes a completely prostrate miniature.

CASCADE WILLOW, *S. cascadensis*, is a very tiny species of the Uintah Mountains and has been found on Trail Ridge in Rocky Mountain National Park.

SANDBAR WILLOW, *S. exigua*, is a variable species on the plains and in the foothills. Ditch banks and streams are often bordered by a growth of slender-stemmed willow shrubs, either greenish or reddish, 1–4m tall, with very narrow leaves 7.5–13cm long. Widely distributed in the United States. In New Mexico and Arizona, its slender stems are used in basketry.

CAPER FAMILY: CAPPARIDACEAE

1. Stamens 6; capsule stipitate.
    2. Capsule short, rhomboidal, 2-horned, or subspherical.

Cleomella p. 124
Cleomella

    2. Capsule long, oblong or linear.

Cleome p. 124
Bee plant

1. Stamens 8–22; capsule sessile.

Polanisia p. 124
Clammy weed

(1) *Cleome.* Bee plant

1. Flowers pink-purple or near white.
   2. Leaflets lanceolate; capsule 10–30-seeded

                                              C. serrulata
                                              Rocky Mountain bee plant

   2. Leaflets linear; capsule 6–8-seeded.

                                                C. multicaulis

1. Flowers yellow.

                                              C. lutea var. lutea
                                              Yellow bee plant

(2) *Cleomella.* Cleomella

1. Stipe not more than twice as long as the capsule.

                                              C. angustifolia

1. Stipe 3 or more times as long as the capsule.

                                              C. plocasperma

(3) *Polanisia.* Clammy weed

1. Stamens long-exserted; flowers large.

                                              P. dodecandra ssp. trachysperma

1. Stamens barely longer than the petals; flowers smaller.

                                              P. dodecandra ssp. graveolens

84. Rocky Mountain Bee Plant, ½ X

Rocky Mountain Bee Plant, *Cleome serrulata* (fig. 84), varies in height from 3 to 9dm, depending on the amount of moisture available to it. Its leaves are compound, of 3 lanceolate or oblong leaflets. The lower leaves are on long petioles, the upper nearly or quite sessile. The numerous rose or purple (sometimes white) 4-petaled flowers are in racemes that elongate as the flowers open. Buds, flowers, and long-stalked pods are usually present at the same time. The flowers secrete much nectar, and fields of these plants provide very good bee pasture. It is found along roadsides and on overgrazed land of the foothill and lower montane zones throughout the Rocky Mountains. *C. lutea*, a yellow-flowered species with 3-7 leaflets, may be seen in western Colorado up to 7,000 feet..

MUSTARD FAMILY: CRUCIFERAE

Members of this family are easily recognized by their flowers, which have 4 petals arranged in the form of a cross. They usually have 6 stamens, 2 being shorter than the others. The leaves are always alternate, and in several species their enlarged bases clasp the stem (see Plate C25). The foliage is often covered with hairs or tiny scales, which make it rough to the touch or silvery in appearance. Most of our species have white or yellow flowers; a few are purple or pinkish. Species that are commonly yellow or white-flowered will sometimes have pink or purple forms, notably the wallflowers, *Erysimum*.

The inflorescence is a raceme which is crowded at first, but, as the outer—or lower—buds open, it elongates. The petal consists of a broad blade attached to a narrow claw (see Plate D39). The distinctive seedpod is 2-celled, short or long, often flattened, and sometimes constricted between the ripening seeds, which gives it a knobby appearance. These pods, called *siliques* in this family, are necessary for accurate identification and should always be looked for when plants are collected for study. Usually, if the plant has been in bloom for a few days, the raceme will have flowers at the top and pods in various stages of development below, so that the characteristic shape of the pod may be seen (measurements given in keys are for mature pods). This is a large family with many genera and species. Only the most conspicuous can be described below, but the key will include many plants found less frequently.

1. Silique (fruit) dehiscent (the two valves separate from the septum at maturity).
2. Silique borne on a long stipe, over 1cm long.

> *Stanleya* p. 139
> Prince's plume

2. Silique either sessile or very short-stipitate, seldom over 3mm.
3. Silique linear or oblong, at least 3 times longer than broad.
4. Mature silique long-linear (more than 2cm).
5. Silique flattened or compressed parallel to the septum (the partition).
6. Pubescence branched or stellate.

> *Arabis* p. 127
> Rockcress

6. Pubescence simple or absent.

> *Streptanthus* p. 140
> Twistflower

5. Silique terete (cylindrical) or 4-angled, only slightly flattened if at all.
6. Flowers yellow or yellowish-white.
7. Silique 4-angled.
8. Plant roughish-pubescent, usually yellow.

> *Erysimum* p. 133
> Wallflower

8. Plant glabrous, yellowish-white.

> *Conringia* p. 130
> Hare's-ear mustard

7. Silique terete.
8. Middle and upper cauline leaves entire (lower leaves may be pinnatifid).

> *Schoencrambe* p. 138
> Plains mustard

8. Middle and upper cauline leaves hastate or pinnatifid.

> *Sisymbrium* p. 138
> Tumble mustard

6. Flowers not yellow or yellowish-white; silique terete or nearly so.

7. Petals with crisp (wavy or curled) margins; stem succulent.

> *Caulanthus* p. 130
> Wild cabbage

7. Petals flat; stem not fleshy.

> *Thelypodium* p. 140
> Thelypody

4. Mature silique (or silicle) short-linear or oblong (under 2cm).

5. Flowers white or purple.

6. Plants glabrous.

> *Cardamine* p. 129
> Brookcress

6. Plants pubescent.

7. Leaves entire or toothed.

> *Halimolobus* p. 135
> Halimolobus

7. Leaves pinnatifid.

> *Smelowskia* p. 139
> Smelowskia

5. Flowers yellow.

6. Leaves dissected.

> *Descurainia* p. 130
> Tansy mustard

6. Leaves lyrately broad-lobed.

7. Silique distinctly beaked by the persistent style.

> *Brassica* p. 125
> Black mustard

7. Silique beakless.

8. Seeds biseriate (in 2 rows), flat.

> *Barbarea* p. 129
> Wintercress

8. Seeds uniseriate (in 1 row), globose or oblong.

> *Rorippa* p. 137
> Watercress

3. Silique not linear, rarely more than twice as long as broad.

4. Silique of twin cells, each much inflated, subglobose.

> *Physaria* p. 136
> Twinpod

4. Silique globose or oblong, not twin.

5. Silique scarcely flattened, nearly circular in cross section.

6. Flower white; silique pear-shaped.

> *Camelina,* p. 129
> Falseflax

6. Flowers white; silique globose or ellipsoid.

> *Lesquerella* p. 135
> Bladderpod

5. Silique distinctly flattened.

6. Silique flattened parallel to the septum.

7. Silique round in face view.

> *Alyssum* p. 127
> Alyssum

7. Silique longer and narrower.

> *Draba* p. 131
> Draba

6. Silique flattened at right angles to the narrow septum.

7. Silique triangular-obovate or obcordate.
    8. Radical leaves pinnatifid; silique cuneate.
        Capsella p. 129
        Shepherd's purse
    8. Radical leaves entire or merely toothed.
        Thlaspi p. 141
        Wild candytuft
7. Silique elliptic or oval.
    8. Seeds several in each cell.
        Hymenolobus p. 135
        Hymenolobus
    8. Seeds solitary in each cell.
        Lepidium p. 135
        Peppergrass
1. Silique indehiscent (the valves not separating from the septum, even at maturity).
    2. Plants perennial; tall, spreading by horizontal roots; petals white; seeds 1 in each cell.
        Cardaria p. 130
        White top
    2. Plants annual; low; seeds many; petals rose or purplish.
        Chorispora p. 130
        Blue mustard

(1) *Alyssum.* Alyssum

1. Silicles (short siliques) glabrous.
        A. desertorum
1. Silicles stellate-pubescent.
    2. Silicles and leaves finely pubescent.
        A. alyssoides
        Pale alyssum
    2. Silicles and leaves coarsely stellate-pubescent.
        A. minus

(2) *Arabis.* Rock cress

The species of this genus in our area have tall, unbranched stems; clasping leaves, which often have little ears (*auricles*) at their bases; small white, pink or purplish flowers; and long, usually flat, pods. The pods are necessary for accurate identification.

1. Siliques erect or ascending.
    2. Stems hirsute at base; styles present and definite.
        3. Cauline leaves usually glaucous.
            A. glabra
            Tower mustard
        3. Cauline leaves not glaucous.
            4. Petals 3—5mm long, white to yellowish-white.
                A. hirsuta var. pycnocarpa
                Hairy rock cress
            4. Petals 5-9mm long, white to pink.
                A. hirsuta var. glabrata
    2. Stems glabrous to pubescent at base; styles absent or very short; petals 7–10mm long, usually white.
        A. drummondii
         Drummond rock cress

1. Siliques not erect, either spreading or reflexed (hanging).
  2. Siliques uniformly bent down at a sharp angle.
    3. Stem leaves not auriculate (without auricles).

A. holboellii var.
pendulocarpa
Holboell rock cress

    3. Stem leaves auriculate, more or less clasping.
      4. Pedicels abruptly bent rather than curved; basal leaves finely pubescent.
        5. Lower stem coarsely hairy.

A. holboellii var. collinsii

        5. Lower stem finely and uniformly pubescent.

A. holboellii var.
retrofracta

      4. Pedicels more uniformly recurved; basal leaves coarsely pubescent.

A. holboellii var.
pinetorum

  2. Siliques spreading or hanging.
    3. Pedicels and fruit arched downward.
      4. Basal leaves dentate, obtuse; petals pink.

A. fendleri var. fendleri
Fendler rock cress

      4. Basal leaves entire, acute; petals white.

A. fendleri var. spatifolia

    3. Pedicels spreading, slightly up or down, but not arched.

A. divaricarpa
Spreading rock cress

85. Drummond Rockcress, ²⁄₅ X

DRUMMOND ROCK CRESS, *A. drummondii* (fig. 85), has the widest pods in our group of species, 3–4mm wide, 4–10cm long, standing erect, parallel to the stem and to each other. The small, 4-petaled flowers are white or pinkish. The foliage is usually smooth and glaucous. Found on rocky slopes of the montane and subalpine zones throughout our range.

HAIRY ROCK CRESS, *A. hirsuta,* has a pubescent stem and thin, oblong, toothed basal leaves; very slender erect pods; white or pink petals.

HOLBOELL ROCK CRESS, *A. holboellii,* has four forms in our range, the stems varying from 1 to 9dm tall. Sometimes one plant has several stems. Its basal and lower stem leaves are usually densely pubescent; flowers white to purple; pods more or less reflexed, often sharply so.

FENDLER ROCK CRESS, *A. fendleri*. The basal leaves are softly hairy, thin, ovate, or obovate; the flowers are pink or white; the pods hang on slender, arching pedicels. SPREADING ROCK CRESS, *A. divaricarpa*, is similar, but the pedicels are loosely spreading or ascending.

(3) *Barbarea*. Wintercress

1. Style 2–3mm, beaklike; upper stem leaves mostly lobed.

> *B. vulgaris*
> Bitter wintercress

1. Style 0.5–1.5mm, not beaklike; basal leaves with 8–10 lateral lobes.

> *B. orthoceras*
> American watercress

(4) *Camelina*. False flax

1. Stems strongly hirsute-stellate near base.

> *C. microcarpa*
> False flax

1. Stems sparsely hairy to glabrous near base.

> *C. sativa*

FALSE FLAX, *Camelina microcarpa*, has pale yellow petals and long racemes of roundish or pear-shaped pods not over 1 cm long.

(5) *Capsella*. Shepherd's purse

One species in our range. SHEPHERD'S PURSE, *Capsella bursa-pastoris,* is an annual and a common weed of gardens. Plants caulescent, the basal leaves in a rosette, pinnately lobed or dissected; fruit triangular, cuneate at base; petals 1.5–2mm long.

(6) *Cardamine*. Bittercress

1. Leaves entire, broad, somewhat cordate or reniform; stems glabrous to hirsute.

> *C. cordifolia var.*
> *cordifolia*
> Brookcress

1. Leaves pinnate.
    2. Leaves and plant small, 1–2dm tall.

> *C. oligosperma*
> Little western bittercress

    2. Leaves and plant larger, 2–4dm tall.
        3. Leaflets ovate to orbicular.
            4. Leaflets, especially terminal one, 7–11-lobed.

> *C. breweri var. leibergii*

            4. Leaflets usually wavy-margined; if lobed, only 3-5.

> *C. breweri var. breweri*
> Brewer's bittercress

        3. Leaflets linear to elliptic.

> *C. pensylvanica*
> Pennsylvania bittercress

BROOK CRESS, *Cardamine cordifolia* (fig. 86), with bright white 4-petaled flowers, in round-topped racemes, is found along cold mountain streams and around springs of the montane and subalpine zones from Montana and Idaho to New Mexico and Arizona.

(7) *Cardaria*. Whitetop

1. Silicles glabrous, cordate at base.

> *C. draba*
> Whitetop

1. Silicles finely pubescent, more acute than cordate at base.

> *C. pubescens*

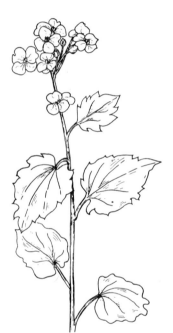

86. Brookcress, ½ X

WHITETOP or WHITEWEED, *Cardaria draba,* is often seen as white patches along roadsides. Its spreading roots make it difficult to eradicate. Stems between 2–7dm tall, with numerous small flowers and tiny, triangular pods.

(8) *Caulanthus*. Wild cabbage

One species in our range. WILD CABBAGE, *Caulanthus crassicaulis,* is a glaucous plant, glabrous below the inflorescence, with a leafy, stout, somewhat inflated stem; sepals greenish purple, 8–12mm long; petals dull purple, 10–14mm long. The pods are 10–13cm long when mature, cylindrical or slightly flattened. Grows on dry soil below 8,000 feet from Wyoming through western Colorado south to Mexico.

(9) *Chorispora*. Blue mustard

One species in North America. BLUE MUSTARD, *Chorispora tenella,* also occurs in patches, often in grain fields. Stems 20–50cm tall, branched, gladular-viscid; petals purple to rose.

(10) *Conringia*. Hare's ear mustard

One species in North America. HARE'S EAR MUSTARD, *Conringia orientalis,* has stems 25–60cm tall; leaves oval to elliptic, cordate and clasping at the base; petals yellowish-white; pods 4-sided, 8–10cm long.

(11) *Descurainia*. Tansy mustard

1. Both upper and lower leaves bipinnate; siliques 20–30mm long.

> *D. sophia*
> Tansy mustard or
> flixweed

1. Upper leaves merely pinnate; siliques less than 20mm long.
    2. Siliques clavate (shaped like a club), widest nearest the apex, tapering to the base; lower leaves usually bipinnate; seeds often in 2 rows, but silique not constricted between seeds.
        3. Fruiting pedicels spreading 65–90 degrees; leaves canescent.

D. pinnata var. halectorum

        3. Fruiting pedicels spreading about 45 degrees; leaves not canescent, but may be pubescent.
            4. Petals 1.5mm long or less; fruiting pedicels 4–6mm long; fruit 4–8mm long.

D. pinnata var. nelsonii

            4. Petals 2–3.5mm long; fruiting pedicels 6–15mm long; fruit 5–15mm long.
                5. Stems glandular and pubescent.

D. pinnata var. branchycarpa

                5. Stems rarely glandular.
                    6. Pedicels 10–15mm long (longer than the fruit).

D. pinnata var. filipes

                    6. Pedicels 6–12mm long (shorter than the fruit).

D. pinnata var. intermedia
Western tansy mustard

    2. Siliques linear or spindle-shaped; lower leaves pinnate; seeds in 1 row.
        3. Pedicels and fruits closely appressed, the pedicels spreading only 5–20 degrees
from the rachis.
            4. Plants canescent.

D. richardsonii var. richardsonii

            4. Plants glabrous to moderately pubescent.

D. richardsonii var. macrosperma

        3. Pedicels and fruit ascending to spreading, the pedicels spreading 40–80 degrees.
            4. Plants glandular-hairy.

D. richardsonii var. viscosa

            4. Plants not glandular-hairy.

D. richardsonii var. sonnei
Mountain tansy mustard

TANSY MUSTARD, *Descurainia sophia,* is a rough-hairy weed found in waste places. Its leaves are twice pinnately dissected into narrow divisions, and its pods are somewhat knobby from being slightly constricted between the seeds.

(12) *Draba*. Draba

There are many kinds of draba. In general they are smaller plants than most members of this family. Their stems are rarely over 15–20cm tall, and some species, especially the alpine ones, are only 2.5–5cm in height. Their stems rise from rosettes of small leaves with racemes of yellow (or white) flowers. Their ovate or lanceolate pods are sometimes twisted, and one often sees old, dry racemes that show the outlines of last year's pods.

1. Annuals (occasionally biennials); petals of yellow or white.
  2. Silique glabrous; flowers yellow (or whitish in drying).
    3. Style absent.
      4.Stems scapose (naked), glabrous.

                                      D. crassifolia var. crassifolia
                                      Thick-leaved draba

      4. Stems scapose, ciliate-hirsute below.

                                        D. albertina
                                      Shiny draba

    3. Style evident.

                                        D. mogollonica

  2. Silique pubescent; stigma sessile.
    3. Flowers white.
      4. Leaves entire; siliques in a terminal cluster.

                                        D. reptans var. reptans
                                      Carolina draba

      4. Leaves toothed or entire; siliques in a somewhat elogated raceme; pedicals ascending or erect.

                                        D. cuneifolia
                                      Wedge-leaved draba

    3. Flowers yellow.
      4. Pedicels longer than the silique.

                                        D. nemorosa
                                      Woods draba

      4. Pedicels shorter than the silique.

                                        D. praealta
                                      Tall draba

1. Perennials; occasionally biennials.
  2. Stems leafless; plants caespitose and matted.
    3. Pubescence of the lower leaf surface mostly 2-rayed (branched).
      4. Pubescence of silicle valves fine.

                                        D. juniperina

      4. Pubescence of silicle valves coarse.

                                        D. oligosperma
                                      Few-seeded draba

    3. Pubescence of leaves branched and densely matted, appearing gray.
                                      D. nivalis
                                      Snow draba

  2. Stems leafy; plants not caespitose or matted (except for *D. exunguiculata*).
    3. Style in fruit mostly less than 1mm long.
      4. Petals white; plants not over 10cm tall.
        5. Style lacking; fruit not over 6mm long.

                                        D. fladnizensis
                                      White arctic draba

        5. Style over 0.2mm long; fruit usually over 6mm long.
          6. Fruit glabrous or sparsley stellate; cauline leaves often absent.
                                      D. nivalis
                                    Snow draba

          6. Fruit densely pubescent; cauline leaves always present.
                                      D. cana
                                    Lance-leaved draba

      4. Petals yellow or lemon yellow; plants often over 10cm tall.
        5. Leaves oblanceolate to broader, over 4mm wide.
          6. Leaves glabrous except for a few cilia.
                                      D. crassa
                                    Thick-leaved draba

          6. Leaves pubescent on both surfaces.
            7. Fruit pubescent.

                                        D. aurea var. aurea
                                      Golden draba

7. Fruit glabrous.

D. stenoloba
Shiny draba
5. Leaves linear to linear-oblanceolate, not over 4mm wide;
leaves ciliate and more or less pubescent on the surfaces.
6. Stems glabrous or sparsely pubescent; fruit 5-14mm long;
plant caespitose.

D. exunguiculata
6. Stems densely pilose (long straight hairs); fruit 4-8mm
long; plant from branched caudex.

D. grayana
3. Style in fruit 1-3.5mm long.
4. Leaves linear-oblanceolate not over 3mm wide; sepals and petals about
equal; stems glabrous or sparsely pubescent.

D. exunguiculata
4. Leaves oblanceolate, lanceolate, or broader, over 3mm wide; sepals at
least 1mm shorter than petals; stems definitely hairy.
5. Style of fruit less than 1.5mm long; fruit pubescent.

D. aurea var.aurea
Golden draba
5. Style of fruit (at least of some) over 1.5mm long; fruit contorted,
glabrous; stems and leaves hairy.

*D.streptocarpa var.
streptocarpa*
Twisted-pod draba

87. Golden Draba, ½X

GOLDEN DRABA, *D. aurea* (fig. 87), usually grows in small clumps of several stems, each 5–15cm tall or rarely taller, and each rising from a rosette of small dark green, spatulate leaves. The stems and leaves are sparsely or densely covered with long hairs. Each raceme has many small, 4-petaled, bright yellow flowers (they become paler as they begin to whither). Found from the montane to the alpine zone, commonly under ponderosa pines in spring, above timberline in early summer, throughout our range. TWISTED-POD DRABA, *D. streptocarpa,* is similar, but its pods are noticeably twisted. SHINY DRABA, *D. stenoloba,* with smooth shining leaves and a long fruiting raceme, occurs in the montane zone.

Several tiny species of draba are found between rocks and on moist ground near and above timberline: THICK-LEAVED DRABA, *D. crassifolia,* which has no leaves on the flower stalks; *D. oligosperma,* which has tiny, narrow leaves on which the hairs grow parallel to the mid-rib, and short pods on which there may be short stiff hairs; and the WHITE ARCTIC DRABA, *D. fladnizensis,* rare on high tundra.

(13) *Erysimum.* Wild Wallflower

This genus includes the most showy members of the yellow-flowered group of mustard relatives. Their flowers are fragrant and their pods 4-angled. The nomenclature below is dated and remains in several cases controversial.

1. Petals generally less than 10mm long; styles rarely more than 1.5mm long.
    2. Pedicels nearly or fully as thick as the fruit; siliques spreading, constricted between the seed; annual.

> *E. repandum*
> Spreading wallflower

    2. Pedicels scarcely half as thick as the fruit; siliques ascending to erect, mostly not constricted between seeds; annual to perennial.
        3. Siliques 1.5–3cm long; petals 3.5–5mm long; sparsely pubescent, mostly greenish; annual.

> *E. cheiranthoides*
> Treacle mustard

        3. Siliques 2.5–5cm long; petals 7–11mm long; densely pubescent, plant grayish; biennial or perennial.

> *E. inconspicuum*
> Small wallflower

1. Petals longer than 11mm; styles longer than 1.5mm.
    2. Plants more or less caespitose; stems under 20cm tall; occurs at high elevations.

> *E. nivale*
> Alpine wallflower

    2. Plants barely caespitose, if at all; stems mostly over 20cm tall; occurs at middle elevations.

> *E. asperum*
> Western wallflower

88. Western Wallflower, ½X

WESTERN WALLFLOWER, *E. asperum* (fig. 88), is a plant 3–6dm tall with bright, orange-yellow, sometimes brownish, 4-petaled flowers in round-topped racemes. Its ripe pods may be ascending or at right angles to the main stalk. Its narrow, grayish leaves may be wavy-margined or have scattered, short teeth. The specific name, *asperum* refers to the harsh pubesence on stem and leaves, which makes it rough to the touch. Blossoming begins in mid-spring on the mesas and lower grasslands. By mid-June it is in full bloom in the mountain parks and open ponderosa pine forests. By the 4th of July, it will be in flower on sunny subalpine slopes. Opinion differs as to whether *E. capitatum* should be separated from *E. asperum*. The plants with deep orange or maroon flowers have been called *E. wheeleri* but they may be either *E. asperum* or *E. capitatum*. Western wallflower is found throughout the Rockies except in northern Montana, where a similar but smaller-flowered species, *E. inconspicuum*, occurs on the eastern slopes in Glacier National Park. It also occurs at low altitudes throughout our entire range.

ALPINE WALLFLOWER, *E. nivale,* is a perennial with several stems up to 20cm tall, a stout woody crown, and lemon yellow (rarely pink or purple) fragrant flowers. It begins to bloom when the stems are still very short. Abundant in high tundra in Colorado and in the high mountains of eastern Utah. Some consider it to be another form of *E. asperum.*

(14) *Halimolobus.* Halimolobus

One species in our range. Twiggy Halimolobus, *H. virgata.* Stems 10–40cm tall, hirsute below, pubescent above; basal leaves narrowly oblanceolate; cauline leaves lanceolate; petals white with narrow claw; sepals 2–3mm long; petals 3–4mm long.

(15) *Hymenolobus.* Hymenolobus

One species in North America. *H. procumbens* is 3-15cm tall; leaves mostly on lower part of stem; racemes without bracts; petals white; silicles sessile, elliptic to elliptic-obovate, 3–3.5mm.

(16) *Lepidium.* Peppergrass

1. Cauline leaves perfoliate (the stem appears to pass through the leaf) or sagittate and clasping at base; annuals.
    2. Petals yellow; upper cauline leaves perfoliate; basal leaves bipinnatifid.
                L. perfoliatum
                Clasping peppergrass
    2. Petals white; upper cauline leaves merely clasping; basal leaves not lobed or merely pinnatifid.
                L. campestre
                Field cress
1. Cauline leaves neither perfoliate nor clasping at base.
    2. Plants perennial.
        3. Silicles ovate-orbicular, sparsely pilose, about 2mm long; long, large basal leaves, entire to dentate.
                L. latifolium
                Broadleaved peppergrass
        3. Silicles ovate to ovate-elliptic, glabrous, 2.5–3mm long; basal leaves pinnate to pinnatifid.
                L. montanum
    2. Plants annual or biennial.
        3. Silicles oblong-obovate to obovate; petals absent or vestigial.
                L. densiflorum
                Prairie peppergrass
        3. Silicles elliptic to oval; petals sometimes conspicuous.
            4. Silicles elliptic, obviously longer than broad.
                L. ramosissimum
                Branched peppergrass
            4. Silicles elliptic-orbicular to orbicular, about as broad as long.
                L. virginicum
                Tall peppergrass

(17) *Lesquerella.* Bladderpod

1. Siliques glabrous.
    2. All leaves narrow, linear or oblanceolate.
                L. fendleri
                Fendler bladderpod

2. Some leaves broad, orbicular to obovate.
   3. Fruiting inflorescence contracted, the pedicels erect or ascending.
                                L. ovalifolia
   3. Fruiting inflorescence elongated, the pedicels somewhat sigmoid (S-shaped).
                                L. pruinosa
1. Siliques stellate-pubescent.
  2. Siliques globose or nearly so, not compressed at apex.
   3. All the leaves narrow, linear or oblanceolate.
                           L. ludoviciana
                           Louisiana or silvery
                           bladderpod
   3. Some of the radical leaves (coming from the root) broad, oval or obovate.
    4. Pedicels uniformly recurved, not straight or sigmoid.
     5. Silique twice as long as the style.
                           L. macrocarpa
     5. Silique as long as the style.
                           L. prostrata
    4. Pedicels sigmoid at maturity.
                           L. rectipes
  2. Siliques ovate or oblong (not globose), often compressed at the apex.
   3. All the leaves narrow, linear or oblanceolate; pedicles sigmoid to straight.
    4. Stems and racemes up to 16cm tall,surpassing the basal leaves.
                           L. alpina var. alpina
                           Alpine bladderpod
    4. Dwarfed, racemes and basal leaf clusters about equal.
                           L. alpina var. condensata
   3. Some of the radical leaves broad, orbicular to obovate; stems 10–20cm long
     pedicels sigmoid.
                           L. montana
                           Mountain bladderpod

MOUNTAIN BLADDERPOD, *L. montana* (fig. 89), has stems 10–20cm long radiating from the crown of the root and turning up at their tips. Its leaves are petioled, oblanceolate or obovate, with entire or irregularly-toothed margins. The flowers are a light yellow on S-shaped pedicels. Found on dry hillsides of the foothill and montane zones from southern Wyoming to northern New Mexico.

(18) *Physaria*. Twinpod

Rosette-type plants with spreading stems, silvery foliage, and comparatively large yellow flowers. They begin blooming when very small and com-

89. Mountain bladderpod, ½X

pact, but the stems elongate later and may become 15–25cm long. They are widely distributed throughout our range on gravelly or sandy foothill slopes or shale banks.

1. Sinuses (the cleft between two lobes) of fruit equal above and below.
　　2. Plants silvery-stellate, densely stellate-canescent.

P. didymocarpa var. didymocarpa
Common twinpod

　　2. Plants white-tomentose, with copious simple pubescence mingled with the long-rayed stellate-canescence.

P. didymocarpa var. lanata

1. Sinuses unequal, the upper deep, the lower shallow or absent.
　　2. Leaves lyrate or fiddle-form.

P. vitulifera

　　2. Leaves oblanceolate, sinuately-toothed.
　　　　3. Cells of silique spherical, small.

P. acutifolia
Alpine twinpod

　　　　3. Cells of silique flattened laterally, large.

P. newberryi

ALPINE TWINPOD, *P. acutifolia* (fig. 90), sometimes called *P. australis*, may be found from 5,000 to 11,000 feet. It has bright yellow flowers and broad leaves, which sometimes have 2 pairs of teeth. Grows in Colorado and Wyoming. *P. newberryi* is found in New Mexico, Arizona, and southern Utah. *P. vitulifera* with fiddle-shaped or runcinate leaves and crinkled pods, occurs on the foothill slopes in Colorado and south-central Wyoming.

90. Twinpod, ½X

(19) *Rorippa*. Yellowcress or Cress

1. Flowers white; plants aquatic perennials.

R. nasturtium-aquaticum
Watercress

1. Flowers yellow.
　　2. Plants perennial by slender rhizomes.
　　　　3. Siliques ovate to oblong, pubescent; plants finely pubescent or papillose.
　　　　　　4. Siliques ovate, 2–5mm long; leaves coarsely-toothed or shallowly-lobed.
　　　　　　　　R. calycina var. calycina

    4. Siliques oblong, 4–7mm long; leaves pinnatifid or lyrate-pinnatifid.

                                        R. calycina var.
                                        columbiae

3. Siliques long, 5–15mm, glabrous; plants mostly glabrous.

    4. Leaves deeply lobed to pinnatifid, the divisions narrow, usually sharply-toothed.

                                        R. sylvestris
                                        Creeping yellowcress

    4. Leaves deeply lobed, the divisions entire to shallowly toothed.

                                        R. sinuata
                                        Spreading yellowcress

2. Plants annual or biennial; no rhizomes.

    3. Stems erect, simple below, branched above.

        4. Hispid on stems.

                                        R. palustris ssp. hispida
                                        Marsh yellowcress

        4. Nearly glabrous throughout.

                                        R. palustris ssp. glabra

    3. Stems branched from the base, low and spreading.

        4. Fruiting raceme normal (bilateral).

            5. Siliques oval to oblong-lanceolate, usually not arching.

                                        R. sphaerocarpa

            5. Siliques linear, usually arching.

                6. Siliques 8–15mm long, curved.

                                      R. curvisiliqua var.
                                      curvisiliqua
                                      Western yellowcress

                6. Siliques 5–7mm long, straight.

                                      R. curvisiliqua var. lyrata

        4. Fruiting raceme unilateral; siliques elongate-ovoid, curved, on curved pedicels.

                                      R. curvipes var. alpina

WATERCRESS, *Rorippa nasturtium-aquaticum,* is frequently segregated as *Nasturtium officinale.* It came from Europe but has become naturalized in cool, running streams throughout the United States. Its stems root at the nodes in shallow water, and its leafy shoots make tangled masses on bright green along stream banks. The 4-petaled white flowers are in racemes, which elongate as the season advances. The peppery young shoots are used in salads, but one should be careful to pick them only in water known to be uncontaminated.

(20) *Schoenocrambe.* Plains mustard

Only one species in North America: *S. linifolia,* which is sometimes placed in the genus *Sisymbrium.* A rhizomatous perennial, the leaves linear, petals yellow; siliques linear, terete, and slightly twisted. Found between 5,000 and 8,000 feet from northwest Wyoming southward.

(21) *Sisymbrium.* Tumble mustard

1. Flowers pale yellow; pedicels stout, nearly as thick as the fruit; siliques widely spreading 5–10cm long; abundant tumbleweed.

                                        S. altissimum
                                        Jim Hill mustard

1. Flowers bright yellow; pedicels slender, not as thick as the fruit; siliques ascending to erect, 2–3.5cm long; less frequent.

S. loeselii
Loesel tumble mustard

JIM HILL MUSTARD, *Sisymbrium altissimum,* is a widely-branched annual plant growing up to 9dm tall, with pinnately dissected leaves and numerous pale yellow flowers, followed by very slender pods. When the Great Northern Railroad was being built across the western states by the tycoon James G. Hill, this plant appeared wherever ground was torn up for track and traveled along behind the advancing railroad. Hence, it became known as the "Jim Hill weed." These dry tumbling weeds scattered their seeds far and wide as winds rolled and bounced them over the high plains, and the freshly disturbed soil of the roadbed provided ideal growing conditions for the young seedlings.

(22) *Smelowskia*. Smelowskia

FERNLEAF CANDYTUFT, *S. calycina* var. *americana* (fig. 92), is the only species of this genus in our range. A small, densely-caespitose perennial with pinnatifid leaves, its caudex is always covered by the bases

92. Fernleaf Candytuft, ½X

of the dead leaves. Its flowering stalks, 5–20cm tall, bear racemes of white to rose flowers. The fruit is 5–12mm long, glabrous, tapering at both ends. Found only above timberline on our highest mountains.

(23) *Stanleya*. Prince's plume

1. Leaves variously pinnatifid.
   2. Plant tomentose or white-villous.

S. tomentosa
Woolly stanleya

   2. Plant glabrous or pubescent, not tomentose.
      3. Flowers pale or cream-color.

S. albescens

     3.Flowers bright yellow.
        4. Leaves mostly bipinnate; siliques twisted.

S. pinnata var. bipinnata

        4. Leaves simple-pinnate; siliques somewhat twisted.

S. pinnata var. pinnata
Golden or bushy prince's plume

1. Leaves entire or nearly so.
   2. Leaves mostly cauline, ovate to ovate-lanceolate; siliques little twisted.
                               S. pinnata var.
                                  integrifolia
   2. Leaves mostly basal; stem leaves reduced, sessile, sagittate-clasping; siliques
      nearly terete.
                                S. viridiflora

GOLDEN PRINCE'S PLUME, *Stanleya pinnata* (fig. 91), is a handsome, gray-leaved plant with long sprays of yellow flowers. The narrow petals, stalked pistils, and long anthers give a fringed appearance to the racemes. Common in desert areas, it occurs in the mountains only on dry foothill slopes below 8,000 feet. This plant usually indicates the presence in the soil of the poisonous mineral selenium, which is absorbed into the plant tissues, making it poisonous to stock. It is said that the Indians of the Southwest used the young leaves as a pot herb after boiling and discarding the first water.

(24) *Streptanthus.* Twistflower

TWISTFLOWER, *Streptanthus cordatus* var. *cordatus,* the only species in our range, is a tall, smooth, glaucous plant, but

91. Golden Princes Plume, ⅖X

with a few short hairs on the basal leaves and at the tip of the sepals. Basal leaves broadly spatulate-obovate, dentate; cauline leaves broadly-oblong, sessile, and cordate-clasping at base. The petals range from yellowish-green to reddish-purple. Easily confused with WILD CABBAGE, *Caulanthus crassicaulis.*

(25) *Thelypodium.* Thelypody

1. Biennials; flowers white or purple.
   2. Stem leaves not clasping, sessile; siliques sharp-pointed, 2–3cm long.
                                T. integrifolium
   2. Stem leaves clasping by sagittate base; slender silique, 4–7cm long.
                                T. sagittatum
                              Slender thelypody
1. Perennials; petals white or purplish; stem leaves sagittate-clasping; siliques 2–3cm long.
                                T. paniculatum

SLENDER THELPODY, *Thelypodium sagittatum* (fig. 93), is a smooth, bluish-green herb with white or purplish flowers about 12mm broad and leaves with little "ears" that clasp the stem. The erect pods are 2.5–5cm long and somewhat knobby. Dry ground of the montane zone in Wyoming, Idaho, and western Colorado.

(26) *Thlaspi.* Wild candytuft

1. Annuals; stems 30cm tall or more, usually branched; silicles orbicular, at least 8mm wide.

> *T. arvense*
> Field pennycress

1. Perennials; stems under 30mm tall; silicles elliptic to nearly oblong, not over 6mm wide.
    2. Petals small, 2.5–3mm generally; sepals 1–1.5mm.

> *T. parviflorum*
> Smallflowered pennycress

    2. Petals 4–6mm; sepals 1.5–4mm.

> *T. montanum var. montanum*
> Mountain candytuft

MOUNTAIN CANDYTUFT, *Thlaspi montanum* (fig. 94), has short stems from rosettes of grayish or dark green, oval or spatulate leaves, often growing in little clumps. It begins to bloom when its stalks are only 2.5–5cm tall, increasing in height as the season advances, to 15–20cm. Each stalk holds a raceme of white flowers. The pods are obcordate. It grows from the foothills, where it is one of the earliest spring flowers in ravines and other moist locations, to the alpine tundra throughout the Rocky Mountains. PENNYCRESS, *T. arvense,* is an introduced weed with smaller flowers and larger, flat, round pods. It is found around settlements.

HEATH FAMILY: ERICACEAE

In the original edition, this family included plants that are here removed to the two succeeding families, Pyrolaceae and Monotropaceae. The segregation eliminates a variety of artificialites that overloaded the lumped family.

93. Thelypody, ½X

1. Fruit a dry capsule.
    2. Anther cells each tipped with a recurved awn (bristle-like appendage); leaves opposite.

                Cassiope p. 143
                Mountain heath

    2. Anther cells not appendaged; leaves alternate.
        3. Corolla of united petals.
            4. Leaves linear, obtuse.

                Phyllodoce p. 144
                Mountain heath

            4. Leaves lanceolate to elliptic.
                5. Corolla urn-shaped.

                Menziesia p. 144
                Menziesia

                5. Corolla saucer-shaped.

                Kalmia p. 144
                Kalmia or laurel

        3. Corolla of separate petals.

                Ledum p. 144
                Labrador tea

1. Fruit more or less fleshy.
    2. Fruit a capsule enclosed in a fleshy calyx; ovary superior.
        3. Depressed undershrub, mostly prostrate; anthers sometimes 2-awned; evergreen.

                Gaultheria p. 143
                Wintergreen

        3. True shrub with erect branches; anthers not awned; our species deciduous.
                Rhododendron p. 145
                Rhododendron

    2. Fruit a berry or drupe, not enclosed in the calyx; anthers usually awned.
        3. Ovary inferior; leaves deciduous.

                Vaccinium p. 145
                Whortleberry,
                Grouseberry,
                Bilberry, Blueberry

        3. Ovary superior; evergreen.

                Arctostaphylos p. 143
                Kinnikinnick

94. Mountain Candytuft, ½X   95. Kinnikinnik, ⅖X

(1) *Arctostaphylos*. Kinnikinnick

1. Leaves withering but persistent; flowers 1–3 in leaf axils; berries red.

    A. alpina var. rubra
    Alpine bearberry

1. Leaves truly evergreen; flowers racemose or paniculate.
    2. Plants prostrate, forming mats, rarely more than 2dm tall; leaves obovate, 1.5–3.5cm.
        3. Leaves rounded (may have a small terminal notch), oblong to obovate to spatulate; berry bright red.
            4. Branchlets more or less viscid.
                5. Branchlets somewhat viscid, minutely tomentulose, soon losing all the hairs.

    A. uva-ursi ssp. uva-ursi
    Kinnikinnick

                5. Branchlets viscid-villous, with pronounced black glands.

    A. uva-ursi ssp.
    adanotricha

            4. Branchlets not at all viscid, with dense canescent tomentum.

    A. uva-ursi ssp. stipitata

        3. Leaves abruptly mucronate (ending in a sharp point), oblong to spatulate; berries brownish-red.

    A. nevadensis
    Pinemat manzanita

    2. Plants erect or somewhat spreading, 8dm or more tall; leaves broad oval; berry brownish.

    A. patula
    Green-leafed manzanita

KINNIKINNICK, *Arctostaphylos uva-ursi* (fig. 95), is probably the commonest evergreen plant, aside from members of the Pine Family, found in the entire Rocky Mountain region. Its leaves are firm, bright green, and shiny. It spreads over rocky or gravelly soil, forming big or little carpets 10–15cm thick, and is important in the soil-building process. Its flowers are dainty pink and white urns on curved pedicels, which are followed in autumn by scarlet berries. Both leaves and berries provide important food for wildlife and were also used by the Indians. A related species, GREENLEAF or UTAH MANZANITA, *A. patula,* is a shrub with smooth reddish branches growing in western Colorado on the Uncompahgre Plateau and in the mountains of Utah.

(2) *Cassiope*. Mountain heath

Only one species is likely to be found in our range. WHITE MOUNTAIN HEATH, *Cassiope mertensiana* var. *gracilis,* is a low shrub with very small, 4-ranked leaves and pure white, urn-shaped corollas, set off by red sepals and pedicels. Found in the northern Rockies and in southern Montana and northern Idaho.

(3) *Gaultheria*. Wintergreen

One species is found in our range. WESTERN CREEPING WINTERGREEN, *Gaultheria humifusa,* is a low plant that embeds itself in moss on moist ground. It is rarely noticed except when its bright red edible berries ripen in late summer. Mat-forming; leaves 1–2cm long, orbicular or oval, entire to crenulate or serrulate; corolla 3–5mm long, white.

(4) *Kalmia*. Kalmia or Laurel

BOG KALMIA, *Kalmia microphylla* var. *microphylla* (fig. 96), is the only species in our range, a diminutive relative of the eastern mountain laurel. Rose-colored flowers hang from erect pedicels and its opposite leaves appear narrow because their margins are involute. Shrubs to 20cm high.

(5) *Ledum*. Labrador tea

TRAPPERS TEA, *L. glandulosum* var. *glandulosum,* the only species in our range, is related to the more northern Labrador tea. It is an evergreen shrub with fragrant leaves, ovate to elliptic, leathery, entire, often revolute, and white beneath. Flowers are white, in terminal racemes or corymbs.

(6) *Menziesia*. Menziesia

96. Bog Kalmia, ½ X

One species occurs in our range: RUSTY-LEAF MENZIESIA, *Menziesia ferruginea* var. *glabella*. A deciduous shrub, 1–2m high; leaves ovate-elliptic, rounded at apex. Flowers appear with the leaves, usually 4-merous, pinkish to reddish-yellow, urn-shaped. The fruits are dry capsules, accounting for one popular name, fool's huckleberry. Found in the northern portion of our range, especially in Yellowstone National Park. The genus was named in honor of Archibald Menzies, the surgeon and naturalist who accompanied Vancouver on his exploration of the Puget Sound area in 1790 to 1795, one of the first biologists to collect in that area.

(7) *Phyllodoce*. Mountain heath

1. Corolla greenish-white, dirty-yellowish, or pink to rose; dwarf evergreen alpine shrubs, mat-forming.
    2. Corolla more than twice as long as calyx, pink to rose, glabrous outside, lobes strongly recurved.
        P. empetriformis
        Pink mountain heath
    2. Corolla narrowly urn-shaped, not twice as long as calyx, greenish-white to dirty-yellowish, strongly glandular-pubescent outside, lobes spreading.
        P. glanduliflora
        Cream mountain heath
1. Corolla pale pink, narrowly campanulate, sparsely glandular-pubescent outside. A hybrid of the two previous species, found only with both parents nearby.
        X P. intermedia
        Hybrid mountain heath

The MOUNTAIN HEATHS are low, evergreen shrubs with small, crowded leaves, which, when the plant is not in bloom, might suggest a juniper or spruce relative. They cover large areas with mats of densely branched stems 30cm or less in height.

(8) *Rhododendron*. Rhododendron

WHITE RHODODENDRON or ROCKY MOUNTAIN AZALEA, *R. albiflorum,* occurs from the Canadian Rockies south into Oregon, but it is also found in one area of moist, rich forest in northern Colorado. A loosely-growing shrub with deciduous, pale green, thin leaves, and saucer or bell-shaped white flowers 10–15mm long, solitary or in pairs from lateral buds.

(9) *Vaccinium*. Blueberry, Bilberry, Whortleberry, Grouseberry

1. Flowers 1 or more per leaf axil, rising directly from a bud on a twig from the previous year; leaves and twigs glabrous; plant 2–6dm tall; corolla pinkish, petals united.
             V. occidentale
             Western huckleberry
1. Flowers usually 1 per leaf axil on twigs of the year.
 2. Plants low and more or less matted, 1–3dm tall.
  3. Much-branched, broom-like; leaves 8–15mm; berry bright red, 3–5mm thick.
           V. scoparium
           Grouseberry or
           whortleberry
  3. Few branches, not broom-like; leaves 10–30mm; berry dark red to bluish, 5–8mm thick.
           V. myrtillus ssp. oreophilum
           Dwarf bilberry
 2. Plants often more than 4dm tall; berry dark blue to blackish.
  3. Corolla narrowly urn-shaped, 5–6mm tall; leaves strongly reticulate beneath, not glaucous, mostly 1–3cm, teeth of leaves gradually disappearing toward the base; mat-forming.
           V. caespitosum
           Dwarf bilberry
  3. Plants 4–20dm tall; leaves sharply serrulate full length; shrubs, not mat-forming.
   4. Leaves oblong-obovate, usually rounded, strongly glaucous.
           V. globulare
           Globe huckleberry
   4. Leaves ovate, ovate-oblong, or elliptic-obovate, usually long-pointed or acuminate, not strongly glaucous.
           V. membranaceum
           Tall whortleberry

DWARF BILBERRY or BLUEBERRY, *V. caespitosum* (fig. 97), has brown, round stems from 5–25cm tall, often tufted. It is usually found in timberline regions, although in Glacier National Park it is common at low altitudes and may reach a height of 4dm or more. MYRTLE-LEAF or DWARF BILBERRY, *V. myrtillus,* is similar but usually slightly taller and not so densely tufted. It grows among rocks in the subalpine zone and, in good seasons, bears quantities of delicious bluish-black berries.

GROUSEBERRY or RED WHORTLEBERRY, *V. scoparium,* has green, angled stems, thin bright green leaves, and small red berries. In many

97. Dwarf Blueberry, ½X

98. Pipsissewa, ½X   99. Wood Nymph, ⅖X

areas it is almost the only plant growing under coniferous trees, especially in engelman spruce forests, where it makes a lacy ground cover.

V. scoparium, V. myrtillus, and V. caespitosum occur throughout our range. From northern Wyoming northward, in moist areas of the foothills and montane, the TALL WHORTLEBERRY, V. membranaceum, occurs, growing 6dm–2m tall and bearing delicious black fruits nearly 12mm in diameter.

WINTERGREEN FAMILY: PYROLACEAE
1. Stems 1-flowered; leaves whorled near the base.

        Moneses p. 147
        Wood nymph

1. Stems more than 1-flowered.
    2. Stems leafy (though short); flowers corymbose; styles very short or lacking.

        Chimaphila p. 146
        Pipsissewa

    2. Stems leafy at base only; flowers in elongated racemes; styles (in most species) over 3mm long.

        Pyrola p. 147
        Wintergreen

(1) *Chimaphila*. Pipsissewa

1. Flowers 1–3; peduncles 2–5cm; filaments hairy.

        C. menziesii
        Little pipsissewa

1. Flowers 5–15; peduncles 5–10cm; filaments ciliate.

        C. umbellata var. occidentalis
        Common pipsissewa

Pipsissewa, *Chimaphila umbellata* (fig. 98). The dark green, stiff, and glossy leaves are in whorls on a stem about 10–15cm long. From this, the flower stalk rises another 5–10 (15)cm, holding several round, pink buds or open flowers. A delightful find on a walk through coniferous forests of the Rockies.

(2) *Moneses*. Wood nymph

One species is found in North America. Wood Nymph, *Moneses uniflora* ssp. *uniflora* (fig. 99), is one of the most charming flowers of the spruce forests. Petals 5, widely spreading, orbicular; the white or rose-tinged flower head hangs down; 1-flowered scape 5–10cm high; cluster of roundish and serrulate thin leaves at base. Found in moist woods of the montane and subalpine zones throughout the Rockies and in boreal forests around the world, usually on rotting wood.

100. One-Sided Wintergreen, ½X

(3) *Pyrola*. Wintergreen

1. Style and stamens curved; anthers apiculate or with short awns.
   2. Flowers white or greenish; plants of dry ground.
      3. Leaf-blades shorter than the petioles, leathery.
         4. Leaves orbicular, not mottled.

            P. chlorantha
            Green-flowered pyrola

         4. Leaves ovate to oblong, mottled or blotched.

            P. picta
            White-vein pyrola

      3. Leaf-blades longer than the petioles, membranous.

            P. elliptica
            White pyrola

   2. Flowers pink or purple, plants in wet ground.

            P. asarifolia
            Bog wintergreen

1. Style straight; anthers connivent (converging) and without projections or awns.
   2. Racemes usually 1-sided; style 3–4mm long; petals white or greenish-white.

            P. secunda
            One-sided wintergreen or
            sidebells

   2. Racemes not 1-sided; style 1–2mm long; petals pinkish.

            P. minor
            Lesser wintergreen

One-sided Wintergreen or Sidebells, *P. secunda* (fig. 100), often occurs in small beds. Its leaves are ovate, and its white or greenish flowers are along one side of the 10–20cm stalk, which is often slightly curved. It sometimes suggests a lily-of-the-valley, to which it is in no way related. Two other species have white or greenish flowers: White-vein Pyrola or Spotted Shinleaf, *P. picta*, occurs in

Yellowstone and Glacier National Parks and throughout the northern part of our range; LESSER WINTERGREEN, *P. minor,* is found in boreal woods around the world.

PINK FLOWERED PYROLA or BOG WINTERGREEN, *P. asarifolia* (fig. 101), has round leaves and a stalk up to 30cm tall, which holds a raceme of several to many pink flowers. Grows around forest springs, along stream banks, and in bogs.

INDIAN-PIPE FAMILY: MONOTROPACEAE

1. Petals united, persistent; plants without chlorophyll.

> *Pterospora andromedea*
> Pinedrops

1. Petals distinct, deciduous; flowers solitary or racemose; plants without cholrophyll.

> *Hypopitys monotropa*
> Fringed pinesap

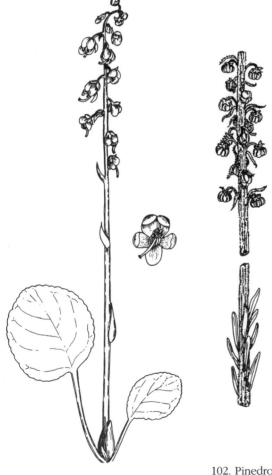

PINEDROPS, *Ptero- -spora andromeda* (fig. 102), is an interesting plant found in dry coniferous woods, often under lodgepole pines. It is a *saprophyte,* a plant that derives its food from dead vegetable matter such as rotting wood. In July it sends up stout, brownish-pink hairy stems, which may be 2.5dm–1m tall. A series of bell-like flowers with whitish petals hangs from each stalk. The entire plant becomes rusty-brown as it matures, and the seed capsules are persistent. These dry, brown stalks are often seen in winter standing up from the pine-needle carpet. PINESAP, *Hypopitys monotropa,* is another leafless plant, yellowish or reddish in color, and shorter and stouter than PINEDROPS.

101. Pink-Flowered Pyrola, ½X

102. Pinedrops, ½X

PRIMROSE FAMILY: PRIMULACEAE

Our plants of this family have united corollas made up of a *tube* and a *limb* (see Plate E46). The place where the tube expands into the limb is the *throat*. Five stamens are attached inside the tube, opposite the lobes of the corolla, and there is 1 pistil. In SHOOTING STAR the tube is very short. The 5-merous flowers distinguish this family from the 4-merous Evening-primrose Family.

1. Corolla present.
    2. Plants scapose with basal leaves.
        3. Corolla lobes erect or spreading; stamens included, distinct.
            4. Corolla tube equaling or exceeding the calyx.
                5. Corolla lobes obcordate or emarginate; capsule many-seeded.
                        Primula p. 152
                        Primrose or cowslip
                5. Corolla lobes entire; capsule few-seeded (1–2).
                        Douglasia p. 151
                        Douglasia
            4. Corolla tube shorter than the calyx.
                        Androsace p. 149
                        Rock jasmine
        3. Corolla lobes reflexed; stamens exserted, connivent in a cone and somewhat monodelphous (united by their filaments).
                        Dodecatheon, p. 150
                        Shooting star
    2. Leafy-stemmed plants; flowers axillary, solitary or in spikes.
                        Lysimachia p. 151
                        Loosestrife
1. Corolla absent; calyx petaloid; leaves opposite.
                        Glaux p. 151
                        Sea milkwort

(1) *Androsace*. Rock jasmine

1. Caespitose perennials; umbels subcapitate; leaves dimorphic.
                        A. chamaejasme ssp. carinata
                        Alpine rock jasmine
1. Annuals (sometimes short-lived perennials); leaves nearly alike; umbels never capitate.
    2. Sepals broadly triangular, each 3-nerved; calyx tube not 5-angled.
                        A. filiformis
                        Slender-stem fairy-candelabra
    2. Sepals narrowly triangular to awl-shaped, not 3-nerved; calyx tube 5-angled.
        3. Involucre bracts ovate to lanceolate-ovate.
                        A. occidentalis
                        Western fairy-candelabra
        3. Involucre bracts narrowly lanceolate to awl-shaped.
            4. Plants usually with 1 well-developed scape; pedicels with glandular hairs.
                        A. septentrionalis var. glandulosa
            4. Plants usually with several to many well-developed scapes; pedicels and scapes glabrous or hairy, not glandular.
                5. Calyx lobes 1/4 to 1/3 the length of the tube; scapes 7–25cm tall.
                        A. septentrionalis var. sublifera
                        Northern rock jasmine

5. Calyx lobes 1/3 to 1/2 the length of the tube; scapes 1–10cm tall.
   6. Plant mostly glabrous, with minute puberulence at base of
      scape and on leaves.

*A. septentrionalis var.
subumbellata*

   6. Plant puberulent all over.

*A. septentrionalis var.
puberulenta*

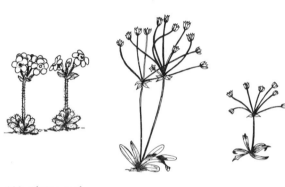

103. Alpine rock
Jasmin, 4X

104. Northern Rock Jasmine, ²⁄₅X

The ROCK JASMINE
are diminutive prim-
rose relatives, white-
flowered but incon-
spicuous   plants.
ALPINE ROCK JASMINE,
*Androsace chamae-
jasme* ssp. *carinata*
(fig. 103), is a tiny,
cusion-type alpine,
5–7.5cm tall, with sil-
very hairs on stem
and leaves and a clus-
ter of fragrant flowers
that might suggest a
forget-me-not. The creamy white corolla has a tiny yellow eye, which turns
pink as the flower ages.

NORTHERN ROCK JASMINE, *A. septentrionalis* (fig. 104), is an annual with
four varieties in our range. At high altitudes the stems will be shorter and the
plants more compact. At lower altitudes they become diffusely and umbel-
lately branched.

(2) *Dodecatheon*. Shooting star

1. Capsule circumscissile; filaments usually less than 1mm long, free or nearly so.

*D. conjungens var.
conjugens*
Slimpod shooting star

1. Capsule dehiscing by valves at the apex; filaments usually longer than 1.5mm.
   2. Staminal tubes reddish-purple to black.

D. pulchellum var.
monanthum

   2. Staminal tube yellow.
      3. Plant dwarfed, 2–5cm tall; 1–2 flowered.

D. pulchellum var.
watsonii

      3. Plants mostly over 6cm tall; mostly several-flowered.

D. pulchellum var.
pulchellum
Western shooting star

WESTERN   SHOOTING
STAR, *Dodecatheon pulchellum* (fig. 105), is easily recognized, with its clusters of
bright pink flowers, the
corolla lobes turned back
from a point of dark-col-
ored anthers, held on slender, leafless stalks above
basal rosettes of smooth,
bright   green   leaves.
Blooms   along   mossy
stream banks in the spring
and in wet montane and
subalpine meadows in
early summer. In some
areas, such as South Park,
Colorado, they are so
abundant that whole mead-
ows are colored pink. The
turned-back petals suggest
their relationship to the
genus *Cyclamen,* also in
this family.

105. Western Shooting Star, ½X

(3) *Douglasia.* Douglasia

1. Flowers solitary; leaves dark green, awl-shaped, imbricated on the shoots.

D. montana var. montana

1. Flowers usually 2–3; leaves pale, closely rosulate on the crowns of the caudex.

D. montana var. biflora

DOUGLASIA, *Douglasia montana*, is a matted, moss-like plant with small, crowded leaves and short-stemmed, phlox-like, lilac or rose flowers. It is found on dry hills and mountain tops in Montana and Idaho.

(4) *Glaux.* Sea milkwort

One species occurs in North America: *G. maritima* Flowers are small, apetalous, single, sessile in leaf axils. Plant rhizomatous; leaves sessile, oval to oblong or oblanceolate; calyx white or pinkish. Found in subsaline soil.

(5) *Lysimachia.* Loosestrife

1. Flowers solitary in the axils; leaves petiolate, the petioles ciliate.

L. ciliate
Fringed loosestrife

1. Flowers in short axillary racemes; leaves sessile.

L. thyrsiflora
Tufted loosestrife

TUFTED LOOSESTRIFE, *Lysimachia thrysiflora*, is a plant of wet ground with small yellow flowers in crowded tufts.

(6) *Primula.* Primrose or cowslip

1. Lower surface of leaves whitish and farinose (mealy); corolla lobes each deeply cleft.

P. incana
Bird's eye primrose

1. No part of the plant farinose; corolla lobes only notched at apex.
  2. Plant over 10cm tall; 3 or more-flowered.

P. parryi
Parry primrose

  2. Plant shorter; flowers solitary, sometimes 2 on a scape.

P. angustifolia
Fairy primrose

106. Fairy Primrose, ½X

FAIRY PRIMROSE, *Primula angustifolia* (fig. 106), is a dainty alpine with stems 5–7.5cm tall, and rose-colored flowers with bright yellow eyes 12–24mm across. It grows singly or in clumps in very exposed alpine locations. Sometimes on mountaintop boulder fields, this plant takes advantage of a sheltered nook between rocks.

PARRY PRIMROSE, *P. parryi* (fig. 107), which has similar but larger flowers, is a much taller plant, up to 4dm in height. The rather stout stalks hold umbels of bright rose, yellow-centered flowers. It is usually found along subalpine and alpine streams, often growing in the edge of the water, though it also occurs in bogs. The flowers and foliage are disagreeably strong-scented. Found from southern Montana and Idaho southward into the mountains of northern New Mexico and Arizona. Frequently seen at high altitudes in Rocky Mountain National Park.

BIRD'S EYE PRIMROSE, *P. incana,* a smaller plant of wet meadows in the foothill and montane regions, has crowded umbels of lilac flowers and white-powdered leaves.

HYDRANGEA FAMILY: HYDRANGEACEAE
1. Shrubs; stamens more than 10; ovary inferior.

Philadelphus p. 154
Syringa or mock orange

1. Shrubs; stamens 10 or fewer; ovary partly inferior.
  2. Cymes many flowered; leaves serrated.

Jamesia p. 153
Cliffbush

  2. Cymes 1–3-flowered; leaves entire.

Fendlera p. 153
Fendlerbush

(1) *Fendlera*. Fendlerbush

Only 1 species occurs in our range. FENDLERBUSH, *Fendlera rupicola* (fig. 108), is a shrub with opposite, rigid branches and pale bark. Its leaves are narrow and grayish. Except when in bloom, it attracts little attention. But when covered with paper-white, 4-petaled blossoms and rose-tinged, squarish buds, it is a handsome sight. It grows from southwestern Colorado to Texas and Arizona on rocky slopes in canyons and on mesas. Particularly abundant and beautiful along the cliffs and canyon rims in Mesa Verde National Park, usually blooming in late May. The genus was named in honor of Augustus Fendler, a German botanist who came to Houston in 1839, working there as a market gardener for a time. He later became an important plant collector in the Southwest, sending his collections to Asa Gray at Harvard. Dr. Gray named several of the new species for the collector.

107. Parry Primrose, ½ X

(2) *Jamesia*. Cliffbush

One species grows in North America. CLIFFBUSH or WAXFLOWER, *Jamesia americana* (fig. 109), is a shrub with clusters of 5-petaled creamy-white flowers. Its buds are sometimes faintly tinged with pink. The velvety and beautifully-veined leaves turn to shades of rose and dark red at the end of the growing season. Leaves ovate, serrate, green and glabrate above,

fruit

108. Fendlerbush, ½ X

109. Jamesia, ½X

white-canescent beneath and on the petioles. In autumn, a patch of red foliage high up on the side of a rock face is almost surely this shrub. It occurs throughout our range and is abundant in the montane zone along the east side of the Front Range in Colorado. The genus was named for Dr. Edwin James, the physician-botanist accompanying Major Long on his exploring expedition to the Rocky Mountains in 1820. The expedition discovered the great peak to the north later named for Major Long. Dr. James and two others made the first ascent by white men of Pikes Peak. This was probably the first time that white Americans had seen the alpine plants of the Rocky Mountains. Dr. James wrote of their beauty and collected several new species, which he later described and named.

(3) *Philadelphus*. Syringa or mock orange

1. Leaves small, 3cm or less, glabrate and shining, subsessile; petals oval, slightly erose-dentate on margins.

P.microphyllus ssp. occidentalis
Littleleaf mock orange

1. Leaves larger, 4cm or more, short-petiolate, rounded at the base with short ciliate incurved hairs on margins.

P. Lewisii
Lewis mock orange

110. Littleleaf Mock Orange, ½X

LITTLELEAF MOCK ORANGE, *Philadelphus microphyllus* (fig. 110), is easily recognized by its white, 4-petaled flowers with many stamens, and its opposite branching. A fine-textured shrub with slender twigs and brown and white bark. Its small pale leaves have a slightly silvered aspect. It grows on rocky slopes between 5,000 and 8,000 feet from Arizona to Texas and northward through southern and western Colorado and Utah into Wyoming. LEWIS MOCK ORANGE, *P. lewisii,* with fragrant flowers and larger leaves, occurs in western Montana and in Idaho, where it is the state flower.

GOOSEBERRY FAMILY: GROSSULARIACEAE

This family is closely related to the Saxifrage Family, and its members are sometimes included within that family. It contains only the genus *Ribes,* the gooseberries and currants. All are shrubs, some of them prickly, and their flowers have 5-lobed tubular or saucer-shaped calyxes, which may be colored. The small petals are inserted on the calyx. The fruits are juicy berries, which are sometimes covered with sticky bristles. The leaves are more or less lobed. These shrubs are browsed by game animals, and the berries furnish food for small mammals and birds.

1. Stems usually armed with spines and often prickles (setose-hispid); bears gooseberries.
  2. Racemes 1–4-flowered; calyx-tube cylindrical or campanulate.
    3. Spines short or unequal; setose-hispid or nearly unarmed.
      4. Leaves glabrate or puberulent (downy), fruit smooth.

            Ribes inerme
            White-stemmed gooseberry

      4. Leaves closely pubescent; fruit hispid.

            R. setosum
    3. Spines long and conspicuous; setae (bristles) few; leaves glabrate; fruit smooth.

            R. leptathum
            Thin-flowered gooseberry
  2. Racemes several-flowered; calyx-tube salverform (with a border spreading at right angles to the slender tube).
    3. Leaves glabrate; fruit glandular-bristly, deep purple or black.
            R. lacustre
            Black prickly gooseberry
    3. Leaves pubescent, somewhat glandular; fruit glandular-bristly, reddish or red.
            R. montigenum
            Red prickly gooseberry
1. Stems unarmed (neither spines nor prickles present); bears currants.
  2. Flowers not yellow.
    3. Berry glandular-bristly.
      4. Leaves glabrous; fruit spherical.
        5. Calyx-tube salverform.

            R. coloradense
            Colorado currant

        5. Calyx-tube campanulate.

            R. wolfii
            Wolf's currant

      4. Leaves pubescent and glandular; fruit oval.

            R. viscosissimum
            Sticky currant

    3. Berry smooth.
      4. Calyx-tube campanulate.
        5. Racemes erect or ascending.

            R. hudsonianum
         5. Racemes drooping.

            R. americanum
      4. Calyx-tube cylindrical, with dilated base.

            R. cereum
            Squaw currant
  2. Flowers yellow.

            R. aureum
            Golden currant

GOLDEN CURRANT, *Ribes aureum* (fig. 111), has smooth foliage that turns rose or red in autumn. The dark red or black berries are edible. It grows in foothill canyons throughout our range. SQUAW CURRANT, *R. cerum* (fig. 112), is a common shrub of sunny, dry foothill and montane slopes. The bright red berries are insipid. Occurs throughout our range.

111. Golden Currant, ½X   112. Squaw Currant, ½X

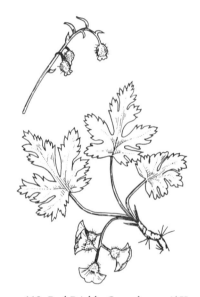

113. Red Prickly Gooseberry, ½X   114. Mountain Gooseberry, ½X

STICKY CURRANT, *R. viscosissimum,* which has sticky stems, leaves, and berries, is found in mountain woods from Montana south to Colorado. Its black berries are almost dry. COLORADO CURRANT, *R. coloradense,* is most frequently seen in the subalpine forests. Its black berries are slightly sticky. Found in western Colorado, Utah, and New Mexico. WOLF'S CURRANT, *R. wolfii,* is similar but a more erect shrub.

RED PRICKLY GOOSEBERRY, *R. montigenum* (fig. 113), occurs on rocky montane and subalpine slopes throughout our range. The BLACK PRICKLY or SWAMP GOOSEBERRY, *R. lacustre,* which occurs in moist, often shaded locations over much of North America, is also found in the montane zone of the Rockies.

WHITE-STEMMED or MOUNTAIN GOOSEBERRY, *R. inerme* (fig. 114), is a prickly shrub of moist, shaded ground found throughout our range. It has smooth flowers, and its dark purplish, very tart berries make excellent jelly. The specific name, *inerme,* which means *unarmed,* is inappropriate, as the stems have both prickles and spines; but is less prickly than the two species above. THIN-FLOWERED GOOSE-BERRY, *R. leptanthum,* is an excessively prickly shrub found on dry mountain slopes of the southern Rockies. It may be distinguished from *R. inerme* by its pubescent flowers.

115. Yellow Stonecrop, ½ X

ORPINE OR STONECROP FAMILY: CRASSULACEAE

The plants of this family are succulents, having fleshy leaves. Many species grow in arid regions, as they are well able to withstand drought. Only one genus, *Sedum,* is likely to be found in our range. The sepals, petals, and pistils are always of the same number, either 4 or 5, and the stamens are twice that number.

1. Flowers pink to purple, or whitish; leaves mostly on the flowering stems.
    2. Petals about 4mm long, usually purple; flowers usually imperfect.
            Sedum integrifolium
            King's crown
    2. Petals 8–10cm long, greenish or white to pink; flowers perfect.
            S. rhodanthum
            Rose crown
1. Flowers yellow; leaves in basal clusters.
    2. Leaves strongly keeled, often persistent on old stems.
            S. stenopetalum
            Yellow stonecrop
    2. Leaves not strongly keeled, often deciduous by anthesis (opening of the flower).
            S. lanceolatum
            Yellow stonecrop

YELLOW STONECROP, *S. lanceolatum* (fig. 115), is a very abundant plant throughout our range, occurring at all altitudes. Its cylindrical leaves are about 1–2cm long and form rosettes. In the winter and spring they are brownish green, making the plants inconspicuous. In early summer, each rosette sends up a stalk 5–13cm tall with many bright yellow star-like flowers.

116. Rose Crown, ½X

ROSE CROWN, *S. rhodanthum* (fig. 116), has leafy stems up to 30cm tall with a cluster of flowers, greenish or white to pink or rose. The leaves are thick but also flattened. It grows in clumps of several stems together, in marshy places or on stream banks of the subalpine zone from New Mexico to Montana. KING'S CROWN, *S. integrifolium,* is similar, but it has purple to dark red flowers, sometimes tinged yellowish or orange, the staminate and pistillate on separate plants. Less restricted to wet situations, it also occurs in gravelly or rocky places of the subalpine and alpine areas.

SAXIFRAGE FAMILY: SAXIFRAGACEAE

The woody shrubs sometimes included in Saxifragaceae are here treated in either Hydrangeaceae or Grossulariaceae. The name *Saxifragaceae* means *breaker of rocks,* and many of these plants will grow in rock crevices.

1. Herbaceous plants; stamens 10.
    2. Ovary 2-celled with axile placentation (the ovules are borne near the center of the compound ovary).
        3. Styles 2, united; petals purple, long-clawed; leaves alternate, petiolate.
                Telesonix p. 166
                Telesonix
        3. Styles 2, distinct; petals commonly white (sometimes spotted with yellow or redish-purple) to greenish, usually entire; leaves opposite, sessile.
                Saxifraga p. 163
                Saxifrage
    2. Ovary 1-celled with parietal placentation (the ovules are borne on the walls); styles or style-branches 3; petals white or pinkish, laciniate; bulblet-bearing on leaf axils.
                Lithophragma p. 161
                Woodland star
1. Herbaceous plants; stamens 4–5 (there may be clusters of staminodia in between).
    2. Clusters of staminodia alternating with fertile stamens; flowers solitary on the stems; leaves entire.
                Parnassia p. 161
                Grass-of-Parnassus
    2. No staminodia present; flowers in clusters; ovary either 1 or 2-celled; leaves toothed or lobed.
        3. Stems leafy; flowers clustered in axils of upper leaves; stamens and sepals 4; petals absent.
                Chrysosplenium p. 159
                Golden carpet

3. Stems scapose; leaves nearly or quite basal; flowers terminal; stamens
    and sepals 5; petals present but may be small or deciduous.
    4. Ovary 2-celled with axile placentation..

>   *Sullivantia* p. 165
>   Sullivantia

    4. Ovary 1-celled with 2 parietal placentae.
        5. Filaments and styles exserted; petals entire.

>   *Heuchera* p. 159
>   Alumroot

        5. Filaments and styles very short; petals pinnatifid or 3-cleft.

>   *Mitella* p. 161
>   Miterwort

(1) *Chrysosplenium*. Golden carpet

One species occurs in our range. *Chrysosplenium tetrandrum* is a low alpine tundra plant. Perennial, with slender stoloniferous rootstock; leaves 4–12mm wide,
alternate, blades thick, reniform, crenate; apetalous. Found in moist or wet places.

(2) *Heuchera*. Alumroot

This genus includes many crevice plants. The petiolate leaves are mostly
basal, usually more or less lobed, and palmately-veined. The calyx tube is
joined to the lower part of the ovary, which develops into a 2-horned capsule.

1. Inflorescence open-paniculate; stamens well exserted.
    2. Peduncles glabrate; petioles somewhat hirsute.

>   *H. rubescens*

    2. Peduncles and petioles both very hirsute-hispid.

>   H. richardsonii
>   Richardson's alumroot

1. Inflorescence spicate; stamens included or nearly so, except for *H. bracteata*.
    2. Stamens exserted at maturity; petals filiform.

>   *H. bracteata*
>   Bracted alumroot

    2. Stamens never exserted; petals spatulate.
        3. Flowers in a narrow one-sided raceme.

>   *H. hallii*
>   Hall's alumroot

        3. Flowers not one-sided.
            4. Scapes glabrous.

>   *H. grossularifolia var.*
>   *grossularifolia*
>   Gooseberry-leaved
>   alumroot

            4. Scapes glandular-puberulent.
                5. Plants usually over 20cm tall; seeds short-spiny.

>   *H. parvifolia var.*
>   *parvifolia*
>   Common alumroot

                5. Plants usually under 20cm tall; seeds verrucose (with wart-like
                    elevations); usually grows above timberline.

>   *H. parvifolia var. nivalis*
>   Alpine alumroot

BRACTED ALUMROOT, *H. bracteata*, forms tufts in rock crevices. Its leaves are bright green, 5–7-lobed and sharply toothed. Looking closely, one can usually see the rust-colored dried leaves of the previous season around the edges. Flowers are small, greenish, and in compact, one-sided clusters on somewhat scaly but leafless stalks only 10–20cm tall. Common in the foothill and montane zones of Colorado and Wyoming.

HALL'S ALUMROOT, *H. hallii* (fig. 117), also grows as tufts and elongated clumps in crevices, but its bell-shaped flowers, only about 6mm long, are not crowded but hang along slender stalks 10–25cm tall. Found only in the mountains of central Colorado at elevations between 7,500 and 10,000 feet.

COMMON ALUMROOT, *H. parvifolia* var. *parvifolia,* usually grows on the ground under trees, with a loose rosette of dark green, long-petiolate, roundish leaves, which are cordate at base and have 5–9 lobes. The margins are toothed. Flowers are tiny in interrupted clusters on stalks 2.5–6dm tall. Grows throughout our range. The similar ALPINE ALUMROOT, *H. parvifolia* var. *nivalis,* is smaller and more compact. Found in rock crevices of the subalpine and alpine zones in Colorado.

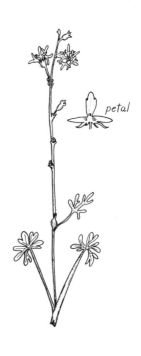

117. Hall's Alumroot, ½X   118. Woodland Star, ½X

(3) *Lithophragma*. Woodland star

1. Stem leaves with clusters of bulblets in the axils; some flowers replaced by bulblets.
    L. glabrum
    Woodland star
1. Bulblets seldom present on stem or inflorescence.
    2. Calyx-tube campanulate, the base somewhat rounded, whole calyx 3–4mm long.
        L. tenellum var. tenellum
        Slender fringecup
    2. Calyx-tube turbinate, cuneate at base, the whole calyx 4–8mm long.
        L. parviflorum
        Small-flowered fringecup

WOODLAND STAR, *Lithophragma glabrum* (fig. 118), is a dainty, slender plant with whitish or pinkish flowers; 5 petals, each cut into 3 or more divisions. The stem is reddish, and there are tiny red bulblets in the axils. Clusters of these sometimes replace the flowers. Grows in shaded places of the montane zone. Found from South Dakota and British Columbia west and south through Colorado. *L. parviflorum,* similar but without the bulblets and the reddish stem color, is found in moist woods from southern Wyoming through western Colorado into New Mexico.

(4) *Mitella*. Miterwort or bishop's cap

1. Stamens opposite the greenish pinnatifid petals.
    M. pentandra
    Bishop's cap
1. Stamens alternate with white, three-cleft (trifid) or entire petals.
    2. Racemes clearly secund (one-sided), 10–48-flowered; calyx 4–6mm.
        M. stauropetala
        var. stenopetala
        Side-flowered miterwort
    2. Racemes weakly if at all secund, 10–20-flowered; calyx 1.5–3.5mm.
        M. trifida
        Three-tooth miterwort

BISHOP'S CAP, *Mitella pentandra* (fig. 119), is a plant of moist, shaded forest banks. Usually found around springs or seepage areas; often grows in moss along with the tiny Tway-blades or One-leaved bog orchid. It has a basal rosette of scalloped leaves and an erect stalk holding a raceme of rather widely spaced, small green flowers. The little calyx cup is shallow, with 5 pinntafid petals inserted on its rim. Grows in subalpine forests from Colorado north to Alaska.

(5) *Parnassia*. Grass-of-Parnassus

1. Petals fringed with a short claw; staminodia short and thick; leaves reniform.
    P. fimbriata var.
    fimbriata
    Fringed parnassia
1. Petals entire; staminodia various; leaves ovate or cordate.
    2. Scape with a single leaf.

3. Staminodia 5–7; leaf near the middle of the scape.

P. parviflora
Small-flowered parnassia

3. Staminodia 7–15; leaf usually below the middle of the scape.

*P. palustris var.
montanensis*
Northern parnassia

2. Scape naked, or with 1 leaf near the base; staminodia slender, 3–5 at the base of each petal.

P. kotzebuei
Kotzebue parnassia

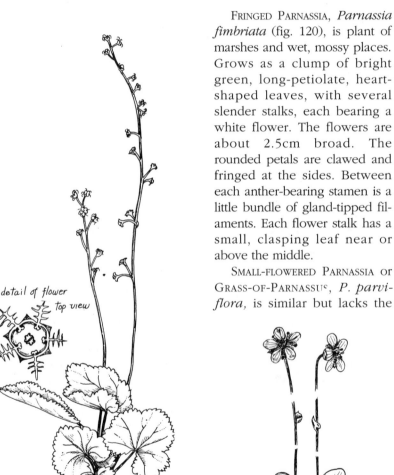

FRINGED PARNASSIA, *Parnassia fimbriata* (fig. 120), is plant of marshes and wet, mossy places. Grows as a clump of bright green, long-petiolate, heart-shaped leaves, with several slender stalks, each bearing a white flower. The flowers are about 2.5cm broad. The rounded petals are clawed and fringed at the sides. Between each anther-bearing stamen is a little bundle of gland-tipped filaments. Each flower stalk has a small, clasping leaf near or above the middle.

SMALL-FLOWERED PARNASSIA or GRASS-OF-PARNASSUˢ, *P. parviflora,* is similar but lacks the

detail of flower
top view

119. Bishops Cap, ½X

120. Fringed Parnassia, ½X

fringe on the petals, and the leaf on the stalk is usually below the middle. It occurs throughout our range, and both species grow in the Colorado mountains and northward into Canada in the wet situations of the montane and sub-alpine regions.

(6) *Saxifraga*. Saxifrage

This genus has many species in the mountains and in the northern parts of the northern hemisphere. Flowers regular with 5 calyx-lobes and 5 petals. The calyx cup is usually joined, at least at the base, to the ovary; and the petals are inserted on the rim of that cup. The petals are often clawed. Stamens 10; ovary 2-celled, developing into a 2-beaked capsule.

1. Leaves all basal; stems scapose.
  2. Leaves reniform-cordate.

> *S. odontoloma*
> Brook saxifrage

  2. Leaves ovate to oblong.
    3. Sepals erect; scape glandular-pubescent above.

> *S. rhomboidea var.*
> *rhomboidea*
> Snowball saxifrage

    3. Sepals reflexed, at least in fruit.

      4. Inflorescence open-paniculate.

> *S. occidentalis var.*
> *occidentalis*

      4. Inflorescence spiciform.
        5. Filaments and sepals greenish.

> *S. oregana var.*
> *montanensis*
> Bog saxifrage

        5. Filaments and sepals purplish-tinged to deeply reddish-purple.

> *S. oregana var.*
> *subapetala*

1. Stems more or less leafy (may have only 1–2 leaves).
  2. Stem leaves opposite.

> *S. oppositifolia ssp.*
> *oppositifolia*
> Purple Saxifrage

  2. Stems leaves alternate.
    3. Petals yellow.
      4. Basal leaves crowded-rosulate.
        5. Plant stoloniferous.

> *S. flagellaris ssp.*
> *flagellaris*
> Whiplash saxifrage

        5. Plant without stolons.

> *S. chrysantha*
> Goldbloom saxifrage

      4. Basal leaves not rosulate.

> *S. hirculus*
> Arctic saxifrage

    3. Petals white, or nearly so.
      4. Leaves as broad as they are long, reniform, lobed or entire.

    5. Calyx turbinate; ovary about 1/2 inferior.

S. debilis
Pygmy or weak saxifrage

    5. Calyx campanulate; ovary hardly 1/4 inferior.

S. cernua
Nodding saxifrage

  4. Leaves longer than broad, entire, cleft, or parted.

    5. Leaves entire, linear or oblong.

      6. Calyx cleft nearly to the base.

S. bronchialis
Dotted saxifrage

      6. Calyx-lobes shorter than the tube.

S. adscendens var.
oregonensis
Wedge-leafed saxifrage

    5. Leaves 3–5-cleft; plants tufted or matted.

S. caespitosa var. minima
Alpine saxifrage

SNOWBALL SAXIFRAGE, *S. rhomboidea* (fig. 121), sends up a stout, leafless stalk from a flat rosette of very short-petiolate leaves. At first the inflorescence is a tight ball-like cluster, but as it develops it becomes an interrupted panicle having 2–3 rather compact clusters of small white flowers. As it goes to seed, it becomes still more open. One of the very early spring flowers of the foothill and montane zones, particularly in north-central Colorado. Widely distributed as an early-

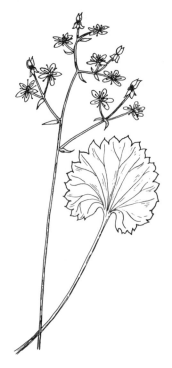

121. Snowball Saxifrage, ½X   122. Brook Saxifrage, ½X

blooming plant of the subalpine and alpine zones throughout our range.

BROOK SAXIFRAGE, *S. odontoloma* (fig. 122), is a plant that grows only in very wet places. Its rosette of shiny, green, rounded, strongly-toothed leaves is usually found among stones at the edges of cold mountain streams. A slender flowering stalk, 2.5–4.5dm tall, bears an open panicle of small white-petaled flowers, their pedicels often reddish. Grows in the subalpine and upper montane zones from Montana south into New Mexico.

DOTTED SAXIFRAGE, *S. bronchialis* (fig. 123), is a plant frequently found in dry coniferous forests, but it also occurs on rocks in the subalpine and alpine regions. Its small, spine-tipped leaves are in tight rosettes. These rosettes often form a mat or carpet, covering several square feet of surface. The slender flowering stalks are 5–15cm tall and bear several white flowers. The 5 oblong petals are distinctly spotted with small orange and red dots. Found throughout the Rocky Mountains.

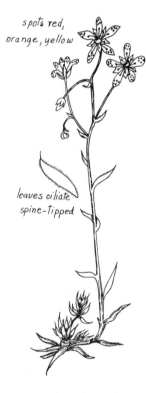

spots red, orange, yellow

leaves ciliate spine-tipped

ALPINE SAXIFRAGE, *S. caespitosa,* is a tiny, white-flowered plant with crowded, 3-lobed leaves, and stems 5–8cm tall, found in high alpine situations. It is circumboreal and comes as far south as Arizona on the highest mountains.

WHIPLASH SAXIFRAGE, *S. flagellaris,* is a yellow-flowered alpine with sticky hairs on its leaves and stems. It sends out 1–2 stolons from the basal rosette, which root at their tips to form new rosettes, as a strawberry plant does. Its capsule is often bright red as it ripens.

GOLDBLOOM SAXIFRAGE, *S. chrysantha* (fig. 124), is a tiny alpine plant that has rosettes of smooth leaves 7–14mm long. The stems are usually 5–8cm tall, with comparatively large flowers. The little round buds are on reflexed pedicels but are lifted some when in bloom. Its bright yellow petals are dotted with orange. As the flower ages, the ripening pistil becomes bright red. Grows in gravelly places on the high mountains of New Mexico and Colorado above 11,000 feet. Another yellow-flowered species, ARCTIC SAXIFRAGE, *S. hirculus,* up to 20cm tall, is found in subalpine bogs. It is a circumboreal species which extends south as far as central Colorado.

123. Dotted Saxifrage, ½X

(7) *Sullivantia*. Sullivantia

One species is found in our range. *S. hapemannii* var. *hapemanni* is a perennial, acaulescent plant. Leaves mostly basal, long-petiolate, lobes sharply

124. Goldbloom
Saxifrage, ½X

dentate, leaves mostly orbicular-reniform; petals white, somewhat exceeding the sepals, spatulate to obovate.

(8) *Telesonix*. Telesonix

1. Petals long-clawed with well-exserted orbicular blades, 3–5.5mm long; styles united.
    T. jamesii var. jamesii
    James saxifrage
1. Petals short-clawed with oblong blades to 3mm in length, scarcely longer than the sepals; styles free.
    T. jamesii var. heucheriformis

125. James Saxifrage, ½X

JAMES SAXIFRAGE, *Telesonix jamesii* (fig. 125). A narrow panicle of bright rose-colored flowers from a tuft of green leaves in a rock crevice is apt to be this species. Its petals are distinctly clawed, making conspicuous spaces between their bases. The petiolate, rounded leaves are distinctly toothed. Found only in granitic rock crevices or, rarely, on very rocky ground, in the mountains of central and north central Colorado from 9,000 to 10,000 feet. This species has sometimes been placed in either *Boykinia* or *Saxifraga*.

ROSE FAMILY: ROSACEAE

A large family, and it includes plants that differ widely. Some botanists have divided it into separate families, but even then great variation is found within the segregated groups. The Rosaceae, as treated here in its broadest sense, includes trees, shrubs, and herbs. Its members have simple or compound leaves; the pistils vary from one to many. In general: the branching is alternate; the leaves have stipules; the flowers are showy, regular, with a more or less united calyx and 5 separate petals; and the stamens are usually many. By careful observation, one may soon learn to recognize combinations of these characters as indicating a member of this family.

The fruits are also variable; apples, plums, raspberries, and strawberries are examples. But dry seed pods and long-tailed achenes are also fruits within the rose families. Many species, especially the shrubs, are important food plants for wildlife. The fruits are eaten, and game animals browse the foliage and twigs.

1. Ovary inferior, adnate to calyx tube; fruit a pome; plants woody.
    2. Leaves pinnately compound.
        Sorbus p. 186
        Mountain ash
    2. Leaves simple, sometimes lobed.

3. Flowers usually solitary (2–3); styles 2–5; leaves narrowly oblanceolate, entire or serrulate, sessile or subsessile.

Peraphyllum p. 175
Squaw-apple

3. Flowers in racemes or corymbs; styles 2–5; leaves lanceolate to orbicular, toothed short- to long-petiolate.
    4. Branches armed with stout spines; flowers in corymbiform cymes; seeds bony.

Crataegus p. 171
Hawthorn

    4. Branches unarmed; flowers racemose; seeds not bony.

Amelanchier p. 169
Serviceberry

1. Ovary superior, not adnate to calyx tube; fruit not a pome; plants woody or herbaceous.
    2. Plants woody, shrubby or tree-like; leaves alternate or fascicled, either entire, toothed, lobed, or compound; 5 sepals; petals various.
        3. Petals absent; pistil 1, becoming an achene with a long plumose style; leaves simple.

Cercocarpus p. 170
Mountain mahogany

        3. Petals present; pistils more than 1 (if 1, no development of plumose style); leaves simple or compound.
            4. Carpels enclosed within globose to urn-shaped calyx (which becomes reddish and fleshy in fruit); leaves pinnately compound; stems more or less prickly; flowers rose-colored.

Rosa p. 183
Wild rose

            4. Carpels not enclosed in fleshy calyx tube (but may be included in a saucer-shaped or campanulate calyx); leaves simple to compound, but if pinnate, prickles absent; petals yellow or white.
                5. Leaves compound, 3 or more leaflets; pistils many.
                    6. Petals yellow; sepals 5 with 5 alternating bractlets; no bristles or prickles; fruit as dry achenes.

Potentilla p. 177
Potentilla

                    6. Petals white; sepals 5, without bractlets; bristles or prickles present; fruit fleshy, of drupelets.

Rubus p. 184
Wild raspberry

                5. Leaves simple (may be deeply lobed); pistils 1 to many.
                    6. Leaves entire or toothed (may be deeply toothed), but not lobed.
                        7. Pistil 1, becoming a fleshy drupe.

Prunus p. 182
Plums and cherries

                        7. Pistils 3 to many, becoming dry follicles or achenes.
                            8. Low, depressed undershrubs; fruits of numerous achenes or of 3–5 follicles.
                              9. Pistils many, becoming achenes with long plumose styles; leaves crenate; petals 8–15mm long.

Dryas p. 171
Mountain dryad

                              9. Pistils 3–5, becoming follicles; leaves entire; petals 1.5–2mm long.

Spiraea p. 186
Spiraea

                            8. Upright shrubs; fruit of 5 achenes.

Holodiscus p. 175
Mountain spray

6. Leaves palmately or pinnately lobed.
7. Calyx-lobes 5, alternating with 5 bractlets (appearing to have 10 sepals); fruit of many achenes with elongated, plumose styles.

Fallugia p. 172

7. Calyx-lobes 5, no bractlets present; fruits various.
8. Pistils many, becoming drupelets, more or less juicy.

Rubus p. 184
Wild raspberry

8. Pistils 1–5, becoming dry achenes or capsules.
9. Fruit a capsule; leaf-blades with broad shallow lobes blade over 3cm long, more or less stellate-hairy, never white-tomentose.

Physocarpus p. 176
Ninebark

9. Fruit of 1–10 achenes; leaf blades deeply lobed or cleft, blade not over 3cm long, white-tomentose below, not stellate-hairy.

Purshia p. 183
Antelope brush

2. Plants herbaceous above ground.
3. Leaves simple, entire or crenate.
4. Pistils many, becoming achenes with long, plumose styles; leaves crenate; petals 8–15mm long.

Dryas p. 171
Mountain dryad

4. Pistils 3–5, becoming follicles, the styles not elongating or plumose; leaves entire; petals 1.5–2mm long.

Petrophytum p. 176
Rockmat

3. Leaves compound or deeply divided.
4. Calyx-tube with hooked prickles above; pistils 1–2.

Agrimonia p. 169
Agrimony

4. Calyx-tube without hooked prickles; pistils 5–many (in *Ivesia* 1–6).
5. Calyx-lobes 5, alternating with bractlets (appearing to have 10 sepals);stamens 5–many; pistils 1–many, becoming dry achenes.
6. Stamens 5; pistils 1–20; petals small, little if any larger than the sepals, not showy.
7. Styles terminal or subterminal on the achenes; pistils 1–6; flowers capitate.

Ivesia p. 175
Ivesia

7. Styles basal or subbasal on the achenes; pistils 5–20; flowers not capitate.
8. Leaves palmately compound with 3 leaflets, these oblanceolate or wider, not lobed; plants 5–15cm tall; petals yellow.

Sibbaldia p. 186
Sibbaldia

8. Leaves 2–4 ternately divided into linear lobes; plants 10–30cm tall; petals white.

Chamaerhodos p. 171
Chamaerhodos

6. Stamens 10 or more; pistils 10 to very many; petals usually longer than the sepals and conspicuous.
7. Basal leaves with only 3 leaflets.
8. Petals white; plants spreading by aerial stolons; plants acaulescent.

Fragaria p. 172
Wild strawberry

      8. Petals light to bright yellow; no aerial stolons; plants
          more or less caulescent.

                         Potentilla p. 177
                         Potentilla

7. Basal leaves with more than 3 leaflets.
      8. Styles articulated at base and deciduous from the
          achenes, neither plumose nor jointed near the
          middle.

                         Potentilla p. 177
                         Potentilla

      8. Styles not articulated at the base or deciduous, but
          forming part of the mature fruit; middle of the style
          often with an abrupt bend or joint, style often
          plumose.

                         Geum p. 173
                         Avens

5. Calyx-lobes 5, no bractlets present; pistils many, becoming fleshy
   drupelets.

                         Rubus p. 184
                         Wild raspberry

(1) *Agrimonia*. Agrimony

1. Calyx-tube bristles erect or ascending; sepals acute; leaves definitely pubescent below.
                         A. striata
1. Calyx-tube bristles strongly reflexed; sepals acuminate; leaves sparsely hairy below.
                         A. gryposepala

AGRIMONY, *Agrimonia striata,* is an erect plant that grows up to 1m tall, with pinnately compound leaves, and a spike-like raceme of small yellow flowers, which develop into burrs with hooked prickles. There are 7–13 leaflets, alternating large and small. Occurs in meadows and open wooded areas of the foothill and montane zones.

(2) *Amelanchier*. Serviceberry

Our species are large or small shrubs. They have racemes of white, fragrant flowers, and little apple-like fruits. Their bark is gray or reddish-brown. The fruits are eaten by wild animals and were an important food for the Indians. The foliage and twigs are browsed by deer, elk, and moose. Their leaves are toothed only from the middle (or from a little below it) around the apex.

1. Leaves glabrous (or with sparse silky hairs beneath); petals 10–22mm.
   2. Top of the ovary glabrous or with few scattered hairs.

                         A. alnifolia var. pumila
                         Smooth or western ser-
                         viceberry

   2. Top of the ovary mostly hairy.

                         A. alnifolia var. alnifolia
                         Saskatoon serviceberry

1. Leaves woolly (especially beneath); petals 5–10mm.

                         A. utahensis
                         Utah serviceberry

126. Saskatoon Serviceberry, ½X

SASKATOON SERVICEBERRY, *A. alnifolia* var. *alnifolia* (fig. 126), with sweet, juicy fruit, is the commonest and most widely distributed species in our range. Its foliage is hairy when young, but becomes smooth and dull green. SMOOTH SERVICEBERRY, *A. alnifolia* var. *pumila,* with entirely smooth foliage, is less common but widely distributed throughout our region in the foothill and montane zone.

UTAH SERVICEBERRY, *A. utahensis,* has grayish hairy leaves and rather dry fruits. It is most conspicuous in the higher sections of the Southwest, but its range extends from Montana to New Mexico. It is prominent between Moab, Utah, and the Arches National Monument, as large, rounded, grayish shrubs growing singly from crevices in the pavement-like sandstone.

(3) *Cercocarpus.* Mountain mahogany

1. Leaves not resinous, obovate or lanceolate, toothed.

C. montanus var. montanus
Mountain mahogany

1. Leaves resinous, linear to elliptic-lanceolate, entire.
    2. Leaves scarcely revolute (rolled backward upon the lower side).

C. ledifolius var. ledifolius
Curl-leaf mountain mahogany

    2. Leaves strongly revolute.

C. ledifolius var. intercedens

127. Mountain Mahogany, ½X

MOUNTAIN MAHOGANY, *Cercocarpus montanus* (fig. 127), is a very common shrub on dry, rocky slopes of the foothills throughout the Rocky Mountains. Commonly 1–2m tall, but may grow taller in favorable locations. Its branches are rigid and the bark gray. Its leaves are 2.5–5cm long, oval or obovate with rounded apex, whitish beneath with a covering of fine hairs. The flowers are apetalous, and the 1 pistil ripens into a pubescent achene with a conspicuously twisted and plumose tail 8–10cm long. Called "mahogany" because the wood is very hard and heavy. Considered an excellent fuel in areas where it is available.

CURL-LEAF MOUNTAIN MAHOGANY, *C. ledifolius,* is a shrub or small round-topped tree, which often has a twisted trunk. Its small leaves are dark green above, light beneath, with revolute margins. They are leathery and evergreen. The flowers and long-tailed fruits are similar to those of *C. montanus.*

(4) *Chamaerhodos.* Chamaerhodos

One species occurs in North America. CHAMAERHODOS, *Chamaerhodos erecta* var. *parviflora,* is erect and branching, glandular-pubesecent, with leaves cleft into many linear segments; calyx campanulate, deeply 5-cleft; stamens and petals 5, white; radical leaves rosulate. Occurs in central Colorado and Utah, extending north to Alaska.

(5) *Crataegus.* Hawthorn

These thorny shrubs or small trees are difficult to distinguish and identify specifically. They have clusters of white flowers in which the conspicuous anthers may be white, pink, rose, or purple. The fruits, called *haws,* are berry-like and may be orange, bright or dark red, or black.

1. Spines short, usually 2cm long or less, seldom over 3cm.
  2. Leaves elliptic to lanceolate, serrate, 1.5cm or less broad.
                                                                    C. saligna
                                                                    Willowy hawthorn
  2. Leaves ovate or obovate, 2–4cm broad.
    3. Leaves serrate, not lobed; petiole without glands.
                                                                    C. douglasii var. rivularis
                                                                    River hawthorn
    3. Leaves lobed or toothed; petiole with scattered glands.
                                                                    C. douglasii var. douglasii
                                                                    Douglas hawthorn
1. Spines long, 3–8cm.
  2. Teeth of leaves gland-tipped.
    3. Leaves small, 1.5cm or less broad.
                                                                    C. saligna
                                                                    Willowy hawthorn
    3. Leaves large, 3–5cm broad.
      4. Leaves thin, shiny above with scattered hairs; fruit mahogany brown.
                                                                    C. erythropoda
                                                                    Red-stemmed hawthorn
      4. Leaves thick, shiny, teeth of leaves dark red-tipped; fruit golden yellow to reddish-orange.
                                                                    C. chrysocarpa
  2. Teeth of leaves not gland-tipped; fruit blood-red.
                                                                    C. succulenta
                                                                    Western hawthorn

(6) *Dryas.* Dryad

One species in our range. MOUNTAIN DRYAD, *D. octopetala* var. *hookeriana* (fig. 128). This dwarf, matted plant is easy to recognize. Its flowers have 8–9 creamy-white petals, and no other flower in our region regularly has 8 petals. Even if the petals have fallen, you can still recognize it by the 8–9 sepals, plus

128. Mountain Dryad, ½X   129. Apache Plume, ½X

the clusters of long, plumy styles. It is an arctic plant that extends south along our mountain tops. Its leaves are 12–25mm long, crenate, white beneath, their veins conspicuously impressed. These leaves are an important source of food for the ptarmigan that live on the tundra.

(7) *Fallugia*. Apache plume

APACHE PLUME, *Fallugia paradoxa* (fig. 129), is the only species of this genus in North America. A finely-branched, white-barked shrub with dainty, dissected leaves. White flowers on long peduncles, solitary or few; petals 5; several separate pistils, each with a long, hairy style. In fruit these styles persist and elongate to produce a cluster of achenes, each with a plumose tail 2.5–5cm long. Common along roadsides around Santa Fe and around the Grand Canyon, it also occurs on the hillsides of northern New Mexico, southern Colorado and adjacent Arizona and Utah.

(8) *Frageria*. Strawberry

1. Leaves very green, not glaucous above, the veins prominent; terminal tooth of leaflets usually projecting beyond the lateral teeth; flower white, usually equalling or surpassing the leaves.

F. vesca ssp. bracteata
Woodland strawberry

1. Leaves glaucous and somewhat bluish-green above, the veins not prominent; terminal tooth of leaflet usually surpassed by adjacent lateral teeth; flower white, usually shorter than the leaves.

F. virginiana ssp. glauca
Wild strawberry

WILD STRAWBERRY, *Fragaria virginiana* (fig. 130). Except for color, the 5-petaled flowers are much like those of potentilla. The compound leaves are made up of 3 toothed, slightly bluish leaflets. The flowers have many separate pistils. As they go to seed, the receptacle to which these pistils are attached enlarges and becomes a strawberry, and each pistil develops into an achene seated on a juicy red receptacle. Strawberry seeds are on the outside of the fruit. In this species, they are in pits. Grows throughout the western part of the United States and Canada in meadows and on open slopes.

WOODLAND STRAWBERRY, *F. vesca* ssp. *bracteata,* has greener leaflets, and the seeds are in very shallow depressions on the surface of the receptacle. Less common than *E. virginiana,* it is often found at the edge of woods, sometimes in the shade of open coniferous forests.

130. Wild Strawberry, ½ X

(9) *Geum.* Avens

1. Stems leafy, leaves of the lower stem almost as large as the basal leaves.
   2. Petals violet to pink; sepals purplish.

   > *G. rivale*
   > Purple or water avens

   2. Petals yellow or white; sepals usually green.
      3. Petals white.

      > *G. canadensis var. camporum*

      3. Petals yellow.
         4. Terminal leaf divisions much larger than lateral divisions; persistent part of style glandular-puberulent.

         > *G. macrophyllum*
         > Large-leaved or bur avens

         4. Terminal leaf divisions only slightly larger than lateral divisions; persistent part of style not glandular, either glabrous or pubescent.

         > *G. aleppicum*
         > Yellow avens

1. Stems mostly subscapose, stem leaves greatly reduced.
   2. Petals reddish-purple to cream, erect; flowers usually nodding.
      3. Larger leaflets of the basal leaves cleft or toothed less than half their length.

      > *G. triflorum var. triflorum*
      > Prairie smoke

      3. Larger leaflets of the basal leaves much more deeply cleft or divided into linear or oblong segments.

      > *G. triflorum var. ciliatum*

   2. Petals yellow, spreading; flowers erect.

   > *G. rossii var. turbinatum*
   > Alpine avens

131. Alpine Avens, ½X

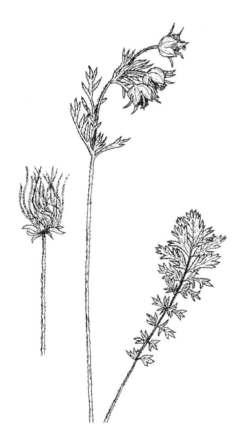

132. Pink Plumes, ½X

ALPINE AVENS, *Geum rossii* (fig. 131). A 5-petaled, yellow-flowered plant with finely dissected leaves, somewhat similar to potentilla, but with its calyx-tube top-shaped (turbinate) so that the flower is somewhat depressed at its center instead of being flat. In many alpine locations in the Rockies, this appears to be the most abundant and conspicuous plant throughout the season. It blooms profusely in late June and early July (and scatteringly during the rest of the summer). Its leaves indicate the coming of autumn by beginning to turn bronze or maroon in late August. Found in the subalpine and alpine zones from Montana to New Mexico and Arizona.

BUR AVENS, *G. macrophyllum,* grows to 9dm in height, has yellow flowers and compound leaves, with the rounded, terminal leaflet much bigger than the others. The styles, which persist on the many achenes, are jointed and bent, so that the fruit becomes a bur. Grows in meadows throughout the northern part of North America. In the mountains, it is found in the foothills and montane zones.

PRAIRIE SMOKE or PINK PLUMES, *G. triflorum* (fig. 132), has finely dissected leaves and nodding flowers, usually 3 on a stalk. The whole inflorescence is rose-colored. The calyx lobes overlap to form a sort of urn, from which whitish or pale yellow petals protrude slightly. In fruit the plumose styles elongate to 2.5–5cm, forming a tuft of rose-colored fluff.

(10) *Holodiscus*. Ocean spray or rock spires

1. Leaf blades over 3cm long, not decurrent; petioles 10–15mm long.
    H. discolor
    Creambush ocean spray
1. Leaf blades 1–2cm long, somewhat decurrent; petioles 2–4cm long.
    H. dumosus var. dumosus
    Mountain spray

MOUNTAIN SPRAY or ROCK SPIREA, *Holodiscus dumosus* (fig. 133), is a medium-sized shrub that bears plume-like panicles of very small, creamy-white flowers in June. As they fade, they turn pinkish and finally become rust-colored, persisting on the branches throughout the summer. Grows on shaded, rocky slopes of foothill and montane canyons of Wyoming, Colorado, Utah, and southward. Its close relative, *H. discolor,* is found in Montana and Idaho.

133. Mountain Spray, ²⁄₅X

(11) *Ivesia*. Ivesia

One species in our range. IVESIA, *Ivesia gordonii* (fig. 134), is also closely related to the potentillas. Leaves pinnately compound, the leaflets more or less uniform, divided to the base into 3 lobes. The small flowers are clustered in a compact, head-like inflorescence, yellow at first, then turning reddish brown. Occurs from the montane to the alpine zone, and from Montana south through Colorado.

(12) *Peraphyllum*. Squaw apple

Only one species in North America, SQUAW APPLE, *Peraphyllum ramosissimum*. The Latin adjective used for the specific name of this relative of the apple means *the most branched* and well describes this rigid, thicket-forming shrub. It grows up to 18dm to 2m tall, has spatulate leaves, which are usually crowded together at the ends of the branchlets. Flowers are pink and white and resemble apple blossoms. The round fruits are yellow-

134. Ivesia, ½X

ish or reddish brown. Occurs on open fields and slopes at altitudes of 5,500 to 8,000 feet in southwestern Colorado and westward to Oregon and California.

(13) *Petrophytum*. Rockmat

One species occurs in our range and is sometimes put in the genus *Spirea*. ROCKMAT, *Petrophytum caespitosum,* is a low plant with creeping, woody stems. Its spatulate, silky-hairy leaves, not over 7.5mm long, are 1-ribbed and are arranged in crowded rosettes. The entire plant is compact and dense, with flowering stalks 2.5–20cm tall holding short, dense spikes of small white flowers. Grows on rocks in the upper foothill and lower montane zones, but is rare in Colorado.

(14) *Physocarpus*. Ninebark

1. Carpels 3–5, united only at the base, swollen, 7–9mm long.
   > *P. opulifolius var. intermedius*
   > Tall ninebark
1. Carpels 1–2, swollen or flattened.
   2. Carpel 1; leaf blades not over 15mm long, stellate; filaments alternately long and short; shrub not over 1m tall.
      > *P. alternans*
   2. Carpels 2; some leaf blades over 15mm long, glabrous or nearly so; filaments virtually alike; shrubs often over lm tall.
      3. Mature carpels flattened and keeled; styles erect.
         > *P. malvaceus*
         > Mallow ninebark
      3. Mature carpels swollen; styles divergent.
         > *P. monogynus*
         > Rocky Mountain ninebark

135. Rocky Mountain Ninebark, ½X

ROCKY MOUNTAIN NINEBARK, *Physocarpus monogynus* (fig. 135). The leaves of this shrub are roundish in outline, 3- to 5-lobed and doubly toothed. The small flowers are white and arranged in corymbs along the arching branches. The leaves turn orange and red and are responsible for much of autumn color on the slopes and in the valleys of the upper foothill and montane regions. Frequently seen in ravines, under open aspen groves, and at the edge of forests. Occurs from southern Wyoming south to New Mexico and Arizona.

The TALL NINEBARK, *P. opulifolius*, is larger in every way. The leaves are only 3-lobed. It is found in some of the deep foothill canyons in Colorado, Montana, Wyoming, and Utah. *P. malvaceus* is similar, occurring from Wyoming northwestward.

(15) *Potentilla.* Potentilla or cinquefoil

This genus has many representatives in western America. Their flowers are so similar that, as soon as one learns one species, the others are easy to recognize as belonging to the same clan. Their flowers have 5 separate, heart-shaped petals, which are usually yellow and have an orange spot at base. The 5 sepals are united into a calyx that resembles a 5-pointed, saucer-shaped star. Five extra, narrower bractlets alternate with the main calyx-lobes to give the calyx a 10-lobed appearance. This type of calyx, which is illustrated on Plate E41, is found only in the Rose Family and mainly in this genus (and in a few closely related ones). It is one positive character by which to distinguish these plants from yellow-flowered plants within other families. The leaves of potentillas are always compound, either palmate or pinnate, and always have stipules attached at the base of the petioles.

The old common name for this genus, cinquefoil, which in French means *5-leaf,* was originally applied to a European species that has 5 leaflets. Among the large number of American species, most have more than 5 leaflets, so the name is inappropriate for the genus as a whole.

1. Flowers red to purple; rhizomatous plant of wet places.
> *P. palustris*
> Marsh or purple potentilla

1. Flowers yellow, cream, or white; not rhizomatous.
  2. Plant shrubby or woody.
> *P. fruticosa*
> Shrubby potentilla

  2. Plant herbaceous.
    3. Styles at or below the middle of the ovary; leaves pinnate; stamens 25–40.
      4. Flowers solitary from leaf axils or on long slender stolons; bright yellow.
> *P. anserina*
> Silverweed

      4. Flowers in clusters, terminal on caulescent stems; petals white to lemon-yellow.
        5. Flowers in open cymes, the branches spreading.
> *P. glandulosa var.*
> *intermedia*

        5. Flowers in dense cymes, the branches short, appressed or ascending.
          6. Leaflets 9–13, regularly decreasing in size from apex to base (rudimentary leaflets often interspersed).
> *P. fissa*
> Leafy potentilla

          6. Leaflets 5–9, rather irregularly decreasing in size from apex to base.
            7. Leaves densely hairy; stem very villous, reddish-tinged.
> *P. arguta var. arguta*
> Sticky potentilla

            7. Leaves sparingly hairy; stem moderately villous or reddish.
> *P. arguta var. convallaria*

    3. Styles terminal, attached at or near the top of the ovary; leaves pinnate or palmate; stamens about 20.
      4. Annuals or biennials; no rosettes of basal leaves; petals seldom surpassing sepals in length.
        5. Achenes swollen on 1 side near base; leaves pinnately 7- to 11-foliate.
> *P. paradoxa*
> Bushy potentilla

5. Achenes not swollen near base; some or all of the leaves 3-foliolate.
6. Lower leaves pinnately 5-foliolate below, 3-foliolate above.
P. rivalis var. rivalis
6. All leaves 3-foliolate, above and below.
7. Stamens 15–20; petals about equal to the sepals; stem hirsute below.
P. norvegica ssp. monspeliensis
Norwegian potentilla
7. Stamens about 10; petals definitely shorter than the sepals; stem not hirsute below.
8. Stems erect or ascending, often glandular-hairy.
P. biennis
Biennial potentilla
8. Stems with weak and decumbent branches, not glandular-hairy.
P. rivalis var. millegrana
River potentilla
4. Perennials; stout crowns, often bearing old leaf bases or a cluster of basal leaves; petals usually surpassing the sepals.
5. Basal leaves with 3 leaflets.
P. nivea
Alpine potentilla
5. Basal leaves with 5 or more leaflets.
6. Basal leaves palmately compound; leaflets 5–7; rachis very short if present.
7. Basal leaves plus a reduced pair of leaflets on the petiole remote from the others.
P. subjuga
7. Remote reduced leaflets absent.
8. Leaflets of basal leaves divided 1/2 or more to the midrib into linear or oblong segments, densely white tomentose beneath.
P. quinquifolia
Snow potentilla
8. Leaflets of basal leaves merely toothed.
9. Leaflets green above, white below.
10. Plant 5–20cm tall.
P. concinna
Early potentilla
10. Plant 10–60cm tall.
P. pulcherrima
Beauty potentilla
9. Leaves greenish both sides.
10. Anthers ovate to lanceolate, 1mm long.
P. gracilis
Goldcup potentilla
10. Anthers oval to round, about 0.5mm long.
P. diversifolia
Blueleaf potentilla
6. Basal leaves pinnately compound, the rachis definite; leaflets often more than 7.
7. Styles shorter than the mature achene.
P. pensylvanica
Prairie potentilla
7. Styles exceeding the mature achene in length.
8. Basal leaves with leaflets deeply incised (pinnatifid or dissected); stems not over 30cm tall.
9. Basal leaves with 5 (7) leaflets, tomentose or gray-silky below.
P. rubricaulis
Red-stemmed potentilla

9. Basal leaves with 7–17 leaflets, glabrous or appressed-strigose below.
P. plattensis
Nelson potentilla
8. Basal leaves coarsely to shallow toothed; stems sometimes over 30cm tall.
9. Leaflets silvery or gray on both surfaces.
10. Basal leaves with 5–7 leaflets; stems 10–20cm tall.
P. diversifolia
Blueleaf potentilla
10. Basal leaves with 7–13 leaflets; stems often over 20cm tall.
11. Bractlets of calyx darker and much smaller than the calyx lobes.
P. effusa
Woolly potentilla
11. Bractlets of calyx usually same color and size of calyx lobes.
P. hippiana
Silvery potentilla
9. Leaflets glabrous or green-strigose, at least above.
10. Leaflets green above, white below, definitely bicolored.
P. pulcherrima
Beauty potentilla
10. Leaflets not definitely bicolored.
11. Leaflets of basal leaves 5–7, glabrous below, except midrib and margins.
P. rupincola
11. Leaflets of basal leaves 9–15, silky villous beneath; stems 40–70cm tall.
P. ambigens

SHRUBBY POTENTILLA, *P. fruticosa* (fig. 136), grows from 3 to 9dm tall and is found on moist hillsides and in meadows. Noticeable during winter as compact, rounded, dark brown or grayish bushes scattered over grassland, in summer these little shrubs are covered with bright yellow, rose-like blossoms 2.5–4cm broad. The plant is widely distributed in arctic and northern regions of the world and extends southward along mountain ranges. Occurs in our area from the foothills to the alpine zone; the higher you find it, the brighter and bigger the flowers seem to be.

ALPINE POTENTILLA, *P. nivea,* is a small, tomentose plant with trifoliate leaves and bright yellow, comparatively large flowers

136. Shrubby Potentilla, ½X

12mm broad. The leaflets are white-felted, at least on their lower surfaces. An arctic-alpine plant that extends southward to the high mountains of Colorado. The very small, 1-flowered forms have been called *P. uniflora*

GOLDCUP POTENTILLA, *P. gracilis,* is usually 3–6dm tall with small flowers in open, flat-topped clusters. Each petal has an orange spot at its base. It grows in meadows and on open slopes of the montane and subalpine zones from Alaska south through Colorado.

BLUELEAF POTENTILLA, *P. diversifolia,* has spreading stems not over 3dm long. Its basal leaves usually have 5 leaflets. Occurs in moist meadows of the subalpine and alpine regions from the Canadian Rockies south to New Mexico.

137. Beauty Potentilla, ½X

SILVERWEED, *P. anserina,* is a plant of wet soil, with long, pinnate leaves that may have as many as 31 leaflets, which are sometimes silvery on both sides but may be green above and white beneath. Spreads by means of runners that root at their tips. Found along ditch banks and in meadows from the plains to the subalpine zone throughout our range. The specific name *anserina* means *of geese,* in reference to the plant being eaten by geese. Its root was used by Indians as both food and medicine. This species is sometimes put in the genus *Argentina*

PRAIRIE POTENTILLA, *P. pensylvanica,* has 5–9 leaflets, which are cut about halfway to the midrib and pinnately arranged. They are sometimes so crowded as to appear almost palmate. The leaflets are more or less hairy on their upper surfaces and usually tomentose beneath. Flowers are about 12mm broad. Found throughout our range on dry hillsides and fields in the foothill and montane zones. NELSON POTENTILLA, *P. plattensis,* is similar, but its leaves are mostly basal and may have as many as 17 leaflets. It is pubescent but never tomentose.

BEAUTY POTENTILLA, *P. pulcherrima* (fig. 137), has stems 3–6dm tall. Most of its leaves are basal on long petioles, and they may be either palmate or pinnate. The leaves usually have 7 leaflets, but there may be 5–11; dark green above, white beneath, obovate or oblanceo-late, with sharp teeth cut less than halfway to the midrib. The flowers are 12–20mm broad. Found in meadows and on mountain slopes throughout the Rockies up to timberline.

EARLY POTENTILLA, *P. concinna,* is a low, silvery plant with 5 leaflets. It grows on dry meadows and rocky hillsides, usually in the subalpine zone.

SILVERY POTENTILLA, *P. hippiana,* is a leafy-stemmed plant. The leaves have 7–13 toothed leaflets, which are gray with hairs on both surfaces. Common on fields and hillsides of the foothill and montane zones throughout our range.

LEAFY POTENTILLA, *P. fissa* (fig. 138), is a very abundant plant, 2–3dm tall and easy to recognize. Its stems are erect; the flowers in a narrow cluster are compara-tively large, up to 2.5cm broad, and usually of a paler shade of yellow than other species. Its short-petiolate leaves have 9–13 broad leaflets, which decrease in size from the terminal one to the basal pair. The plant is rather shaggy with brown, sticky hairs. Grows on rocky slopes and along roads and trails from the foothills to the subalpine zone in South Dakota, Wyoming, Colo-rado, and New Mexico. *P. glan-dulosa* is a related species that seems to replace the leafy poten-tilla in Montana and Idaho. It is taller, sometimes has reddish sta-mens, and its petioles are long and slender.

STICKY POTENTILLA, *P. arguta,* is a tall, rather weedy plant with sticky, hairy foliage. The crowded flowers are cream-col-ored or white. At first the petals

138. Leafy Potentilla, ½ X

are slightly longer than the calyx-lobes, but in fruit the calyx-lobes enlarge and bend together over the cluster of achenes. Grows in meadows throughout our range except in the alpine zone.

MARSH POTENTILLA, *P. palustris,* may be recognized by its red petals, which are shorter than the reddish sepals. Its leaves are pinnately compound with sharply-toothed leaflets, which are pale beneath. Grows in swamps and bogs from Alaska to southern Wyoming.

(16) *Prunus.* Plum, cherry, chokecherry

The flowers in this genus have 5 separate petals; many stamens, which are inserted on the edge of the calyx-tube; and 1 superior pistil, which develops into a fleshy fruit called a drupe, containing a hard stone. In general, the stones of plums are flattened; those of cherries are round.

1. Flowers in narrow racemes on leafy branches, many in a cluster; fruit dark-purple or black.

P. virginiana var. melanocarpa
Chokecherry

1. Flowers in umbellate clusters, 3–6, the clusters sessile (or nearly so), axillary.
  2. Calyx-lobes hairy inside and on margins; branches usually spiny; fruit a plum.

P. americana
Wild plum

  2. Calyx-lobes glabrous on both sides; branches never spiny; fruit a cherry.
    3. Low or prostrate shrubs, mostly under 1m tall; petals glabrous; fruit 12–18mm in diameter.

P. besseyi
Sand cherry

    3. Upright shrub, usually over 1m tall; petals hairy outside near base; fruit 6–7mm in diameter.

P. pensylvanica
Pin or bird cherry

139. Wild Plum, ²⁄₅ X

WILD PLUM, *P. americana* (fig. 139), is a rigid, much-branched, thicket-forming, somewhat spiny shrub or small tree. Bark grayish; few-flowered umbels of white, fragrant flowers, which open before the leaves; petals rounded; leaves 5–10cm long, taper-pointed at apex. The plums are about 2.5cm long, orange to red or purplish when ripe and much sought after by birds and people. They have a fine flavor for eating raw or making into jelly. Widely distributed from the Atlantic coast to the Rocky Mountains. Found in our area in foothill canyons and along streams on the eastern side of the Continental Divide.

140. Pin Cherry, ²⁄₅ X

PIN or BIRD CHERRY, *P. pensylvanica* (fig. 140), is less rigid than the plum, and its bark, except on old trunks, is glossy brown with horizontal markings. Its flowers are similar, but the fruits are small, bright red cherries. They also are very desirable for jelly. Blooms as the leaves are unfolding. Widely distributed east of the Rockies. A small tree in canyons at elevations between 6,000 and 7,000 feet, but it becomes a shrub at higher altitudes. In Colorado it is found as a low shrub even as high as the subalpine zone.

CHOKECHERRY, *P. virginiana* (fig. 141), is the common chokecherry of the United States. In our region, where it is called *P. virginiana* var. *melanocarpa,* it rarely exceeds 3.5m in height. The bark is reddish-brown, the leaves taper to both ends, and the creamy-white flowers hang in racemes. Its abundant dark red or black fruit is an important food for bears, small animals, and birds. The cherries are somewhat puckery in flavor but make delicious syrup and can be used for jelly by combining them with apple juice or other pectin. These shrubs form thickets in valleys and on hillsides of the foothill and montane zones throughout our range.

141. Chokecherry, ½X

(17) *Purshia.* Antelope brush or bitterbrush

Only one species is likely to be found in our range. ANTELOPE BRUSH, *P. tridentata* (fig. 142), is a shrub with 3-toothed leaves. It becomes covered with small, pale yellow flowers in late May and early June. Stamens about 25; carpels 1; fruit pubescent. Livestock and game animals browse this extensively. Usually seen as a low shrub spreading on banks or among rocks. Found throughout the Rocky Mountains in the foothill and lower montane zones.

142. Antelopebrush, ½X

(18) *Rosa.* Rose

The wild roses are all easily recognized as roses, but they are difficult to differentiate as species.

1. Stems bristly with slender prickles; the infrastipular prickles (1 or 2 prickles at or just below the nodes) not differentiated from other prickles.
   2. Usually 1 flower (or 2) on the new lateral branches; leaflets mostly 5–7.
                                      Rosa acicularis ssp. *sayi*
                                        Prickly rose
   2. Flowers cymose, either on lateral shoots or at the end of the main shoot; leaflets mostly 9–11.
                                        R. arkansana
                                        Arkansas rose

1. Stems with infrastipular prickles that are larger and differentiated from the other
      prickles.
      2. Flowers small and clustered; petals 1.2–2.5cm long; sepals 1–2cm long.
            3. Up to 1m tall; leaflets 1–2 cm long, the teeth sometimes gland-tipped.
                                          R. woodsii var. woodsii
                                          Plains woods rose
            3. Usually 1–2m tall; leaflets to 5cm long, the teeth not gland-tipped.
                                          R. woodsii var.
                                          ultramontana
                                          Mountain woods rose
      2. Flowers larger and often solitary; petals 2.5–4cm long; sepals 1.5–4cm long.
                                          R. nutkana var. hispida
                                          Nootka rose

ARKANSAS ROSE, *Rosa arkansana,* is usually under 6dm tall, and its stems
often die back to the ground during winter. It has 9–11 leaflets, and the flowers
are usually clustered. Widely distributed in the western United States, in our
area it is most frequently found on railroad embankments, ditch banks, and
rocky hillsides of the foothill zone.

WOODS ROSE, *R. woodsii,* can be the tallest of our roses (see key above). Its
blossoms are usually clustered, 2–4, and its fruits are round and hard.

PRICKLY ROSE, *R. acicularis,* is usually 3–9dm tall but may grow to be 18dm.
Its blossoms are usually solitary, occasionally 2. Its fruits are often elongated
and soft when ripe, and its leaves turn beautiful shades of orange, rose, and
maroon in the fall. Spreads by underground stems and forms patches. Both it
and *R. woodsii* are found from the foothills to timberline throughout our area.

(19) *Rubus.* Raspberry; blackberry

1. Stems with prickles or stiff bristles.
      2. Stems with stiff bristles; inflorescence racemose; fruit red.
                                          R. idaeus var. sachaliensis
                                          Red raspberry
      2. Stems with prickles; inflorescence corymbose; fruit black or purple.
            3. Pedicel prickles usually straight, not stout or broad-based; fruit black.
                                          R. occidentalis
                                          Black raspberry
            3. Pedicel prickles usually hooked, very stout and broad-based; fruit purple.
                                          R. leucodermis
                                          Black raspberry
1. Stems entirely unarmed.
      2. Leaves compound, leaflets 3; stems herbaceous.
            3. Petals white; stems prostrate.

                                          R. pubescens
                                          Dwarf red blackberry

            3. Petals rose-colored; stems erect and tufted.

                                          R. acaulis
                                          Nagoonberry

      2. Leaves simple (only lobed); stems woody.
            3. Flowers in clusters, 2–9; leaves 6–30cm wide.

                                          R. parviflorus
                                          Thimbleberry

            3. Flowers solitary; leaves 4–6cm wide.
                  4. Peduncles and petioles glandular-hairy.

                                          R. exrubicundus

4. Peduncles and petioles not glandular-hairy.

R. deliciosus
Boulder raspberry

BOULDER RASPBERRY or ROCKY MOUNTAIN THIMBLEBERRY, *R. deliciosus* (fig. 143), is a handsome shrub with rounded and more or less lobed, bright green leaves, and white, rose-like flowers 5–8cm across. Grows from 6–9dm tall, with arching branches clothed with light brown shreddy bark. The young tips and sprouts are bright rose during the growing season. Its raspberry-like insipid fruit (the specific name is a misnomer) is eaten by birds and animals. Grows on rocky slopes and banks of the foothill and montane zones in Colorado and southern Wyoming, mostly on the eastern slope. One of the best of our native shrubs for garden use.

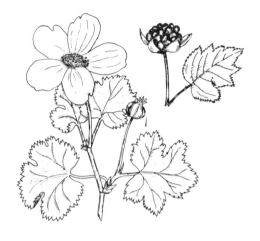

143. Boulder Raspberry, ½X

THIMBLEBERRY, *R. parviflorus* (fig. 144), is a related shrub with smaller, more cup-shaped flowers, and large, angular, 3- to 5-lobed leaves. Occurs over a wide area from Alaska to Mexico. Found in a few moist, shaded canyons on the eastern slope in Colorado, but much nore abundantly in the north central mountains.

WILD RED RASPBERRY, *R. idaeus,* is a bristly shrub with

144. Thimbleberry, ⅖X

compound leaves of 3 or 5 leaflets, which are usually white on their under surfaces and sharply toothed. The white flowers, which are in terminal and axillary clusters, are followed by typical raspberry fruits of excellent flavor. Grows along trails, roadsides, and on rocky slopes from the foothills to the alpine zone throughout mountainous areas of the United States. Recently-burned areas are often invaded by it.

(20) *Sibbaldia*. Sibbaldia

Only one species in North America. SIBBALDIA, *Sibbaldia procumbens,* is a dwarf, tufted, cushion plant with bluish, trifoliate, compound leaves, the leaflet 3-toothed at apex. The calyx is 5-lobed with 5 alternating bractlets, and the 5 minute yellow petals are shorter than the sepals. The plant is circumpolar and abundant at high altitudes, especially on gravelly tundra.

(21) *Sorbus*. Mountain-ash

MOUNTAIN ASH, *Sorbus scopulina* var. *scopulina* (fig. 145), is easily recognized at any time by its symmetrically patterned leaves made up of 11–15 evenly and sharply toothed leaflets. In early summer it bears large, flat-topped clusters of white flowers, which are followed by orange or red berries. The berries are like tiny apples, each one showing the tips of the 5 calyx-lobes. Found on moist hillsides of the montane and subalpine zones from New Mexico northward. Grows as a large shrub. Its main stems are often decumbent where they have been bent down by the heavy snow of the subalpine region.

145. Mountain Ash, ½X

(22) *Spirea*. Spirea

1. Shrubs 4–8dm high; leaves alternate, larger above than below, pale and glaucescent beneath, 5–8cm long; flowers in corymbs, white.

> *S. betulifolia var. lucida*
> Meadowsweet

1. Shrubs rarely 4dm tall, stems usually decumbent and branching; leaves alternate, glabrous 2–4cm long; flowers in corymbs, rose color.

S. densiflora
Pink meadowsweet

MEADOWSWEET, *S. betulifolia* var. *lucida* (fig. 146), has stems rising singly from a long, woody, underground stem. The upright stems scarcely seem woody. Each bears a flat-topped corymb of many small, white, fragrant flowers. The PINK MEADOWSWEET, *S. densiflora,* has brown, shreddy bark and a rounded panicle of pink flowers.

PEA FAMILY: LEGUMINOSAE

The members of this family are easily recognized by their distinctively irregular corollas. All species commonly found within our range have *papilionaceous* corollas (see Plate E49). The flowers have a united calyx; 5 petals, including a *banner,* 2 *wings,* and 2 lower petals joined to form the *keel*; 10 stamens; and 1 pistil, which develops into a 1-celled, usually several-seeded pod called a *legume*. The stamens and pistil are usually enclosed by the keel. (Sweet pea and clover blossoms are typical papilionaceous flowers, and bean and pea pods are typical legumes.) Many plants of this

146. Meadowsweet, ½X

family have 9 of the stamens joined together by their filaments and 1 separate. Most species have compound leaves, which usually have stipules.

1. Stamens 10, entirely distinct.
    2. Leaves digitately 3-foliolate; flowers yellow.

Thermopsis p. 204
Golden banner

    2. Leaves odd-pinnate; flowers white.

Sophora p. 204
Sophora

1. Stamens (some of them) united by their filaments, at least at base (i.e. monadelphous or diadelphous).
    2. Anthers of 2 forms; stamens monadelphous; leaves digitately 5 or more-foliate.

Lupinus p. 197
Lupine

    2. Anthers all alike, reniform.
        3. Leaves even-pinnate, terminated by a tendril or bristle.
            4. Style slender, with a tuft of hair near the apex.

Vicia p. 207
Vetch

            4. Style flattened, hairy on the inner side.

Lathyrus p. 195
Peavine

        3. Leaves odd-pinnate, without tendrils.

4. Pod a loment (separates transversely into joints at maturity).

Hedysarum p. 194
Sweetvetch

4. Pod not a loment, either 2-valved or indehiscent.
  5. Foliage not glandular-dotted.
    6. Digitately 3-foliolate, rarely 5-foliolate.
      7. Leaflets entire.

Lotus p. 196
Deervetch

      7. Leaflets serrulate or denticulate.
        8. Flowers in racemes.

Melilotus p. 199
Sweetclover

        8. Flowers capitate or in loose spikes.
          9. Pods curved or coiled.

Medicago p. 199
Alfalfa, lucerne, or medic

          9. Pods straight, membranous.

Trifolium p. 205
Clover

    6. Pinnately 5- or more-foliolate (rarely simple or 3-foliolate in *Astragalus*).
      7. Shrubs.

Robinia p. 203
Locust

      7. Herbs, or rarely with a woody base.
        8. Keel of the corolla blunt.

Astragalus p. 188
Milkvetch

        8. Keel of the corolla acute.

Oxytropis p. 199
Loco

  5. Foliage glandular-dotted.
    6. Pod with hooked prickles.

Glycyrrhiza p. 194
Licorice

    6. Pods not prickly.
      7. Leaves digitately 3–5-foliolate.

Psoralea p. 202
Scarf pea

      7. Leaves pinnately 5 to many-foliolate.
        8. Stamens 10.

Dalea p. 194
Prairie-clover

        8. Stamens 5.

Petalostemon p. 202
Prairie-clover

(1) *Astragalus*. Milkvetch

This large genus includes many species that differ widely in general appearance, and many others that are so similar to each other that identification depends on small technical features. Botanists consider it a difficult genus. Accurate determination is usually impossible without well-developed pods. The flowers vary in size but are quite constant in form and structure. At flowering time the calyx is a sleeve-like tube with 5-pointed teeth. It enlarges or is split open as the legume develops. The tip of the keel is rounded (this character distinguishes *Astragalus* from the locoweeds of the genus *Oxytropis,* which have a sharply-pointed keel). A few species are cushion or mat plants with their leaves reduced to one or few leaflets; a few others are of low habit and more or less caespitose; but most have erect or ascending, branched stems and odd-pinnate leaves with numerous leaflets.

Many species grow on alkali, or "badland" type, soils where *selenium* is present, and they absorb the poisonous mineral into their foliage. They become dangerously poisonous to livestock if eaten and are sometimes called locoweeds by stockmen. But that name is used more properly for plants of the genus *Oxytropis*. Most of the seleniferous species are found at elevations below our range. Anyone needing accurate identification will do well to send specimens, including flowers and well-developed legumes, to an experiment station or herbarium.

1. Leaves not pinnate, either unifoliate blades, trifoliate, or 5–7-foliate.
  2. Leaves mostly unifoliate (some 3–5-foliate) with silky-canescent appressed hairs; leaves tufted.
                    A. spatulatus
                    Tufted milkvetch
  2. Leaves not unifoliate or tufted.
    3. Leaves with 5 or more leaflets, spine-tipped, rigid, not densely pubescent.
                    A. kentrophyta var.
                    implexus
                    Matvetch
    3. Leaves not spine-tipped, either 3-foliate or reduced to phyllodia (broadened petioles that appear to be linear leaves).
      4. Leaves 3-foliate, densely white pubescent.
                    A. tridactylious
                    Three-fingered milkvetch
      4. Leaves reduced to phyllodia.
        5. Calyx tube 2–3.5mm long; pods oblong or elliptic, much inflated.
                    A. ceramicus var. filifolius
                    Painted milkvetch
        5. Calyx tube 3.2–5.4mm long; pods linear to oblong, compressed.
                    A. convallarius var.
                    convallarius
                    Rushy milkvetch
1. Leaves pinnate.
  2. Pubescence dolabriform (pick-shaped).
    3. Stipules clasping opposite the petiole (i.e. connate).
      4. Flowers declined, pale-greenish to ochroleucous; stems rising singly or few together, from rhizomes.
                    A. canadensis var.
                    brevidens
                    Canada milkvetch
      4. Flowers erect or ascending at a narrow angle; stems arising from root crown or forking caudex; flowers purplish.
                    A. adsurgens var.
                    robustior
                    Standing milkvetch
    3. Stipules decurrent but not connate opposite the petiole; flowers purple, ascending.
                    A. missouriensis
  2. Pubescence basifixed, not dolabriform.
    3. Stipules fully clasping the stems (connate) opposite the petiole.
      4. Banners 15mm or more long.
        5. Flowers erect, clustered in ovoid heads; pods sessile, erect, hirsute.
                    A. agrestis
                    Field milkvetch

5. Flowers mostly spreading, in elongate racemes; pods stipitate or
sessile pendulous.
6. Calyx-tube white-hairy, 6–9mm long; leaflets 2–3cm long, with
straight appressed hairs; pods 2–3.5cm long, straight.
A. racemosus
6. Calyx-tube grayish- to black-hairy.
7. Pedicels 1–3mm long; pubescence spreading; pods stipitate.
8. Calyx teeth nearly equal to the tube; pods 15–22mm
long, nearly straight.
A. bisulcatus
Two-grooved milkvetch
8. Calyx teeth about 1/2 as long as the tube; pods
20–40mm long, nearly straight.
A. drummondi
Drummond milkvetch
7. Leaves sessile; leaflets linear or filiform, grayish with few
appressed hairs; pods 10–20mm long, fleshy when
green, woody at maturity; a malodorous plant.
A. pectinatus
Tine-leaved milkvetch
4. Banners 15mm or less long.
5. Racemes with 25-80 flowers.
A. bisulcatus
Two-grooved milkvetch
5. Racemes with 2–20 flowers.
6. Plants with widespread rootstocks (rhizome-like) or with
buried, branched crowns, with many slender stems.
7. Stems from rhizome-like caudex branches; flowers pale
lilac to purplish; pods black-pilose, 8–12mm long.
A. alpinus
Alpine milkvetch
7. Stems from underground root crowns; flowers pink to
yellowish; pods sparsely pubescent, 12–21mm long.
A. flexuosus
Wire milkvetch
6. Stems rising from more limited root crowns or taproots.
7. Plants freely branched from the base.
8. Petals bicolored, whitish with keels purple or purple-
tipped.
9. Calyx 3–4mm long, the teeth triangular; keel
purple-tipped; pod 20–30mm long.
A. miser
Weedy milkvetch
9. Calyx 4–8mm long, the teeth linear, often black-
hairy the keel purple; pod 15–30mm long.
A. aboriginum
Indian milkvetch
8. Petals purple to reddish-lilac; calyx 4–5mm long; pod
about 10mm long.
A. bodinii
Bodin milkvetch
7. Plants simple and without leaves at the base.
8. Lower stipules with dark bands; pods 7–15mm long,
glabrous or with white hairs.
A. tenellus
Loose-flowered milkvetch
8. Lower stipules uniformly green or brown; pods
8–12mm long, gray-to blackish-hairy.
A. eucosmus
Elegant or arctic
milkvetch

3. Stipules decurrent but not connate opposite the petiole.
    4. Banners under 15mm long.

> *A. eucosmus*
> Elegant or arctic
> milkvetch

    4. Banners 15mm or more long.
        5. Ovary and pod glabrous.
            6. Stems and leaves with short, straight, appressed hairs; ovary and pod sessile; pods thick and fleshy, subglobose.

> *A. crassicarpus*
> Groundplum

            6. Stems and leaves villous-hirsute; ovary and pod stipitate; pods linear and pendulous.

> *A. drummondii*

        5. Ovary and pod pubescent.
            6. Plants caulescent; stems 10–30cm long; pods 10–20mm long, villous.

> *A. parryi*
> Parry milkvetch

            6. Plants subacaulescent; stems under 10cm long.
                7. Pods shaggy-villous, ovoid.

> *A. purshii*
> Woolly-pod milkvetch

                7. Pods with short, appressed hairs, 25–45mm long, not ovoid.

> *A. shortianus*
> Short's milkvetch

MATVETCH, *A. kentrophyta* var. *implexus,* is a very low cushion or mat-forming plant with tiny, stiff, spine-tipped leaflets and small purple flowers, which are partly hidden among the leaves. Occasionally found on moist, granitic soils in openly wooded areas from Montana and Idaho south to Colorado. The TUFTED MILKVETCH, *A. spatulatus,* is somewhat similar in habit of growth, but it has softer, spineless leaves, often with one long, slender leaflet, and numerous clusters of purple or cream-colored flowers held above the foliage on slender stalks. Grows on dry, sandy, exposed areas at altitudes of 5,000–9,000 feet from Canada along the mountains into northern Colorado.

EARLY PURPLE MILKVETCH, *A. shortianus* (fig. 147), is a plant of the mesas and foothills, especially along the eastern slope. It has stems about 10cm tall, each bearing a crowded cluster of rose-purple flowers 2.5cm long. The rounded leaflets are covered with silky hairs, and the pods are 2.5–5 cm long and curved. Occurs from southern Wyoming to northern New Mexico.

147. Early Purple Milkvetch, ½ X

148. Parry Milkvetch, ²⁄₅ X

149. Groundplum, ½ X

PARRY MILKVETCH, *A. parryi* (fig. 148), is a tufted, gray-hairy plant with creamy-white blossoms that have a purple-tipped keel and are sometimes tinged with pink. Its pods are about 2.5cm long. Found along the eastern slope on dry fields and open areas of the foothill and montane zones from southern Wyoming to central Colorado. PURSHES or WOOLLY-POD MILKVETCH, *A. purshii,* is similar, but it has very woolly pods and is frequently found in the western part of our range.

GROUNDPLUM, *A. crassicarpus* (fig. 149), is another low-growing species, with stems 15–30cm long. Its flowers are white with blue or purple tips. The fruits are round, smooth, and fleshy. When ripe they are hard and do not split open. This species is found on high plains and fields of the foothill zone from Canada to Colorado.

INDIAN MILKVETCH, *A. aboriginum,* has 10–30cm stems in tufts that grow from large woody taproots. The flowers are whitish with purple tips, and the linear or oblong leaflets are usually covered with curly hairs. The smooth, thin pods, 15–30mm long, hang from the calyx on slender stalks. Its roots were used for food by Indians. Occurs from the Yukon south into Nevada and Colorado in open woods and on hillsides, often on limestone, from 7,000 to 10,000 feet, and is found in Yellowstone and Glacier National Parks.

ARCTIC MILKVETCH, *A. eucosmus,* is a tufted plant with several erect or ascending stems that grow from a woody taproot. The leaves are green with oblong leaflets that have appressed hairs on their undersides. The dark purple or blue flowers are 8mm long, in a loose, often one-sided, raceme. The pendant pods, which are appressed-hairy, are comparatively short and thick, not over 2cm long. It grows in partly shaded situations on moist stream banks and borders of lakes in the montane and subalpine zones from arctic America through Montana and Wyoming into the Colorado mountains.

Several of the larger species of the genus *Astragalus* become conspicuous in early summer in fields and pastures and along roadsides on the mesas, high plains, and mountain valleys. These plants may be from 3 to 6dm tall, with several leafy stems forming clumps. One of the most frequently seen is the Two-grooved Milkvetch, *A. bisulcatus,* with its erect stems and numerous rose-purple flowers, which are usually bent downward and arranged in compact racemes. The pods are 15–22mm long with 2 grooves along the upper surface. This is one of the worst of the stock-poisoning plants. It grows on alkali and seleniferous soils, usually below 7,000 feet, from Canada southward into New Mexico.

Drummond Milkvetch, *A. drummondii* (fig. 150), is another species with erect, bunched stems. It has conspicuously gray-hairy foliage, whitish or cream-colored flowers, and pods 20–40mm long, which are stipitate and protrude from the more or less black-hairy calyx. Grows on dry fields of the foothills and lower montane zone from Canada to northern New Mexico, mostly east of the Continental Divide. *A. tenellus* is another species with cream-colored flowers. Its leaflets are green, narrow, and usually smooth. The stems, 25–45cm long, are often reddish at least toward the base, and more or less decumbent. The thin-walled pods are 7–15mm long. Occurs from Canada southward to Colorado and Nevada.

150. Drummond Milkvetch, ½X

Alpine Milkvetch, *A. alpinus* (fig. 151), is a dainty plant with conspicuously two-toned flowers. The banner and wings are pale bluish, violet or more rarely white, and the keel is purple. The stems are 10–30mm in height; the leaflets are oval and notched at apex. The calyx is covered with black hairs, and the stipitate, slightly-inflated pods are

151. Alpine Milkvetch, ½X

pendant. Grows in shaded locations from the upper montane through the subalpine zone. This species is circumboreal in distribution, occurring throughout the northern and mountainous regions of America, Europe, and Asia.

Wire Milkvetch or Limber Vetch, *A. flexuosus,* is a sprawling plant with stems 15–50cm long, racemes of small pink or yellowish blossoms, and narrow leaflets. The calyxes are covered with short, appressed hairs, some of which are black. The slender, cylindric pods are usually under 2.5cm long. Grows on fields and hillsides of the foothill and montane zones from Montana into Utah and New Mexico.

A. convallarius has upper leaves so reduced that the plant appears rush-like. Its whitish flowers are about 10mm long with an erect banner. It has creeping rootstocks with stems 15–45cm long. Grows on dry slopes, often among sagebrush, in the foothill and lower montane zones from Montana to Colorado and northeastern Arizona.

(2) *Dalea*. Prairie Clover

1. Annual; inflorescence a densely-flowered, oblong spike; leaves odd-pinnately compound leaflets 13–41 per leaf; white or rose-tinged.
   D. leporina
1. Perennial; spike densely flowered, corolla yellow when fresh; leaves palmately compound with 3 leaflets.
   D. jamesii
   James dalea

JAMES DALEA, *Dalea jamesii,* has numerous, silky, somewhat decumbent stems 5–25cm long, which are tufted at the top of a thick, woody root. Its leaves are of 3 silky leaflets, under 2.5cm long, and the yellow flowers are in dense, woolly clusters. They turn rose or purple as they dry. Grows on dry plains and hillsides of the foothills and lower montane zones from Kansas and Colorado south into Mexico.

(3) *Glycyrrhiza*. Licorice

Only 1 species found in our range. WILD LICORICE, *Glycyrrhiza lepidota* var. *lepidota,* is a stout, weedy, widely-distributed plant of ditch banks and moist, sandy soil, found in the foothills. Leafy stems; leaves odd-pinnate, many leaflets, entire, glandular-dotted; flowers greenish-white in compact, axillary clusters; fruits sessile, brown, covered with hooked prickles. The common licorice used to flavor candy is a different species, but the root of our species is equally sweet and was used as food by the Indians.

(4) *Hedysarum*. Sweetvetch

1. Flowers yellowish to nearly white.
   H. sulphurescens
   Yellow sweetvetch
1. Flowers pink or reddish to purplish.
   2. Upper calyx-lobes slender, about equal to the lower lobes and to the calyx-tube; loments (pods) cross-corrugated (obviously folded).
      H. boreale
      Northern sweetvetch
   2. Upper-calyx lobes broader but shorter than the lower 3; pods not obviously cross corrugated.
      3. Flowers 11–18mm long; loment 3.5–6mm broad.
         H. alpinum
      3. Flowers 16–22mm long; loment 7–12mm broad.
         H. occidentale
         Western sweetvetch

Plants of this genus look very much like the milkvetches but may be distinguished by their pods, which are pendant and made up of separable segments. NORTHERN SWEETVETCH, *H. boreale* (fig. 152), has racemes of rose or purple flowers. The pod is divided into 2–4 round segments. Grows in rocky places throughout North America and in the higher mountains as far south as Utah and New Mexico. Its roots are eaten by Indians and Eskimos. WESTERN SWEETVETCH, *H. occidentale,* is similar, but the segments of the fruit are elliptic and sometimes narrowly winged. Occurs in Wyoming, Idaho, and southwestern Colorado.

152. Northern Sweetvetch, ½ X

(5) *Lathyrus.* Peavine

Plants of this genus have their leaves tipped with either tendrils or bristles. Our common native species are usually more bushy and less inclined to. trail or climb than those of *Vicia,* and their flowers are usually larger. Some species are very fragrant. The cultivated sweetpea belongs to this genus.

1. Flowers purple.
   2. Tendrils not prehensile (not adapted for seizing or grasping), but bristlelike.
      3. Leaflets glabrous.

      L. branchycalyx

      3. Leaflets villous-canescent.

      L. polymorphus ssp. incanus
      Hoary peavine
   2. Tendrils prehensile.
      3. Calyx sparsely hairy, teeth shorter than the tube; calyx 10–15mm long; 6–88 leaflets, oblong to oblong-elliptic.

      L. eucosmus
      Purple peavine

      3. Calyx glabrous (occasionally pubescent).
         4. Calyx teeth longer than the tube; calyx 8–18mm long; flowers 18–27mm long; 6–13 leaflets, linear to ovate.

         L. pauciflorus var. utahensis

         4. Calyx teeth shorter than the tube; calyx 5–8mm long; flowers 8–16mm long; leaflets 4–10, linear to oblong-elliptic.

         L. lanszwertii
         Thick-leaved peavine
1. Flowers white or yellowish.
   2. Tendrils never prehensile; leaflets 2–6.

   L. arizonicus

2. Tendrils usually prehensile; leaflets 6 or more.
   3. Calyx 7–9mm long, the teeth shorter than the tube; flowers 14–22mm long.
                                                        L. leucanthus var.
                                                        leucanthus
                                            White-flowered peavine
   3. Calyx 8–10mm long, the teeth obviously unequal in length; flowers
      12–16mm long.
                                                     L. ochroleucus
                                                Cream-flowered peavine

153. White-Flowered Peavine, ½X

WHITE-FLOWERED PEAVINE, *L. leucanthus* (fig. 153), has cream-colored flowers that turn light-brown as they fade. Its leaves sometimes end in short tendrils, sometimes only in bristles. Occurs in large patches on banks of gulches and on hillsides from the foothills to timberline, from Wyoming and Utah through Colorado into New Mexico and Arizona.

PURPLE PEAVINE, *L. eucosmus,* has climbing tendrils, at least on the upper leaves. Grows in the foothills from Colorado and Utah to Mexico.

(6) *Lotus.* Deervetch

1. Plants annual; flowers whitish, 4–7mm long; leaflets mostly 3; fruit glabrous.
                                                  L. purshianus
                                                  Spanish clover
1. Plants perennial; flowers yellow to orange, sometimes reddish-tinged, 8–15mm long.
   2. Leaflets 5; pod 20–40mm long, glabrous.
                                                  L. corniculatus
                                                  Bird's foot trefoil
   2. Leaflets 3–6; pod 20–25mm long, strigose (with appressed short hairs).
                                                  L. wrightii
                                                Wright deervetch

WRIGHT DEERVETCH, *Lotus wrightii,* is a plant with spreading, branching stems covered with grayish hairs; yellow or orange flowers, which turn reddish as they fade; and 3–6 crowded leaflets. Grown on rocky slopes, often in shade, from southwestern Colorado into Utah, Arizona, and New Mexico. Its seeds and foliage are a valuable source of food for wildlife.

(7) *Lupinus.* Lupine

All plants in this genus may be recognized by their distinctive leaves: 5-9 long, narrow leaflets, all evenly attached at the tip of the petiole, and always more or less hairy. The pea-form flowers are usually blue, sometimes two-toned, and in one case dingy white.

1. Plants annuals.
    2. Flowers in dense racemes, mostly less than 2cm long.
        3. Leafy stem hardly 1cm long (subacaulescent), the leaves mostly basal.

> *L. brevicaulis*
> Short-stemmed lupine

        3. Plants caulescent, the leafy stem at least 3cm long and branched.

> *L. kingii*
> King's lupine

    2. Flowers in loose racemes, mostly over 3cm long.

> *L. pusillus*
> Low lupine

1. Plants perennials.
    2. Plants dwarf and caespitose, 5–15cm tall.

> *L. caespitosus*
> Cushion lupine

    2. Plants medium to large, 2dm or more tall.
        3. Flowers conspicuously bicolored; leaflets green, appressed-pubescent beneath.

> *L. plattensis*
> Nebraska lupine

        3. Flowers often bi- or tri-colored, but not strikingly so.
            4. Herbage very canescent or rough-hoary.
                5. Flowers 10–12mm long; pedicels 4–7mm long; racemes 10–15cm long.

> *L. sericeus*
> Silky lupine

                5. Flowers 8–10mm long; pedicels 1–3mm long; racemes 15–30cm long.

> *L. leucophyllus*
> Velvet lupine

            4. Herbage green or greenish, at least in part.
                5. Pubescence of stems and petioles appressed.
                    6. Flowers 5–7mm long.

> *L. argenteus var.*
> *parviflorus*
> Lodgepole lupine

                    6. Flowers 9–11mm long.

> *L. argenteus var.*
> *argenteus*
> Silvery lupine

                5. Pubescence of stems and petioles spreading or reflexed, or the plant nearly glabrous.
                    6. More or less rough-hirsute.
                        7. Raceme about equal in length to the leaf-bearing part of the stem.

> *L. ammophilus*

                        7. Raceme relatively short, the stems tall.
                          8. Flowers ochroleucous (to purple).

> *L. barbiger*

                          8. Flowers blue (or pinkish).

> *L. wyethii*
> Wyeth's lupine

6. Plant glabrous or nearly so.

L. polyphyllus var. burkei
Large-leaved lupine

KING'S LUPINE, *L. kingii,* is a low annual or biennial, usually having dark blue flowers. It is found on dry ground of the foothills in Utah, New Mexico, Arizona, and west of the Continental Divide in Colorado. LOW LUPINE, *L. pusillus,* is a similar annual, with pale or almost white flowers. It is found on sandy soil of the foothills and plains along the eastern side of the Rockies from Canada south into New Mexico.

CUSHION LUPINE, *L. caespitosus,* is a mat-forming perennial with leaflets not over 2.5cm long and very short clusters of pale blue or lavender flowers, which are partly buried in the leaves. Grows on dry hills in the montane zone from Montana to central Colorado and is often seen in Yellowstone National Park.

NEBRASKA LUPINE, *L. plattensis,* is a showy plant between 3 and 6dm tall, found on the high plains and in the foothills of Nebraska, Wyoming, western Kansas, and eastern Colorado. Frequently associated with sagebrush, it is easily recognized by the large, bicolored corolla.

154. Silvery Lupine, ½X

SILVERY LUPINE, *L. argenteus* var. *argenteus* (fig. 154), is a common blue-flowered, sometimes bicolored, lupine of the mountains. Found from the foothills through the subalpine zone, from Montana to New Mexico and Arizona. Occurs in all the national parks of the Rocky Mountain region. It is frequently found at the edge of the woods or other partly shaded places, but it grows in the open at higher altitudes.

LODGEPOLE LUPINE, *L. argenteus* var. *parviflorus,* has long racemes of bluish to dingy white flowers, which turn brown as they fade. This is the commonest small-flowered lupine found in the montane zone throughout our range.

SILKY LUPINE, *L. sericeus,* is a large, branched plant usually having several stems 3–6dm tall. A very silky and conspicuous species found on fields and open slopes, often with sagebrush, throughout the western parts of the Rocky Mountains and particularly in Glacier and Yellowstone National Parks.

(8) *Medicago*. Alfalfa, lucerne, or medic

1. Perennials, deep-rooted; mostly erect.
    2. Flowers blue; leaflets 2–4cm.

M. sativa
Alfalfa or lucerne

    2. Flowers mostly yellow; leaflets 1–2cm.

M. falcata
Yellow lucerne

1. Annuals, shallow-rooted; stems prostrate to somewhat ascending; yellow-flowered.
    2. Racemes ovoid, many; fruit pubescent but not prickly.

M. lupulina
Hop clover

    2. Racemes subcapitate, few; fruit glabrous but prickly on the edges.

M. hispida
Bur clover

(9) *Melilotus*. Sweet clover

1. Corolla white; fruit ovoid, glabrous, not rugose (not wrinkled).

M. alba
White sweet clover

1. Corolla yellow; fruit ovoid, rugose.

M. officinalis
Common yellow sweet clover

YELLOW SWEET CLOVER, *Melilotus officinalis* (fig. 155), is an escape from cultivation and is widely distributed, especially along roadsides and in fields below 8,000 feet. It is usually a tall, branched, bright green plant with many short racemes of heavily sweet-scented yellow flowers. But sometimes after the original upright stems have been cut by road machinery, the plant sends out lateral branches near the ground and goes on blooming. WHITE SWEET CLOVER, *M. alba,* grows in similar locations. Both furnish good forage for livestock, and the flowers provide excellent bee pasture.

155. Yellow Sweet Clover, ½X

(10) *Oxytropis*. Loco

The plants of this genus are often confused with the milkvetches. Some species of each are poisonous to stock and are indiscriminately called *locoweeds*. Both have pinnately compound leaves, usually of numerous leaflets, and papilionaceous flowers. In general, the milkvetchs are leafy-stemmed plants (caulescent), and the oxytropes are stemless (acaulescent) with basal leaves, but there are exceptions in both groups.

The best character by which to distinguish the two genera is the shape of the keel: the tip of the keel is blunt in the milkvetches and pointed in the oxytropes (Plate E49). The poisonous element in oxytropes is said to be derived from soils that contain the mineral *barium*. It acts as a narcotic on animals who eat considerable amounts of it. They become addicted and, at times, "go loco." Several of the species are inclined to hybridize, producing color variations in the hybrid populations.

1. Plants caulescent; stipules slightly adnate by their bases to the petioles; pods pendulous.
　　　　　　　O. deflexa
　　　　　　　Drop-pod loco
1. Plants acaulescent; scapose; stipules adnate to the petiole; pods erect or spreading.
　　2. Inflorescence glandular-viscid.
　　　　　　　O. viscida var. viscida
　　　　　　　Sticky loco
　　2. Inflorescence not glandular-viscid.
　　　　3. Leaves with whorled leaflets.
　　　　　　　O. splendens
　　　　　　　Showy loco
　　　　3. Leaves only pinnate, never whorled.
　　　　　　4. Racemes 1- to 5-flowered; corolla purple to pinkish.
　　　　　　　　5. Fruiting calyx inflated, completely enclosing the ripe fruit.
　　　　　　　　　　O. multiceps
　　　　　　　　　　Tufted loco
　　　　　　　　5. Fruiting calyx not inflated, nor enclosing the ripe fruit.
　　　　　　　　　　6. Pods papery and inflated, ellipsoidal.
　　　　　　　　　　　　O. podocarpa
　　　　　　　　　　　　Stalked-pod loco
　　　　　　　　　　6. Mature pods leathery, not inflated;, cylindrical or oblong.
　　　　　　　　　　　　O. parryi
　　　　　　　　　　　　Parry loco
　　　　　　4. Racemes 6- to many-flowered; corolla purple, lavender, white, or ochroleucous.
　　　　　　　　5. Corolla purplish, pinkish, or bluish.
　　　　　　　　　　√6. Pods inflated, ovoid, with abrupt beak.
　　　　　　　　　　　　O. besseyi var. obnapiformis
　　　　　　　　　　　　Bessey loco
　　　　　　　　　　6. Pods not inflated, oblong or cylindrical, gradually narrowing into a beak.
　　　　　　　　　　　　7. Calyx with spreading hairs; pubescence of foliage basifixed.
　　　　　　　　　　　　　　O. lagopus
　　　　　　　　　　　　　　Rabbit-foot loco
　　　　　　　　　　　　√7. Calyx with appressed hairs; some of the foliage hairs dolabriform (2-armed or pick-like).
　　　　　　　　　　　　　　O. lambertii
　　　　　　　　　　　　　　Lambert or Colorado loco.
　　　　　　　　5. Corolla white or ochroleucous, the keel often purple-tinged.
　　　　　　　　　　6. Corolla ochroleucous, flowers 10–20mm long; foliage green; pods papery.
　　　　　　　　　　　　7. Stipules very hairy; scapes mostly over 15cm tall.
　　　　　　　　　　　　　　O. campestris var. gracilis
　　　　　　　　　　　　　　Field loco
　　　　　　　　　　　　7. Stipules glabrous or glabrate; scapes rarely over 15cm tall.
　　　　　　　　　　　　　　O. campestris var. cusickii
　　　　　　　　　　√6. Corolla white, flowers 18–25mm long; foliage gray or silvery; mature pods leathery.
　　　　　　　　　　　　O. sericea var. sericea
　　　　　　　　　　　　Silky or Rocky Mountain loco

LAMBERT LOCO, *O. lambertii* (fig. 156), is one of the most conspicuous and beautiful wildflowers of the Rocky Mountains. Its foliage is silvery with appressed silky hairs, and the bright rose-purple flowers are held erect in racemes on leafless stalks. *O. parryi* is a somewhat similar but smaller, alpine species found at high altitudes throughout the Rocky Mountains.

SHOWY LOCO, *O. splendens* (fig. 157), is a handsome plant with silvery, silky foliage, rose-colored flowers, and numerous leaflets arranged in whorls. Grows in the montane zone.

DROP-POD LOCO, *O. deflexa,* has many-flowered, crowded racemes of whitish, sometimes purple-tipped, flowers that elongate in

156. Lambert Loco, ½X

157. Showy Loco, ½X (leaf only)

fruit. Some black hairs are mixed with light ones on the calyx. As the pods ripen, they hang. The leaflets are so spaced as to give a ladder-like appearence. Grows on hillsides and especially in lodgepole pine forests from Canada to New Mexico.

TUFTED LOCO, *O. multiceps* (fig. 158), often grows in a low circle. Its leaves are gray with silky hairs. Each short stem bears only a few rose-colored flowers. As the petals fade, the calyx turns red and becomes inflated so that the plant appears as a mat of gray, rose, and red. Found on granitic gravel from northwestern Nebraska and Wyoming to central Colorado in the foothill and montane zones, and occasionally into the subalpine zone.

STICKY or YELLOWHAIR LOCO, *O. viscida,* is an erect plant with sticky green foliage and blue flowers. A rare alpine in Colorado, but it occurs more abundantly farther north, especially in Yellowstone National Park.

FIELD LOCO, *O. campestris* var. *gracilis,* is a taller plant than most of the other locos, the stalks often 30–50cm high, and it bears crowded racemes of

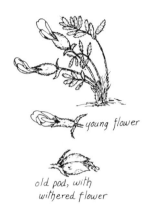

young flower

*old pod, with
withered flower*

158. Tufted Loco, ½ X

yellowish flowers. Grows on rocky or gravelly slopes of the montane and subalpine zones from Canada southward to Colorado.

ROCKY MOUNTAIN LOCO, *O. sericea,* is a handsome plant that usually grows in clumps with numerous flower stalks 25–45cm tall. Its leaves are silvery-pubescent. Occurs in Wyoming, Colorado, New Mexico, and Utah on gravelly slopes and open fields of the montane and subalpine zones. In the true species, the flowers are white with a purple-tipped keel; but *L. lambertii* and *L. sericea* do hybridize. Their seedlings retain the vigor and clump habit of *sericea* but take on various amounts of color from *lambertii.* Where these two species grow together, one often finds beautiful fields of plants with flowers ranging from white through lavender and pink to deep rose. This occurs noticeably along U. S. Highway 24 near Divide, Colorado, west of Colorado Springs; and also in the Centennial Valley west of Laramie, Wyoming.

(11) *Petalostemon.* Prairie clover

1. Flowers white.
    2. Calyx and spike glabrate.

          P. occidentale
          White prairie clover

    2. Calyx and spike densely silky-villous.

          P. compactum

1. Flowers pink to purple.
    2. Leaflets 3–7.

          P. purpureum
          Purple prairie clover

    2. Leaflets 13–17.

          P. villosa

PURPLE PRAIRIE CLOVER, *Petalostemum purpureum* (fig. 159), usually has several erect or ascending leafy stems 20–50cm tall. The leaflets are usually folded, and there are small dots on their lower surfaces. The small purple flowers are in dense terminal spikes, 2.5–5cm long. Grows on the plains and foothills from the mountains eastward.

(12) *Psoralea.* Scurf pea or breadroot

1. Plants low or almost stemless; roots tuberous or thickened.
    2. Foliage canescently-hirsute.

          P. mephitica

    2. Foliage greenish, but rough-hirsute.
        3. Stem short but divaricately-branched; pod glabrous.

          P. esculenta
          Indian breadroot

        3. Stemless; pod somewhat hirsute.

          P. hypogaea
          Subterranean breadroot

1. Plants taller, the stem leafy and branched.
  2. Roots tuberous.

                                                   P. cuspidata

  2. Roots not tuberous.
    3. Densely silvery-silky throughout.
      4. Calyx-lobes about equal.

                                                   P. digitata

      4. Calyx with one elongated lobe.

                                                   P. argophylla
                                                  Silver-leaved scurf pea

    3. Greenish, glabrate, or minutely canescent.
      4. Pods nearly globose, very glandular; leaves glabrate.

                                                  P. lanceolata
                                                Lance-leafed scurf pea

      4. Pods ovoid to ovoid-oblong.
        5. Leaves obovate to oblong-cuneate; pod glandular but glabrous.

                                        P. tenuiflora
                                        Slender-flowered scurf pea

        5. Leaves narrowly linear; pod glandular-punctute but glabrous.

                                        P. linearifolia

(13) *Robinia*. Locust

1. Corolla white; leaflets, twigs, and fruit glabrous; an escaped cultivar.

                                        R. pseudoacacia
                                        Black locust

1. Corolla rose-pink; leaflets with appressed hairs; twigs puberulent; fruit hairy, usually glandular.

                                        R. neomexicana
                                        New Mexico locust

Our only native species is the NEW MEXICAN LOCUST, *R. neomexicana* (fig. 160), which grows as a large shrub on foothill slopes and along roads from southern Colorado into Texas, New Mexico, Arizona, and Nevada.

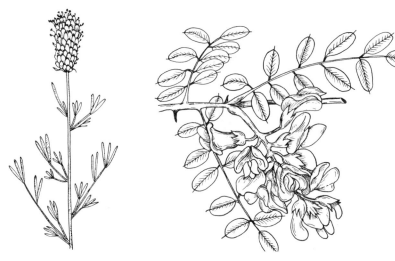

159. Purple Prairie
Clover, ½ X

160. New Mexico Locust, ½ X

(14) *Sophora*. Sophora

One species in our range. *S. nuttalliana* has white flowers and silky-pubescent leaves. Stamens all free; pods constricted between the seeds. Easily mistaken for an *Astragalus*.

(15) *Thermopsis*. Golden pea

These plants have compound leaves of 3 leaflets with large, leaf-like stipules, showy yellow-pea blossoms, and bean-like pods. Because of their vigorous-spreading, underground stems, they tend to occur in patches and often take over banks and roadsides. For accurate identification of the three species in our range, pods are necessary.

1. Mature pods straight or only slightly curved.
  2. Pods erect and nearly straight.

      T. montana
      Mountain thermopsis
  2. Pods spreading, somewhat curved, but not recurved or deflexed.
      T. divaricarpa
      Golden banner
1. Mature pods strongly curved and often recurved.

      T. rhombifolia
      Round-leaved golden pea

The earliest to bloom is *T. rhombifolia*. Rarely over 30cm tall, it grows on sandy soil on the plains from North Dakota to Nebraska and westward to the foothills of the Rockies in Montana, Wyoming, and Colorado. Its pod is curved, sometimes forming a half circle, and is finally bent down.

The other two species are 3–6dm tall, bloom in late spring and early summer, occur from the foothills up to timberline, and can be distinguished only by the position of the pod and the locality where found growing. GOLDEN BANNER, *T. divaricarpa* (fig. 161), has slightly hairy pods that spread away from the

161. Golden Banner, ²/₅X

stalk almost at right angles. Abundant from central Colorado northward through Wyoming on the eastern side of the Continental Divide, it may also be found occasionally in New Mexico and Utah, and throughout the mountainous parts of Colorado. MOUNTAIN THERMOPSIS, *T. montana,* has pods that are very hairy and are held erect, close to the stalk. Occurs abundantly in the mountains of southern Colorado, and through Utah and Nevada to Montana and Oregon.

(16) *Trifolium.* Clover

1. Plants acaulescent.
   2. Flowers subtended by true involucres; calyx silky with brownish hairs.
         T. dasyphyllum
         Whiproot clover
   2. Flowers not subtended by true involucres.
      3. Leaves pubescent, entire; calyx strigose.
         T. gymnocarpon
         Hollyleaf clover
      3. Leaves glabrous.
         4. Head of several flowers, pedunculate.
            5. Leaflets sharply denticulate.
               T. haydenii ✓
               Hayden clover
            5. Leaflets entire.
               T. brandegei
               Brandegee clover
         4. Head of 2–3 flowers on short, radical peduncles.
            T. nanum
            Dwarf clover
1. Plants caulescent.
   2. Flowers subtended by true involucres.
      3. Involucre green; flower 10–13mm long.
         T. wormskjoldii
         Springbank clover
      3. Involucre white-scarious; flower 13–20cm long.
         T. parryi ✓
         Parry clover
   2. Flowers not subtended by true involucres, but upper leaf stipules may seem to be involucral.
      3. Flowers 5–10mm long; heads axillary.
         4. Calyx glabrous; stems creeping and may root at the nodes.
            T. repens
            White clover
         4. Calyx sparsely pubescent at the sinuses between the teeth; stems ascending or erect. ʊｃ𝓉 area
            T. hybridum
            Alsike clover
      3. Flowers 11–20mm long; heads terminal.
         4. Plants 0.5–3dm tall, often stoloniferous or rhizomatous; flowers yellowish-white to purplish.
            T. longipes var. reflexum
            Long-stalked clover
         4. Plants 3–10dm tall; not stoloniferous or rhizomatous; flowers deep red.
            T. pratense
            Red clover

162. Whiproot
Clover, ½X   163. Dwarf Clover, 1X

164. Parry Clover, ½X

Three of the clovers are cushion-type species. The commonest and most widely distributed is WHIPROOT CLOVER, *T. dasyphyllum* (fig. 162). It has 2-toned flowers and hairy leaflets, which are about 2.5cm long, and is found at high altitudes from southern Wyoming southward. Grows in exposed rocky situations of the subalpine zone and on gravelly areas above timberline. The DWARF or DEER CLOVER, *T. nanum* (fig. 163), is a compact, mat-forming, alpine plant with leaflets usually not over 8mm long. Each head is composed of 2–3 comparatively large, rose-colored flowers, which turn brown as they age.

BRANDEGEE CLOVER, *T. brandegei,* is another alpine species of the high mountains of Colorado and New Mexico. It has several purple flowers in a loose head. The lower ones soon become conspicuously reflexed. J. S. Brandegee, a railroad engineer with the Santa Fe Railroad and topographer with the Hayden Survey in 1875, made botanical collections in Colorado.

Among the taller, native species PARRY CLOVER, *T. parryi* (fig. 164), is the showiest. The compact heads of bright rose-pink flowers are raised on long peduncles above the green, glabrous leaves and surrounded by papery involucres. Occurs from Wyoming and Utah southward through the mountains of Colorado, especially in the timberline region of the subalpine zone, but also in moist alpine meadows.

LONG-STALKED or RYDBERG CLOVER, *T. longipes,* a plant of the montane and subalpine regions, has leafy stems up to 4.5dm long, sharply-toothed and strongly-veined leaflets, the veins ending in little, sharp teeth. The white or rose-colored flowers become reflexed as they age. Often there are silky hairs on one half of the lower surface of a leaflet.

SPRINGBANK CLOVER, *T. wormskjoldii,* has leafy stems up to 4.5dm tall and rounded, long-stalked heads of white or rose flowers surrounded by green involucres. Occurs in montane meadows from central Colorado into New Mexico, Arizona, and Utah.

The common cultivated species RED CLOVER, *T. pratense*; WHITE or DUTCH CLOVER, *T. repens*; and ALSIKE CLOVER, *T. hybridium,* have become established in meadows and along roads and trails throughout our area and may be identified by the characters given in the key.

(17) *Vicia*. Vetch

This genus is easily recognized by the fine tendrils that tip its pinnately-compound leaves and by which it attaches itself to other plants.

1. Racemes many-flowered, 20–80, one sided.
    2. Plant densely hirsute-villous; flowers over 15mm long; annual or biennial.
        V. villosa
        Hairy or winter vetch
    2. Plant glabrous to appressed-pubescent; flowers 10–15mm long; perennial.
        V. cracca
        Tufted vetch
1. Racemes of 1–20 flowers, sometimes only 1–2 in leaf axils; perennials.
    2. Flowers 1–3 in leaf axils.
        V. sativa
        Common vetch

    2. Flowers 4–20 in pedunculate racemes.
        3. Leaflets thin, 4–14mm wide, ovate to oblong.
            V. americana var. americana
            American vetch
        3. Leaflets thick, 1–4mm wide, linear to linear-oblong.
            V. americana var. minor

AMERICAN VETCH, *V. americana* var. *americana* (fig. 165), is our most common species. It has few-flowered racemes of showy, purple pea-blossoms, which are followed by pods 2.5cm long. Widely distributed over the United States. Found trailing on the ground or clambering over bushes in the foothills and montane zones of the Rockies.

165. American Vetch, ½ X

OLEASTER FAMILY: ELEAGNACEAE
The native members of this family are shrubs with simple, entire leaves. The young twigs and leaves are scurfy with silvery or rusty scales. The fruit is a dry or juicy berry.

1. Flowers perfect; stamens 4; leaves alternate.
        Eleagnus p. 208
        Silverberry

1. Flowers dioecious; stamens 8; leaves opposite.

Shepherdia p. 208
Buffaloberry

(1) *Eleagnus.* Silverberry

1. Trees; young branches silvery-scurfy; leaves narrow, not over 12mm wide; not native.
E. angustifolia
Russian olive
1. Large shrubs; young branches brown-scurfy; leaves elliptic-ovate to oblong-ovate, often over 12mm wide; native.
E. commutata
Silverberry

(2) *Shepherdia.* Buffaloberry

1. Leaves silvery-scurfy on both sides; thorny shrubs or small trees; berry red.
S. argentea
Silvery buffaloberry
1. Leaves green above, somewhat brown-scurfy below; thornless shrubs; berry red or yellowish.
S. canadensis
Bitter buffaloberry

166. Bitter Buffaloberry, ²/₅X

BITTER BUFFALOBERRY, *Shepherdia canadensis* (fig. 166), is the most commonly seen. A low shrub of shaded, rocky hillsides in the montane zone; leaves opposite, ovate, dark green. In the autumn it bears orange-red, juicy, but very bitter, berries. The buds and under-leaf surfaces are distinctly rusty, as are the small axillary, clustered, round flower buds, which are formed in late summer.

SILVER BUFFALOBERRY, *S. argentea,* is a taller, erect shrub of bottom lands and streamsides in the foothills, often on alkaline soil, especially on the western slope, with silvery leaves and thorny branches. The female flowers produce quantities of oval, red or orange berries, which are sour but edible.

SILVERBERRY, *Eleagnus commutata,* is a stout shrub with mahogany-colored stems and oval or ovate silvery leaves. Its fruit is an olive-shaped, dry, silvery berry, which furnishes food for many birds. Closely related is the small tree called RUSSIAN OLIVE, *E. angustifolia,* which is much planted throughout our area and sometimes escapes.

WATER-MILFOIL FAMILY: HALORAGACEAE

Only 1 genus in our range, *Myriophyllum.* These are smooth leafy herbs with leaves whorled in threes or fours. The upper flowers are usually staminate, the lower pistillate, and the intermediate are perfect.

1. Floral leaves (bracts) shorter than the flowers; a common species.

> *M. spicatum var.*
> *exalbescens*
> Spiked water milfoil or
> parrot feather

1. Floral leaves (bracts) longer than the flowers; very uncommon.

> *M. verticillatum*

PARROT FEATHER, *Myriophyllum spicatum* var. *exalbescens,* with leaves dissected into thread-like divisions, occurs in the quiet water of ponds and slow streams almost throughout the United States and in the Rockies up to about 8,500 feet.

LOOSESTRIFE FAMILY: LYTHRACEAE

Only one species is likely to be found in our range. WINGED LOOSESTRIFE, *Lythrum alatum,* is an erect, slender, perennial herb with angled stems. Flowers are axillary, mostly solitary, and dimorphus. Leaves oblong-ovate to lanceolate, upper leaves alternate; calyx striate, 4–7-toothed; filiform style, the stigma capitate; petals purple.

EVENING-PRIMROSE FAMILY: ONAGRACEAE

The plants of this family are easy to recognize as our species have 4 petals and an inferior ovary (the mustards, also 4-petaled, have a superior ovary). The ovary is usually long, slender, and often 4-sided; the stigma is often 4-lobed; and there are usually 8 stamens. It is a large family, with many native representatives in this region, including several cultivated species and many weeds. Only one genus, *Circaea,* does not follow these general rules.

1. Flowers 2-merous; stamens 2; fruit with hooked hairs.

> *Circaea* p. 210
> Enchanter's nightshade

1. Flowers 4-merous; stamens usually 8; fruit without hooked hairs.
    2. Fruit a many-seeded capsule, opening by valves.
        3. Calyx-tube not prolonged beyond the ovary.
            4. Annuals; seeds without a tuft of hair.

> *Gayophytum* p. 213
> Groundsmoke

            4. Perennials; seeds comose (with a tuft of hair).

> *Epilobium* p. 211
> Willowherb

        3. Calyx-tube prolonged beyond the ovary.
            4. Seeds comose.

> *Epilobium* p. 211
> Willowherb

            4. Seeds naked, sometimes tuberculate.
                5. Stigma deeply cleft into 4 linear lobes.

> *Oenothera* p. 213
> Evening-primrose

                5. Stigma entire (discoid or capitate).
                    6. Stigma discoid; calyx-tube funnelform above.

> *Calylophus* p. 210
> Sundrops

6. Stigma capitate; calyx-tube cylindrical or widened above.
> *Camissonia* p. 210
> Camissonia

2. Fruit indehiscent, nut-like.

> *Gaura* p. 212
> Butterflyweed

(1) *Calylophus*. Sundrops

1. Copiously tufted plant, stems rarely over 1dm high; leaves linear or linear-spatulate; petals yellow, 13–22mm long; plant woody below, herbaceous above.
> *C. lavandulifolius*
> Puckered sundrops or
> galpinsia
1. Stems usually copiously branched at base, the branches canescent, 1–5dm high; leaf blades spatulate to linear-oblong, serrate; petals yellow, 3–13mm long; plant mostly herbaceous.
> *C. serrulatus*
> Sundrops or merilyx

In this genus, the flowers remain open during the day. SUNDROPS or MERILYX, *C. serrulatus,* is a small, bushy plant of the high plains and foothills, found on dry ground and mostly east of the Continental Divide. GALPINSIA or PUCKERED SUNDROPS, *C. lavandulifolius,* is a plant of similar habitats. Its larger, paler-yellow flowers have a squarish outline. The petals are usually crinkled.

(2) *Camissonia*. Camissonia

1. Plants acaulescent; flowers axillary, yellow or white; style usually adherent to the calyx-tube; stigma capitate; stamens 8, erect.
   2. Leaves pinnatifid.
> *C. breviflora*
   2. Leaves sub-entire or undulate-denticulate.
> *C. subacaulis*
1. Plants caulescent; stamens 8, stigma capitate or globose; flowers yellow, yellowish, or whitish.
   2. Capsule more or less clavate (club-shaped); styles often longer than the petals; capsule pedicellate and obtuse.
> *C. scapoidea ssp.*
> *scapoidea*
   2. Capsule sessile or nearly so, cylindrical.
      3. Flowers yellow when young, in leaf axils.
         4. Plants with several naked, fine stems; leafy inflorescence at apex; petals 1.5–2.5mm long; capsules nearly straight, 5–8mm long.
> *C. andina*
         4. Plants with leafy stems; petals 2.5–3.5mm long; capsule strongly curved, 15mm long or more.
> *C. parvula*
      3. Flowers white when young, in terminal spikes; petals and sepals about 2mm long; capsules somewhat enlarged near base, tapering to apex, 18–25mm long.
> *C. minor*

(3) *Circaea*. Enchanter's nightshade

1. Racemes with bracts; leaves sharply dentate; plants not over 20cm tall.
> *C. alpina var. alpina*
1. Racemes bractless; leaves undulate-denticulate; plants usually over 20cm tall.
> *C. alpina var. pacifica*

(4) *Epilobium*. Willowherb

The botanical name of this genus means *on a pod* in Greek, describing the elongated ovary bearing the other flower parts on its tip. Its common name refers to the tuft of hairs at the end of each seed, which is similar to that on willow seeds. A plant in flower will usually have at least a few pods beginning to burst, so that this character may be seen. A difficult genus in our range.

1. Calyx-tube not prolonged beyond the ovary; stigma 4-lobed; petals 1–2cm long; leaves 5–20cm long.
   2. Style longer than the stamens, pubescent at base.

                     E. angustifolium ssp.
                     circumvagnum
                     Common fireweed
   2. Style shorter than the stamens, glabrous; leaves 2–6cm long.

                     E. latifolium
                     Alpine or broad-leaved
                     fire-weed
1. Calyx-tube prolonged beyond the ovary.
   2. Stigma 4-lobed; flowers showy, scarlet.

                     E. canum ssp. garrettii
                     Firechallis
   2. Stigma more or less entire; petals less than 1cm long.
      3. Plants annual; leaves mostly alternate; stem epidermis peeling; petals violet; dry habitat.

                     E. paniculatum
                     Panicled willowherb
      3. Plants perennial; rhizomatous; leaves opposite or nearly so; stem epidermis not peeling; wet habitat
         4. Densely caespitose, the stems 5–15cm high, curved and nodding at apex; petals lilac to violet.

                     E. anagallidifolium
                     Alpine willowherb
         4. If caespitose, only loosely so; stems 1–10dm high, not curved.
           5. Stems glabrous, pubescent, or glandular-pubescent, but without decurrent lines of hairs from the leaf bases.
              6. Sepals 3–5mm long; petals 6–10mm long; leaves (3) 5–12cm long.
                 7. Stems branched, glabrous to glandular-pubescent above; leaves petiolate.

                     E. ciliatum ssp. ciliatum
                     Common willowherb
                 7. Stems unbranched, glabrous (or sometimes with decurrent lines of hairs from the base of leaves); leaves sessile.

                     E. ciliatum ssp.
                     glandulosum
              6. Sepals 1–2mm long; petals 4–7mm long; leaves 1–3cm long.
                     E. oregonense
                     Oregon willowherb
           5. Stems often glandular-pubescent; decurrent lines of hairs from leaf.
              6. Leaves petiolate; capsules not glandular.
                 7. Petals white to pink; stems usually well over 3dm tall; seed coma (tuft of hairs) white.

                     E. ciliatum ssp.
                     glandulosum
                 7. Petals purple to rose; stems mostly under 4dm tall; seed coma dingy.

                     E. hornemannii ssp.
                     hornemannii
                     Hornemann willowherb
              6. Leaves sessile, often clasping the stem; capsules usually glandular-pubescent.

7. Leaves lance-linear, the margins often irregularly dentate; seed coma white.

E. halleanum
Hall willowherb

7. Leaves mostly ovate, the bases rounded, the margins entire; seed coma dingy.

E. saximontanum
Rocky Mountain willowherb

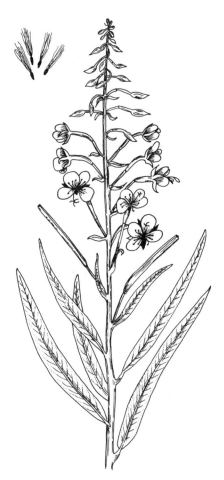

167. Fireweed, ½X

COMMON FIREWEED, *E. angustifolium* (fig 167), is one of our most common and widespread plants due to the ease with which its seeds are distributed by the wind. It often grows in masses on burned areas, hence the name fireweed, or along roadsides. It has tall stems bearing spire-like racemes of bright rose-purple flowers. Occurs in mountainous areas throughout the western United States and into Canada and Alaska. ALPINE or BROAD-LEAVED FIREWEED, *E. latifolium,* has similar but larger flowers in which the crimson sepals contrast with the rose petals. It has shorter stems and grayish leaves. Found on moist, gravelly slopes and along stream beds in the subalpine and alpine regions of the Rockies, becoming more abundant northward.

FIRECHALLIS or HUMMINGBIRD-TRUMPET, *E. canum* ssp. *garrettii,* has spikes of showy scarlet flowers with long tubes, which are inflated just above the ovary. Grows in the mountains of Utah, southern Wyoming, and Idaho.

Most of the smaller species of willowherb are found in moist places, a few on dry, waste ground. ALPINE WILLOWHERB, *E. anagallidifolium* (fig. 168), with small pink flowers and S-shaped stems, occurs in subalpine and lower alpine meadows. *E. hornemannii,* about the same size, is more common in spruce forests. PANICLED WILLOWHERB, *E. paniculatum,* with shreddy bark on the stems and small pink or lilac blooms, grows on dry ground in the foothill zone.

(5) *Gaura.* Butterflyweed

1. Anthers oval; fruits 6-10 mm long, somewhat fusiform, 8-ribbed; flowers pink to rose.

G. parviflora
Small-flowered gaura

1. Anthers linear or nearly so; fruits strongly 4-angled, at least above.
    2. Plants tall, erect, herbaceous; flowers pink to rose.
    G. neomexicana
    2. Plants low, stems gradually rising from a near-woody base; flowers red to scarlet.
    G. coccinea
    Scarlet gaura

(6) *Gayophytum*. Groundsmoke

1. Capsules sessile or subsessile, only slightly constricted between the seeds (or not at all); plants branched mostly at the base.
    2. Seeds usually fewer than 10, about 1mm long.
    G. racemosum
    Groundsmoke
    2. Seeds 15–20, 0.5–0.75mm long.
    G. humile
    Dwarf groundsmoke
1. Capsules pedicelate, usually constricted between the seeds; plants generally branched above.
    2. Petals 0.5mm long; capsules 3–6mm long, the pedicels sharply reflexed.
    G. ramosissimum
    Hairstem groundsmoke
    2. Petals 1–5mm long; capsules 4–15mm long, the pedicels spreading or erect.
    G. diffusum
    Spreading groundsmoke

GROUNDSMOKE or BABYS BREATH, *Gayophytum racemosum* (fig. 169), is a delicate, much-branched plant with tiny pinkish flowers. Found on dry fields and slopes of the foothill and montane zones.

(7) *Oenothera*. Evening-primrose

This large genus has many species in the Rockies. The 4-petaled flowers are white or yellow; the white ones fade to pink, the yellow ones to orange or reddish. The flowers open in the late afternoon or evening, usually withering when the sum becomes bright the following day. An individual flower does not reopen, but there are usually buds ready to open the next evening. They all have a tube called a *hypanthium,* which is really part of the calyx, though easily mistaken for a stem. Through the hypanthium, the style (which in some species may be from 10 to 15cm long) connects the stigma to the ovary.

168. Alpine
Wilowherb, ½X

1. Stamens equal in length.
    2. Flowers yellow; seeds in 2 rows.
        3. Petals 1–2cm long, pure yellow; plant strigose to villous above.
        O. villosa ssp. *strigosa*
        Common evening-primrose
        3. Petals 2.5–4cm long, usually yellow but occasionally orange; plant some what hirsute.
        O. hookeri
        Hooker evening-primrose

2. Flowers white or pink; seeds in 1 row.
  3. Leaves entire.
    4. Plants perennial.
      5. Capsules straight, ascending; leaves pubescent beneath.

O. nuttallii
Nuttall evening-primrose

      5. Capsules divaricate (very divergent), some contorted or twisted; leaves glabrous.

O. pallida
Pale evening-primrose

    4. Plants annual, from winter rosettes; capsule erect or ascending; stems pubescent.

O. albicaulis
Prairie evening-primrose

  3. Leaves pinnatifid; stems somewhat pubescent.

O. coronopifolia
Cutleaf evening-primrose

flower

capsule

split
capsule

169. Groundsmoke, ½X   170. Common Evening-Primrose, ½X

1. Stamens unequal, the alternate ones longer.
    2. Capsules not winged, the angles rounded.
        3. Plant entirely glabrous, or nearly so.

    O. caespitosa ssp. *caespitosa*

        3. Plant either pubescent or hirsute.
            4. Capsule sessile; plant acaulescent.

    O. caespitosas ssp. *macroglottis*
White stemless evening primrose

           4. Capsule stipitate; plant caulescent.

    O. caespitosa ssp. *marginatus*

  2. Capsules either winged or sharply-angled.
        3. Plants acaulescent.
            4. Capsule beaked; flowers 3–4cm broad.

    O. flava
Yellow evening-primrose

           4. Capule obtuse; flowers 8–10cm broad.

    O. brachycarpa
Yellow stemless evening-primrose

      3. Plants with diffuse, wiry stems; petals white or pink, spotted or striped with red.

    O. canescens
Hoary evening-primrose

YELLOW STEMLESS EVENING-PRIMROSE, *O. bruchycarpa*. This large, bright yellow flower, appearing in a rosette of long, green leaves, has a very long hypanthium. Grows on shale banks in the foothills and on the mesas throughout the Rockies.

HOOKER EVENING-PRIMROSE, *O. hookeri*, may have a stem 3dm to over 1m tall. Its flowers are in a terminal, elongated raceme interspersed with leafy bracts. The hypanthium and sepals are reddish, and the yellow petals turn red or purplish as they age. Grows in New Mexico, Arizona, southern Colorado, and Utah to Idaho, in foothills and montane zones. The COMMON EVENING-PRIMROSE, *O. villosa* ssp. *strigosa* (fig. 170), is similar but has smaller flowers and is common throughout the western United States, often occurring as a weed.

171. White-Stemless Evening Primrose, 1/4 X

WHITE STEMLESS EVENING-PRIMROSE, *O. caespitosa* ssp. *macroglottis* (fig. 171), is sometimes called gumbo lily, but its 4 petals and 4-parted stigma tell the observer right away that it is not a true lily. Grows on clay banks, often on red shale or on gravel, from the foothills to the subalpine regions throughout our range.

172. Nuttal Evening-Primrose, ½X

173. Cutleaf Evening-Primrose, ½X

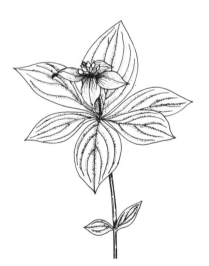

174. Bunchberry, ½X

NUTTALL EVENING-PRIMROSE, *O. nut-tallii* (fig. 172), is a leafy-stemmed plant that grows from a spreading underground root and is often found in patches. It has a shiny-white stem from which the outer bark peels off in shreds. The tips of its stems with unopened buds are usually somewhat bent down. Occurs in the montane zone. Another white-flowered species, the PRAIRIE EVENING-PRIMROSE, *O. albicaulis*, has similar flowers but is different in habit as it grows from a winter rosette. Its stems branch and often form a bushy plant. Found on the high plains and foothills.

CUTLEAF EVENING-PRIMROSE, *O. coronopifolia* (fig. 173), is a white-flowered species of smaller size than the last two. It has leaves that are sharply toothed or divided, and it always seems to show some pink withered flowers. Found on foothill and montane fields, often along roadsides, throughout our region.

DOGWOOD FAMILY: CORNACEAE

Only one genus and two species of this family occur in our range. Most members of the family are trees or shrubs, but one of ours is a low herbaceous plant.

1. A low herb, from a creeping rootstock; fruit red.

Cornus canadensis
Bunchberry

1. Shrubs over 25cm tall; fruit white.

Cornus stolonifera
Red-osier dogwood

BUNCHBERRY, *Cornus canadensis* (fig. 174), with stems only 10–25cm tall, may form carpets in moist forests. The small, clustered flowers are surrounded by 4 white, petal-like bracts, like the flowering dogwood trees of our eastern states and California. The fruit is a cluster of red berries. This species is rare in Colorado and in most of our range. It becomes more abundant in the Northwest and in the Canadian Rockies.

RED-OSIER DOGWOOD, *C. stolonifera* (fig. 175), is an oppositely-branched shrub usually from 6dm to more than 2m tall, with smooth, red stems and ovate, pinnately-veined leaves. Its small, white flowers occur in flat-topped clusters, followed by whitish berries (rarely bluish), which provide good food for birds. Grows along streams of the foothill and montane zones and is easily recognized in winter by its red bark.

175. Red-Osier Dogwood, ²⁄₅X

SANDALWOOD FAMILY: SANTALACEAE

The only member of this family likely to be found in our range is PALE BASTARD TOADFLAX, *Comandra umbellata* ssp. *pallida*. Its leaves are glaucous, oblong to nearly linear, and usually acute. There is a cluster of small, whitish, star-shaped flowers at top of the 15–30 cm tall stem. The plant has subterranean rootstocks and is parasitic on other plants that grow on mesas and foothills. Its root is blue when cut.

MISTLETOE FAMILY: LORANTHACEAE

MISTLETOE is common on the evergreen trees of our forests and is seen as clusters of smooth, yellowish-brown or greenish stems that erupt from the branches of the host tree. Each species of tree usually has its own species of mistletoe. These plants are parasites, which sap the strength of the host tree, eventually causing its death. The one most commonly seen is *Arceuthobium vaginatum* (fig. 176), on ponderosa pines.

1. Anther a single globose cell; berry compressed, fleshy.

Arceuthobium p. 217
Dwarf mistletoe

1. Anthers 2-celled; berry globose, pulpy.

Phoradendron p. 218
Mistletoe

(1) *Arceuthobium*. Dwarf mistletoe

1. Staminate flowers on peduncle-like joints in a paniculate cluster; host is usually lodge pole pine, *Pinus contorta*.

A. americanum
American dwarf mistletoe

1. Staminate flowers in the axils of the scales of a simple or compound spike.
  2. Spikes short and stout, 3–4mm in diameter; brownish-yellow; host is usually *Pinus ponderosa*.

A. vaginatum
Dwarf mistletoe

  2. Spikes slender, 1–2mm in diameter.
    3. Dark, usually greenish-brown; host is piñon pine, *Pinus edulis*.

A. divaricatum
Dwarf mistletoe

    3. Pale, usually yellowish-green.
      4. Flowers few; host is Douglas-fir, *Pseudotsuga menziesii*.

A. douglasii
Douglas dwarf mistletoe

      4. Flowers few; host is limber pine, *Pinus flexilis*.

A. cyanocarpum

(2) *Phoradendron*. Mistletoe

CEDAR MISTLETOE, *P. juniperinum*, occurs as yellowish-green parasites on various species of *Juniperus* in our range.

STAFF-TREE FAMILY: CELASTRACEAE

This is the family to which bittersweet and the deciduous and evergreen species of euonymous belong. Our only native representative is MOUNTAIN-LOVER, *Pachystima myrsinites* (fig. 177), a low shrub with neat, opposite, evergreen leaves, short-petioled and serrulate. Flowers 4-merous, solitary or in few-flowered axillary cymes, greenish-brown or dark-reddish. Grows in moist forests of the montane and sub-alpine zones from Canada to New Mexico and California.

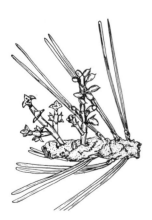

176 Pine Mistletoe, ½X

177. Mountain-Lover, ⅖X

SPURGE FAMILY: EUPHORBIACEAE

This is an interesting family of curious plants. Many are inconspicuous, some are showy, such as *Poinsettia*, and some are strange cactus-like succulents from Africa. The small flowers are unisexual, both kinds generally gathered in few-flowered clusters.

1. Flowers with a true calyx.
   2.Pubescence stellate.

                                     Croton p. 219
                                     Croton

  2. Pubesence simple.
     3. Flowers with petals.

                                     Argythamnia p. 219
     3. Flowers apetalous.

                                     Tragia p. 220
                                     Noseburn

1. Flowers in an involucre, no true calyx.

                                     Euphorbia p. 219
                                     Spurge

(1) *Argythamia*. Argythamia

1. Flowers in axillary clusters; leaves petiolate; stems 10–30cm long.
                                      A. humilis
1. Flowers in elongated axillary racemes; leaves sessile; stems 10–60cm tall.
                                      A. mercurialina

(2) *Croton*. Croton

One species occurs in our range. *C. texensis* is annual; 20–60cm tall; canescent-stellate; plants dioecious; petals apparently lacking; staminate flowers in racemes, pistillate flowers 2–4 together. Found on prairies and plains.

(3) *Euphorbia*. Spurge

1. Glands of the involucre with horns (petaloid appendages) (sometimes placed in *Chamaesyce*).
  2 Leaves not equilateral, oblique at base.
     3. Leaves entire.
       4. Seeds smooth.
         5.Horns conspicuous; seeds terete (cylindrical).
                                E. missurica
         5. Horns small; seeds obtusely 4-angled.
                                E. serpens
       4. Seeds somewhat roughened.
         5. Glands broader than long, with fan-shaped appendages.
                              E. albomarginata
         5. Glands longer than broad, with crescent-shaped appendages.
                                E. fendleri
     3. Leaves serrate or dentate.
       4. Leaves glabrous.
         5. Seeds pitted, sometimes faintly rugose (wrinkled).
                              E. serpyllifolia
                                Sidewalk-weed
        5. Seeds rugose, not pitted.
                                E. glyptosperma
                                Sidewalk-weed
      4. Leaves pubescent.

                                E. stictospora

2. Leaves with both sides similar at the base.
   3. Horns large and conspicuous; leaves alternate.
        E. marginata
        Snow-on-the-mountain
   3. Horns small; leaves opposite.
        E. hexagona
1. Glands of the involucre naked, without horns.
   2. Leaves opposite, ovate or orbicular-oblong, coarsely dentate (sometimes placed in *Poinsettia*).
        E. dentata
   2. Leaves alternate or scattered (sometimes placed in *Tithymalus*).
      3. Leaves entire.
         4. Leaves 0.5–2cm long, thick and somewhat fleshy; upper leaves and bracts green; plants taprooted.
           E. robusta
           Rocky Mountain spurge
         4. Leaves 2–6cm long, not thick; upper leaves and bracts yellowish-green; plants rhizomatous.
           E. esula
           Leafy spurge
      3. Leaves serrate or dentate.
         4. Annuals.
           5. Stems simple below.
              E. spathulata
           5. Stems branched from the decumbent base.
              E. crenulata
         4. Biennials; stem branched from the base; base either erect or decumbent.
           E. commutata

(4) *Tragia*. Noseburn

One species is likely to be found in our range. *T. ramosa* is perennial, light green, bristly with stinging hairs; stems erect, slender, often freely branched, 2–4dm tall.

Our most conspicuous species in this family is SNOW-ON-THE-MOUNTAIN, *Euphorbia marginata* (fig. 178), which scarcely qualifies as a mountain plant. It grows from 3 to 6dm tall, with an umbel-like inflorescence in which the showy parts are the numerous, white-margined, floral leaves. Abundant on the plains; may be found along the eastern foothills up to 7,000 feet. It is occasionally grown in gardens.

ROCKY MOUNTAIN SPURGE, *E. robusta* (fig. 179), is an entirely green plant. Its small flowers are surrounded by involucres, and each involucre contains 4 crescent-shaped glands. Stems 15–30cm tall, usually clustered, sometimes reddish at base, and each bears a 3- to 5-rayed umbel. Occurs with ponderosa pine throughout our range.

BUCKTHORN FAMILY: RHAMNACEAE

A family of woody plants; most of ours are low shrubs. Their small, white petals are clawed and inserted on the calyx tube.

1. Petals short-clawed or absent; fruit fleshy.
        Rhamnus p. 221
        Buckthorn

1. Petals long-clawed; fruit dry, with three dehiscent nutlets.

Ceanothus p. 221
Buckbrush

(1) *Ceanothus*. Buckbrush

1. Shrubs with spines.

C. fendleri
Buckbush

1. Shrubs unarmed.
   2. Leaves entire, rarely over 2cm long.

C. martinii

   2. Leaves toothed, over 2cm long.
      3. Leaves leathery, shining, strongly aromatic, evergreen.

C. velutinus var. *velutinus*
Mountain-balm or
snowbrush

      3. Leaves thin, softly pubescent, not strongly aromatic, deciduous.

C. herbaceous
Redroot ceanothus or
New Jersey tea

(2) *Rhamnus*. Buckthorn

1. Petals present; nutlets smooth.

R. smithii

1. Petals absent; nutlets grooved.

R. alnifolia
Dwarf buckthorn

178. Snow-on-the-mountain, ²/₅ X

179. Rocky Mountain Spurge, ²/₅ X

180. Mountain-balm, ²⁄₅X

MOUNTAIN-BALM or SNOWBRUSH, *Ceanothus velutinus* (fig. 180), has oval, leathery leaves that appear as though varnished. Its twigs are olive green; its evergreen leaves are dark green above, lighter beneath, strongly 3-veined, and fragrant. Flowers are small, creamy-white, in dense clusters; the small, dry, 3-lobed capsules persist for 2–3 seasons. The shrubs are 6dm-2m tall and form large patches on montane hillsides. Found from South Dakota to British Columbia, and south through Wyoming and Colorado to California.

REDROOT CEANOTHUS, *C. herbaceous* (fig. 181), is a shrub 3–6dm tall with oblong to ovate-lanceolate leaves 2.5–5cm long. Small white flowers are crowded in corymbose clusters. Found on sandy soil from New England to the foothills of Colorado and New Mexico.

FENDLER BUCKBRUSH, *C. fendleri,* is a low, thorny shrub with white flowers in umbel-like clusters, arranged in terminal racemes. Its small leaves are pale beneath. A plant of the foothill and montane zones, much browsed by deer. Forms patches under pines in Colorado, Utah, Texas, Mexico, and Arizona.

SMITH BUCKTHORN, *Rhamnus smithii,* is a shrub with alternate or nearly opposite, elliptical to oblong-lanceolate leaves, yellowish beneath, 2.5–4cm long. The small, unisexual flowers are 2 or 3 together in the leaf axils. The fruits are smooth, black berries about 1cm long. Grows on hillsides and in valleys of the upper foothill and lower montane zones in western Colorado

181. Redroot Ceanothus, ²⁄₅X

182. Fox Grape, ²⁄₅X

and New Mexico. DWARF BUCKTHORN, *R. alnifolia,* grows in northwestern Wyoming, Montana, and Idaho.

GRAPE FAMILY: VITACEAE
1. Leaves simple; hypogynous disk around the ovary or its base.

Vitis riparia
Fox grape

1. Leaves compound, palmately 5- to 7-foliolate; disk not present.

Parthenocissus inserta
Virginia creeper

Members of this family have palmately-lobed or palmately-compound leaves, and most of them have tendrils. All the varieties of cultivated grapes and the many wild species belong to this family. In our range, there is only one native species of grape, FOX GRAPE, *Vitis riparia* (fig. 182). It grows on banks and slopes of the foothills. Its relative, VIRGINIA CREEPER or WOODBINE, *Parthenocissus inserta* (fig. 183), is more inclined to climb and is found on trees, fences, or old buildings. It has palmately-compound leaves of 5–7 long, pointed leaflets; the berries resemble small grapes.

183. Virginia Creeper, ²⁄₅ X

FLAX FAMILY: LINACEAE
This family has regular, 5-merous flowers. Its petals are very fragile, its leaves are narrow and small, and its stems are very tough. Only one genus occurs in our range, *Linum*.

1. Petals blue.
    2. Annual plants; inner sepals ciliate or toothed on margins; escapes from cultivation.

Linum usitatissimum
Common cultivated flax

    2. Perennial plants; inner sepals entire; true native plant.

L. lewisii
Wild blue flax

1. Petals yellow or yellowish.
    2. Leaves glabrous.
        3. Leaves imbricated (overlapping).

L. kingii

        3. Leaves small and sparse.
            4. Flowers orange-yellow; styles united to the top.

L. rigidum var. *berlanieri*

            4. Flowers yellow; styles distinct at top.

L. rigidum var. *rigidum*
Yellow flax

2. Leaves and stem puberulent.
    3. Filaments dilated below; styles united to the top.

                    L. puberulum

    3. Filaments linear; styles distinct at top.

                    L. rigidum var.
                    compactum
                    Yellow flax

184. Wild Blue Flax, ½ X

WILD BLUE FLAX, *Linum lewisii* (fig. 184), is a perennial, but similar in general appearance to the annual species cultivated for oil and fiber. Nothing else in our region has such a symmetrical, clear sky-blue blossom. Early morning on a sunny day is the time to see it, because the petals wither and fall under a hot sun. It usually has several ascending stems clothed up to the inflorescence with numerous, narrow leaves. Usually, only one flower is in bloom at a time on each stem, but a fresh one opens each morning during the blooming season. As this progresses, a row of round seed pods, each on a curved pedicel, lines the stalk. This species was named in honor of Meriwether Lewis.

YELLOW FLAX, *L. rigidum,* is a plant of the plains and foothills, occurring from Canada south to Colorado and Texas. *L. kingii,* another yellow-flowered flax, has a narrow, crowded inflorescence and very leafy, tufted stems with many narrow leaves. Grows up to 3dm tall from a woody base. Occurs in the montane zone of the high mountains of Wyoming, southeastern Idaho, and south to Colorado and Utah.

MAPLE FAMILY: ACERACEAE
    This is a family of trees and shrubs native to the cool regions of the northern hemisphere. All North American species belong to the genus *Acer,* and there are 3 species in our range.

1. Leaves simple, palmately-lobed.
    2. Corymbs pedunculate; fruit wings broad.

                    Acer glabrum
                    Rocky Mountain maple

2. Corymbs nearly sessile; fruit wings narrow.

Acer grandidentatum
Wasatch maple

1. Leaves compound.

Acer negundo var.
interius
Boxelder

ROCKY MOUNTAIN MAPLE, *A. glabrum* (fig. 185), is a large shrub or small tree with gray bark. Its young twigs are smooth and dark red; its winter buds are bright red. The leaves are palmately 3- to 5-lobed, sharply-toothed. Its small, chartreuse-colored flowers are fragrant. In the center of the female flowers you can see the young ovary, already in the shape of the double "key" that is the maple fruit. As these ripen, they are often tinged with red. The egg clusters of a frequent insect pest appear as crimson blotches on the leaves. *A. glabrum* always grows as several trunks in a clump. Found from Wyoming and Idaho south to Utah and New Mexico, and it is common in the canyons and on hillsides of the foothills and montane zones throughout Colorado.

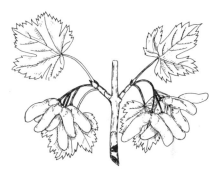

185. Rocky Mountain Maple, ²/₅ X

WASATCH MAPLE, *A. grandidentatum* (fig. 186), is a small tree that grows in groves or thickets on moist canyon-sides and along streams. Its leaves are 3-lobed with rounded sinuses and bluntly-pointed tips. Found from western Wyoming through the mountainous parts of Utah to Arizona, New Mexico, and west Texas.

186. Wasatch Maple, ²/₅ X

BOXELDER, *A. negundo,* is a tree found along streams of the high plains and foothill canyons. It differs from our other maples in having a pinnately-compound leaf and flowers and fruits in pendant clusters.

SUMAC FAMILY: ANACARDIACEAE

A family of shrubs, trees, or woody climbers, with compound leaves and small flowers in axillary or terminal panicles. Traditionally, only one genus was recognized in our range, but it is now segregated into two genera.

1. Leaves pinnately compound, 11–21 leaflets; inflorescence terminal.
Rhus glabra
Smooth sumac
1. Leaves 3-foliolate.
   2. Leaflets obovate, less than 3cm long, crenately-lobed; fruit red or orange-red, hairy.
   R. trilobata var. *trilobata*
   Three-leaf sumac
   2. Leaflets ovate to rhombic-ovate, over 3cm long, serrate or dentate, sometimes entire fruit usually yellowish-white, usually glabrous.
   Toxicodendron rydbergii
   Poison ivy or poison oak
   [Rhus radicans]

187. Smooth Sumac, ⅖X

188. Three-leaf Sumac, ½X

SMOOTH SUMAC, *R. glabra* (fig. 187), is a shrub up to 2m tall with stout stems, few branches, and pinnately-compound leaves; the leaflets are lanceolate, toothed, green above, whitish beneath. The small, cream-colored flowers, in dense, pointed clusters, become dark red, velvety, berry-like fruits. Grows on banks, dry hillsides, and along roads in the foothill zone throughout our range and in much of North America.

THREE-LEAF SUMAC, *R. trilobata* (fig. 188), is a much-branched, rounded shrub with bright green 3-foliolate leaves and light brown bark. Its flowers appear before the leaves, in crowded catkin-like clusters, and are followed by flattened, velvety, red or orange berries. They have an acid flavor and have been used to make a substitute for lemonade, and the plant is sometimes called *lemonadebush*. Because of the strong odor of its foliage, it is also called *skunkbush*; and it gets the name *squawbush* because the long, slender young branches were split by Indian women and used in basketry.

Poison Ivy or Poison Oak, *Toxicodendron rydbergii* (fig. 189), occurs throughout the United States and is occasionally found on hillsides and foothill canyons of the Rocky Mountain region. It is either a low, single-stemmed shrub or a woody climber clinging to tree trunks. It may be recognized by its bright green, compound leaves of 3 shiny, ovate leaflets; and by clusters of small, yellowish-white flowers, and white to yellowish-white smooth berries about 7mm in diameter. The foliage of this plant is poisonous to some people, but a thorough washing with strong soap after exposure will usually prevent bad results. Immediate application of household ammonia stops the itching.

189. Poison Ivy, ½ X

CALTROP FAMILY: ZYGOPHYLLACEAE

Herbaceous or shrubby plants; leaves opposite, pinnately compound; flowers regular, 4–5-merous; stamens twice the number of petals; style 1; ovary 2–12-carpellate, superior. Four genera represented in our range.

1. Woody shrubs; plants strong-scented; fruit woolly-hairy.

> *Larea tridentata*
> Creosote bush

1. Herbaceous plants, annuals or perennials; plants not strong-scented; fruit glabrous to hairy not ever woolly.
   2. Leaflets 2, succulent; perennials with creeping roots.

> *Zygophyllum fabago* var.
> *brachycarpum*
> Syrian bean-caper

   2. Leaflets 6–10, not succulent; plants annual.
      3. Fruits separating into 5 nutlets, very spiny.

> *Tribulus terrestris*
> Ground bur-nut

      3. Fruits separating into 8–12 nutlets, not spiny.

> *Kallstroemia hirsutissima*
> Caltrop

WOOD-SORREL FAMILY: OXALIDACEAE

One genus, *Oxalis,* in our range. Leaves palmately 3-foliolate, leaflets mostly obcordate; flowers perfect, sepals 5, often unequal; petals 5; stamens 10–15; ovary 5-celled.

1. Plant acaulescent; petals violet or purple.

> *Oxalis violacea*
> Violet wood-sorrel

1. Plant caulescent; petals yellow.
   2. Hairs of stem and petioles in part septate (provided with partitions); plant rhizomatous.

> *O. stricta*
> Yellow wood-sorrel

2. Hairs of stem and petioles nonseptate; plant more stoloniferous than
   rhizomatous.

> O. dillenii
> Dillen-wood sorrel

YELLOW WOOD-SORREL, *O. stricta,* is the only common species in our range. The erect or decumbent stems bear 2–3 yellow flowers; the leaves are acid-flavored. Occurs as a weed in gardens throughout most of North America and is occasionally found in woods or along roads of the foothill and lower montane zones.

GERANIUM FAMILY: GERANIACEAE

Only two genera occur within our range. Flowers are regular with 5 sepals, united at base and usually awn-tipped; petals 5, separate; stamens 5–10. The pistil is compound, with 5 1-seeded carpels, which separate at maturity. The 5 styles, which are united into an elongated beak, separate as the seed ripens.

1. Leaves pinnately divided; stamens 5; carpel-tails hairy on the inner side.

> *Erodium cicutarium*
> Storksbill or filaree

1. Leaves palmately lobed or divided; stamens 10; carpel-tails naked on the inner side.
   2. Plants pubescent but not glandular.

> *Geranium caespitosum*
> James geranium

   2. Plants more or less glandular-pubescent.
      3. Plants single or scarcely tufted.
         4. Pubescence tipped with purple glands

> *G. richardsonii*
> Richardson geranium

         4. Pubesence viscid-glandular.
            5. Lower petioles and stems hirsute.

> *G. viscosissimum* var.
> *viscosissimum*
> Sticky geranium

            5. Lower petioles and stems glabrous.

> *G. viscosissimum* var.
> *nervosum*

      3. Plants caespitose-tufted.
         4. Pubesence short-glandular, mostly above only.

> *G. fremontii*
> Fremont geranium

         4. Pubesence viscid-villous as well as short-glandular.

> *G. parryi*
> Parry geranium

STORKSBILL or FILAREE, *Erodium cicutarium* (fig. 190), is one of the earliest flowers to bloom on the mesas and in the lower foothills, especially on over-grazed fields. It has pinnately compound, much dissected leaves, and small bright pink flowers. It is easily recognized, both by these characters and by the corkscrew-like tails of the fruits. These are coiled when dry; but when moist, they uncoil, an action that drives the pointed tip containing the seed into the soil. The Spaniards introduced this plant (which they called *Alfilaria*) from

Europe. It is now widely distributed as a weed in the United States, and is a valuable forage plant throughout the Southwest.

In the genus *Geranium* the styles separate at the base and recoil at maturity, but they are not twisted. The house plants commonly called geraniums belong to a related genus, *Pelargonium,* of African origin.

FREMONT GERANIUM, *G. fremontii* (fig. 191), is common in the foothill and montane zones of the central Rockies. The petioles of its basal and lower leaves are hairy but not sticky; flowers are usually a deep rose-purple. Occurs from southern Wyoming southwestward through Colorado into Utah and Arizona.

JAMES GERANIUM, *G. caespitosum,* is pubescent but not at all sticky. It may be erect, nearly 1m tall, but sometimes its stems are decumbent and rooting at the nodes. The flowers are usually deep rose-purple. Grows abundantly in southwestern Colorado and adjoining states in ponderosa pine forests.

STICKY GERANIUM, *G. viscosissimum,* is a stout, many-flowered plant of the northern part of our range. Its flowers are bright rose-purple, and the plant is very sticky. Conspicuous in the montane zone from Wyoming north.

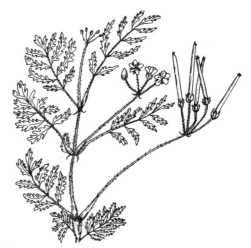

190. Storksbill, ½X

191. Fremont Geranium, ⅖X

√ RICHARDSON GERANIUM, *G. richardsonii,* has white or very pale lavender flowers; its stems are usually single. It grows in moist, usually shaded locations throughout our range, from the foothills to timberline.

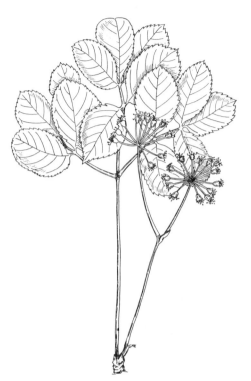

192. Wild Sarsaparilla, ²/₅ X

GINSENG FAMILY: ARALIACEAE

Only one species is likely to be found in our range: WILD SARSAPARILLA, *Aralia nudicaulus* (fig. 192), is about 3dm tall. Stem somewhat woody, short, scarcely rising out of the ground; single long-stalked compound leaf, the leaflets finely toothed; a shorter, naked scape bearing greenish flowers in 2–7 ball-like umbels; petals 5, ovate; stamens 5; fruit a dark berry.

CARROT FAMILY:
UMBELLIFERAE

A large and difficult family. Plants have small, usually numerous, flowers in umbels (see Plate F52), rarely in heads. Stems are commonly hollow; leaves mostly alternate and compound, often decompound (several times compounded or divided). The umbels are usually compound, forming umbellets; the bracts under the general umbel form an involucre, under an umbellet an involucel. Petals 5, generally white or yellow; stamens 5; ovary inferior; styles 2, often forming a *stylopodium* (a swelling at the base).

Correct identification usually requires the fruit. The fruit (a schizocarp splitting along the commissure) is flattened either laterally (at right angles to the commissure) dorsally (parallel to the commissure), or it is nearly terete (not flattened). A schizocarp splits into 2 1-seeded mericarps; the fruit wall has 1 or more oil tubes in the interval between the ribs or wings on the commissural side.

The family includes several common vegetables and culinary herbs, such as carrots, parsnips, parsley, caraway seed, and angelica. The thick, starchy roots of some wild species were used as food by the Indians and early settlers, but several species are poisonous, so accurate identification is very important..

1. Basal leaves simple, toothed, cordate at base; stem leaves mostly trifoliate.

Zizia p. 240
Zizia

1. Basal leaves compound.
  2. Most leaves with well-defined leaflets, not dissected into small, narrow segments.
    3. Leaflets usually 3, mostly 1–4dm long and wide.

Heracleum p. 235
Cow parsnip

3. Leaflets usually more than 3, usually under 1dm wide.
    4. Plants from fibrous or fleshy-thickened, fascicled roots; no taproot or developed caudex.
        5. Involucre absent; plants from fleshy-thickened, fascicled roots.
            6. Involucel absent; leaflets crenate on margins, rounded or obtuse at tip; fruit flattened dorsally; flowers white.

> *Oxypolis* p. 240
> Cowbane

            6. Involucel poorly developed; leaflets linear-lanceolate to lance-ovate, serrate on margins, sharply acute at tip; fruit flattened laterally; flowers white to greenish.

> *Cicuta* p. 233
> Water hemlock

        5. Involucre and involucel present; plants from fibrous roots; leaflets lance-linear, sharply serrate; fruit slightly compressed laterally.

> *Sium* p. 240
> Water parsnip

    4. Plants from a taproot or stout caudex.
        5. Roots usually crowned with many stringy fibers; leaflets usually many left to near the midrib; fruit slightly flattened laterally.

> *Ligusticum* p. 236
> Lovage

        5. Roots usually not crowned with many stringy fibers; leaflets toothed or few-cleft; fruit flattened dorsally or subterete (almost cylindrical).
            6. Stems usually more than 1cm thick; primary rays (branches of the umbel) usually 10 or more in some flower clusters; petals never yellow; fruit flattened dorsally.

> *Angelica* p. 232
> Angelica

            6. Stems usually less than 1cm thick; primary rays less than 8; if more then petals yellow; fruit subterete.

> *Osmorhiza* p. 239
> Sweet cicely

2. Leaves without well-defined leaflets, mostly dissected into small, narrow segments, the ultimate segments mostly less than 3(10)mm wide.
    3. Plants taprooted, the taproot sometimes fleshy-thickened but elongate, often surmounted by a stout, branching caudex, or with deep, creeping rootstocks.
        4. Flowers white, purple, or pink.
            5. Stylopodium absent; fruit flattened dorsally.
                6. Dorsal ribs prominent, at least one of them winged; petals white to purplish.

> *Cymopteris* p. 234
> Cymopteris

                6. Dorsal ribs filiform or obsolete, not at all winged; petals purple.

> *Lomatium* p. 237
> Biscuit-root or desert-parsley

            5. Stylopodium present; flowers white, rarely pink; fruit subterete, or flattened laterally or flattened dorsally.
                6. Taproot usually crowned with many stringy fibers; fruit oblong, mostly subterete.

> *Ligusticum* p. 236
> Lovage

                6. Taproot usually not crowned with many stringy fibers; fruit oval or ovoid and flattened.
                    7. Fruit oval, flattened dorsally.

> *Conioselinum* p. 234
> Hemlock parsley

7. Fruit ovoid, flattened laterally; stems purple or purple-spotted at base.

Conium p. 234
Poison hemlock

4. Flowers yellow.
   5. Plants with leaves all basal.
      6. Plants under 1dm tall.

Oreoxis p. 238
Alpine parsley

      6. Plants about 3.5dm tall.

Cymopteris p. 234
Cymopteris

   5. Plants with some stem leaves.
      6. Taproots usually crowned with many stringy fibers; fruit flattened laterally.

Harbouria p. 235
Whisk-broom parsley

      6. Taproots usually not crowned with many stringy fibers; fruit flattened dorsally.
         7. Caudex simple; leaf segments remote and few; bractlets much shorter than the fruits.

Lomatium p. 237
Biscuit-root or desert parsley

         7. Caudex branched, remnants of old stems evident; leaf segments many and crowded; bractlets equaling or exceeding fruits in length.

Cymopteris p. 234
Cymopteris

3. Plants with fibrous or fleshy-thickened roots.
   4. Leaves all basal, petioles filiform; plants under 1.5dm tall, the root globose.

Orogenia p. 239
Indian potato

   4. Leaves not all basal, petioles wider; plants more than 2dm tall, with a cluster of fleshy roots or an enlongate fleshy root.
      5. Leaflets few, linear to linear-lanceolate, leaves often early deciduous; petals about 1mm long; fruit 2–4mm long, subterete to somewhat flattened laterally.

Perideridia p. 240
Yampa

      5. Leaflets wider, leaves persistent; petals about 1.5mm long; fruit 4–6mm long, flattened dorsally.

Conioselinum p. 234
Hemlock parsley

(1) *Angelica.* Angelica

1. Stems over 1m tall; whole umbels globular in shape.

A. ampla
Giant angelica

1. Stems less than 1m tall; whole umbels rather flat-topped.
   2. Leaves once or twice pinnate.
      3. Involucels of conspicuous bractlets.

A. grayi
Gray's angelica

      3. Involucels none.

A. pinnata
Pinnate-leaved angelica

   2. Leaves ternate, then pinnate.

3. Fruit glabrous; flowers white.

A. arguta
Sharptooth angelica

3. Fruit scabrous; flowers greenish or purplish.

A. roseana
Rose angelica

GIANT ANGELICA, *Angelica ampla* (fig. 193), has stout stems often more than 1.5cm tall and large, pinnately compound leaves. Both the big compound umbels and the small ones are globular. This character plus its size makes the plant easy to recognize. The flowers are white, but after the petals drop and the fruits develop, the umbels look greenish or brown. Grows on moist soil in shade in montane meadows of Wyoming and Colorado.

GRAY'S ANGELICA *A. grayi,* is a stout plant not over 6dm tall, but it is large for the conditions under which it grows. Flowers are greenish and the umbels globular. Found in the subalpine and alpine zones on high mountains, often among rocks, in Wyoming and Colorado.

ROSE ANGELICA, *A.*

193. Giant Angelica, ²/₅ X

roseana, a stout plant that has or pink flowers and less symmetrical umbels than the last, is common in the northern part of our range. PINNATE-LEAVED ANGELICA, *A. pinnata,* is a comparatively slender plant with stems up to 9dm tall and flattened umbels of white or pinkish flowers. Grows in the montane and subalpine zones from Montana south to New Mexico.

(2) *Cicuta*. Water hemlock

One species grows in our range: WATER HEMLOCK, *Cicuta douglasii,* a deadly poisonous plant. It is a smooth marsh perennial, 6–12dm tall, with an open inflorescence of compound umbels. Pinnately compound leaves with serrate leaflets; involucre of few bracts or none; involucels present; fascicled tuberous roots. In old plants, cross partitions are developed in the stem above

194. Mountain Parsley, ½X

the roots (in cross section). The secondary veins of the leaflets end in the notches between the teeth. Occurs on wet ground from Alaska south in the mountains to Mexico. This plant resembles WATER PARSNIP, *Sium suave,* which is not poisonous. The veins of water parsnip leaves run toward the tips of the teeth, a character helping to distinguish it from water hemlock.

(3) *Conioselinum*. Hemlock parsley

HEMLOCK PARSLEY, *C. scopulorum*, is the only species in our range. Plants caulescent; glabrous except in inflorescence; from a taproot or a cluster of fleshy roots; 3–9dm tall; leaves 10–20cm long, lanceolate or ovate in shape, 1- or 2-pinnate, or 1- or 2-ternate-pinnate; leaflets 20–65mm long; flowers white; involucre of several narrow bracts or absent; involucels present.

(4) *Conium*. Poison Hemlock

One species in North America. POISON HEMLOCK, *Conium maculatum,* has branched, spotted stems from 50cm to 3m tall; erect, caulescent, glabrous, biennial plants with stout taproots; involucre present; involucels present. Grows on waste, moist soil along roadsides and on ditch banks over much of the world. Found in western Colorado at elevations of 5,000-9,000 feet. Beware of any such plant that has a spotted stem.

(5) *Cymopteris*. Cymopteris

1. Fruit strongly flattened dorsally.
    2. Dorsal ribs very prominent or slightly winged.
        3. More or less caulescent; flowers yellow.

                C. lemmonii
                Mountain parsley

        3. Acaulescent; flowers white.

                C. bipinnatus

    2. Dorsal ribs filiform.

3. Peduncles shorter than (not exceeding) the leaves; bracts usually absent; flowers white.

C. acaulis
Plains cymopteris

3. Peduncles longer than the leaves; bracts present; flowers usually yellow.
    4. Leaflets about as wide as long; without pseudoscapes (a false scape between roots and leaves, underground or just above ground).

C. newberryi

    4. Leaflets definitely longer than wide; pseudoscapes 4–20cm high.

C. fendleri

1. Fruit not strongly flattened dorsally, usually more or less flattened laterally.
  2. Seed face grooved or concave.
    3. Plants with pseudoscapes, 5–16cm long.
      4. Flowers purple; pedicels 2–3mm long.

C. planosus

      4. Flowers usually yellow; pedicels 3–8mm long.

C. longipes

    3. Psuedoscapes inconspicuous, not over 2cm long.

C. purpureus

  2. Seed face plane or nearly so.
    3. Peduncles shorter than (not exceeding) the leaves; flowers pink to purplish.

C. montanus

    3. Peduncles exceeding leaves at maturity.
      4. Bractlets of involucels many-nerved, often purplish; pedicels under 1mm long or absent.

C. multinervatus

      4. Bractlets few-nerved if any, never purplish; pedicels 2–21mm long.
        5. Fruit broadly ovoid, the wings 2–3 times as wide as the body.

C. purpurascens

        5. Fruit ovoid-oblong to oblong, the wings seldom as wide as the body.

C. bulbosus

MOUNTAIN PARSLEY, *Cymopteris lemmonii* (fig. 194), is a yellow-flowered plant that grows in moist meadows and aspen groves from the foothills to the subalpine zone. Its stems may be from 20cm to 60cm tall. Occurs from Wyoming and Utah southward in the mountains. Sometimes segregated as *Pseudocymopteris montanus*.

(6) *Harbouria*. Whisk-broom parsley

Harbouria is a monotypic genus (includes only 1 species). WHISK-BROOM PARSLEY, *Harbouria trachypleura* (fig. 195), also has yellow flowers, but its stems are usually in tufts, and it is found on dry hillsides on the montane zone. It is a caulescent perennial, but the leaves are mostly basal, pinnately decompound, the ultimate divisions linear and distinct; involucre usually absent; involucel of several linear bractlets.

(7) *Heracleum*. Cow parsnip

One species in our range. COW PARSNIP, *Heracleum sphondylium* (fig. 196), has large, stout, hollow stems 1.8–2.5m tall; leaves large, ternately compound (some simple), the leaflets coarsely toothed; upper cauline leaves with con-

195. Whisk-broom
Parsley, ½ X

196. Cow Parsnip, ⅖ X

spicuous sheaths at base of petioles. The flowers at the edge of the umbels are larger than the inner ones. The umbels are flat-topped, the flowers white.

(8) *Ligusticum*. Lovage

1. Leaflets ovate to oblong or lanceolate, mostly over 3mm wide.
    2. Fruit 5–8mm long; woods and meadows.

L. porteri var. *porteri*
Porter's lovage

    2. Fruit 3–5mm long; wet places.

L. canbyi
Canby lovage

1. Leaflets linear, 1–3mm wide.
    2. Plants caulescent; pedicels 8–12mm long.

L. filicium
Fern-leaved lovage

    2. Plants subcaulescent (only 1 cauline leaf); pedicels 3–8mm long.

L. tenuifolium
Slender-leaved lovage

PORTER'S LOVAGE, *Ligusticum porteri* (fig. 197), has stems from 48cm to 1m tall, with white-flowered, flattened, compound umbels. Grows in meadows of montane and subalpine zones of Wyoming south and west through our region.

(9) *Lomatium*. Biscuit-root, desert parsley

1. Ultimate leaf-segments relatively large, some at least 1cm long or more, the leaflets few, not lace-like.
  2. Involucel usually absent.

> *L. ambiguum*
> Swale desert-parsley

  2. Involucel present.
    3. Flowering stalk puberulent.
      4. Mature fruit twice as long as wide; leaf segments linear.

> *L. triternatum* ssp.
> *platycarpum*
> Nine-leafed lomatium

      4. Mature fruit 3 times as long as wide; leaf segments elliptic to oblanceolate.

> *L. triternatum* ssp.
> *triternatum*

    3. Plant glabrous.
      4. Leaves fleshy; fruits 5–9mm long.

> *L. nuttallii*

      4. Leaves not fleshy; fruits 10–13mm long.

> *L. graveolens*

1. Ultimate leaf-segments mostly less than 1cm long; the leaves very dissected, the segments numcrous or lace-like.
  2. Plants usually over 25cm high.

> *L. dissectum* var.
> *multifidum*
> Indian balsam

  2. Plants mostly under 25cm tall.
    3. Plants glabrous or slightly scabrous.
      4. Involucel bractlets oblanceolate, obovate, or elliptic.

> *L. cous*
> Cous biscuit-root

      4. Involucel bractlets linear or lanceolate.
        5. Ultimate leaf-segments mostly 4mm or less long, 1mm or more wide.

> *L. attenuatum*

        5. Ultimate leaf-segments from nearly 0.5 to 1mm wide, longer than 4mm.
          6. Taproot usually bulbous near top.

> *L. bicolor*
> Bicolor biscuit-root

          6. Taproot not bulbous near top.

> *L. grayi*
> Gray's lomatium

    3. Plants very puberulent.
      4. Ovaries and fruit completely hairy.

> *L. foeniculaceum*
> Fennel-leaved lomatium

      4. Ovaries and fruit mostly glabrous.
        5. Involucel bractlets tomentose to villous.

> *L. macrocarpum*

        5. Involucel bractlets glabrous or nearly so.
          6. Petals white; leaves slightly glaucous.

> *L. orientale*
> Biscuit-root or salt-and-pepper

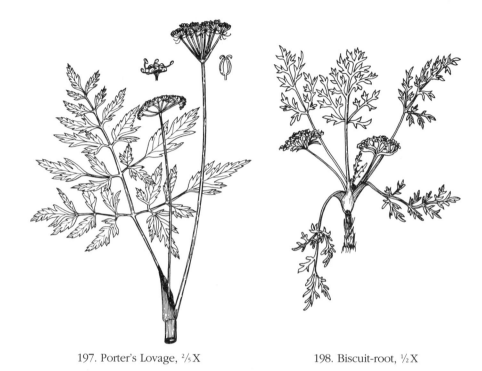

197. Porter's Lovage, ²/₅X   198. Biscuit-root, ½X

6. Petals yellow; leaves not glaucous.

L. juniperinum

BISCUIT-ROOT or SALT-AND-PEPPER, *Lomatium orientale* (fig. 198), with whitish flowers and finely divided grayish leaves, is one of the first plants to appear on dry plains, mesas, and foothills in the spring. To begin with, its leaves spread flat against the ground, and its speckled flower clusters seem almost stemless. Later the stems may become 5cm or more long. Found from Minnesota to Montana and south to Missouri and northeastern Colorado. The thick starchy roots were used by Indians and pioneers for food.

INDIAN BALSAM, *Lomatium dissectum,* is a stout species with stems 3–9dm tall. Its flowers are usually yellow but may be purplish. Grows in the upper foothill and lower montane zones throughout our range.

(10) *Oreoxis.* Alpine parsley

1. Plants glabrous (sometimes slightly puberulent just below the umbel); alpine.
    2. Involucels linear, distinct bractlets.

O. humilis

    2. Involucels with broad bractlets, toothed at the apex, sometimes purplish.

O. bakeri

1. Plants usually puberulent (rarely glabrous); broad bractlets more or less united at base; subalpine to alpine.

O. alpina

ALPINE PARSLEY, *Oreoxis alpina* (fig. 199), is a low, mat-forming plant found on gravel slopes and among rocks of the subalpine and alpine zones. The yellow-flowered umbels are compact, rarely over 7mm in diameter, on stalks from 2.5–10cm tall; and the leaves are finely dissected. Found on high mountains from Wyoming and Utah south to New Mexico and Arizona.

199. Alpine Parsley, ½X

(11) *Orogenia*. Indian potato

One species in our range. INDIAN POTATO, *Orogenia linearifolia,* is a small, smooth plant 2–12 cm tall, rising from a deep-seated round tuber; leaves 2–3, once or twice ternate, the leaflets entire. The rays of the umbel are few and of unequal length, each bearing a small tight cluster of white or pinkish flowers. Sometimes found blooming on mountain slopes as soon as the snow melts.

(12) *Osmorhiza*. Sweet cicely

1. Fruit with bristly ribs; petals greenish-white to purplish.
   2. Involucel present.

   O. longistylis

   2. Involucel absent.
      3. Fruit obtuse at apex.

   O. depauperata
   Sweet cicely

      3. Fruit with a sharp beak.

   O. chilensis
   Western sweetroot

SWEET CICELY, *Osmorhiza depauperata* (fig. 200a and 200b), 15–16cm tall, is a plant of moist, shaded woods. It hardly looks like a member of this family. Its umbels have only 2–5 rays, which are long and spreading or reflexed. The club-shaped fruits are about 7–8mm long.

WESTERN SWEETROOT, *O. occidentalis,* may grow to 9dm tall and has 5–12 rays to an umbel. The flowers are yellowish, and the linear, erect fruits are 7–20mm long. When nearly ripe, each pair becomes detached and separate at the base, while still attached at the tip.

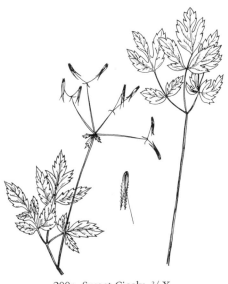

200a. Sweet Cicely, ⅖X

(13) *Oxypolis.* Cowbane

FENDLER OXYPOLIS or COWBANE, *Oxypolis fendleri,* is the only species in our range. It is a slender, smooth plant with stems 3–10dm tall. The small, dainty umbels are white-flowered, and the pinnate leaves, with coarsely-toothed leaflets, are somewhat suggestive of celery. Involucre and involucel are absent. A plant of wet, shaded locations, such as streamsides and bogs, from the montane to subalpine zones. Occurs from Wyoming and Utah south to New Mexico.

(14) *Perideridia.* Yampa

1. Principal leaves dissected, ultimate segments numerous; plants 2–6dm tall.
    P. bolanderi ssp.
    bolanderi
    Bolander's yampa
1. Principal leaves once or sometimes twice pinnate or ternate, ultimate segments few; plants
    4–12dm tall.
    P. gairdneri ssp. *borealis*
    Yampa

200b. Sweet Cicely, in fuit, ²⁄₅X

YAMPA, *Perideridia gairdneri,* is a tall slender plant with compound leaves of narrow, grass-like leaflets 2.5–15cm long, which soon wither. The small white flowers are in compound umbels. Grows from tubers like small sweet potatoes. One of the best wild food plants of the Rocky Mountains, it was an important food used by the Indians and the early explorers. Be sure you have the correct identification of the plant if you wish to eat it.

(15) *Sium.* Water parsnip

One species occurs in our range, *Sium suave,* which resembles the poisonous *Cicuta douglasii.* Stems 50–100cm tall; leaves pinnately compound; involucre and involucel present; petals white. The veins of the leaves run toward the tips of the teeth.

(16) *Zizia.* Zizia

Only one species is likely to be found in our range: *Zizia aptera.*

Stems 30–60cm tall; basal leaves simple, crenate-dentate; cauline leaves ternately divided; involucel present; involucre absent.

GENTIAN FAMILY: GENTIANACEAE

The most conspicuous members of this family found in our range have bright blue or purplish-blue flowers; a few have white or greenish corollas. The united corollas may be saucer-shaped (rotate), funnel-form, or tubular with more or less spreading lobes; the lobes may be either 4 or 5. In general, the leaves are opposite, whorled, or all basal; the foliage and stems are always smooth. The blossoms of several species close when shaded, and they can be overlooked on cloudy days. There has been substantial modification of the traditional nomenclature of the gentians in recent years.

1. Style filiform, deciduous; anthers recurved or twisted in age.
  2. Corolla-tube exceeding the calyx.

          Centaurium p. 241
          Centaury

  2. Corolla-tube shorter than the calyx.

          Eustoma p. 242
          Prairie gentian

1. Style short and persistent, or absent; the stamens remaining straight.
  2. Corolla rotate, with 1–2 nectar-bearing pits or glands at the base of each lobe.
    3. Leaves verticillate (whorled); style evident.

          Frasera p. 242
          Green gentian

    3. Leaves not verticillate; style absent.

          Swertia p. 246
          Star gentian

  2. Corolla without nectar-bearing pits or glands.
    3. Corolla rotate, with the stamens on its base.

          Lomatogonium p. 246
          Felwort

    3. Corolla campanulate, funnelform, or tubular, the stamens on its tube.
      4. Corolla plicate (folded into plaits) in the sinuses, the folds forming teeth or smaller lobes; corolla lobes entire, not fringed at base inside.

          Gentiana p. 242
          Pleated gentian

      4. Corolla not plicate in the sinuses, no teeth or smaller lobes between the main lobes; corolla lobes fringed or with irregular margins, or fringed at the base inside.
        5. Corolla about 2cm long, conspicuous, the lobes fringed or irregular, but not fringed in the throat at the base of the lobes; 4-merous.

          Gentianopsis p. 245
          Fringed gentian

        5. Corolla less than 2cm long, inconspicuous, the margins of the lobes mostly entire, but fringed in the throat at the base of the lobes; 4–5-merous.

          Gentianella p. 244
          Fringed gentian

(1) *Centaurium*. Centaury

1. Corolla salverform; corolla lobes 7–8mm long, at least 3/4 as long as tube.
          C. calycosum
1. Corolla salverform; corolla lobes 3–5mm long, about 1/2 as long as the tube.
          C. exaltatum

(2) *Eustoma*. Prairie gentian

One species in our range: *E. grandiflorum*. Annual to perennial; plants with vertical taproots and rosettes; leaves entire, ovate to lance-elliptic; flowers deep-purple or, rarely, white.

(3) *Frasera*. Green gentian

1. Tall, unbranched; floral branches not exceeding the leaves.

F. speciosa
Monument plant

1. Low, much-branched stem; floral branches much exceeding the leaves; leaves white-margined.

F. albomarginata

basal leaves
½ actual size

201. Monument Plant, ½X

MONUMENT PLANT or GREEN GENTIAN, *Frasera speciosa* (fig. 201), is a stout, pale green plant, which sometimes becomes very conspicuous in middle and late summer on montane hillsides and at the edges of meadows. It is occasionally found in the sub-alpine zone. Stems are from 3 to 18dm tall from a basal rosette of long smooth leaves; stem leaves are in whorls, progressively smaller upward. There are numerous flowers on short stalks from each of the middle and upper whorls. Corollas are 4-lobed, pale green with purplish markings and fringed appendages. A biennial plant, its rosettes of large, pale (sometimes purple-tinged), oblong or spatulate leaves are conspicuous during its first summer. In the second season, when the flower stalk begins to develop, the basal leaves gradually wither. After flowering, the plant dies, and the dry stalks persist through the winter.

(4) *Gentiana*. Pleated gentian

1. Flowers solitary and terminal, not over 18mm long; plants rarely over 10cm tall; annual or biennial.
  2. Corolla greenish outside, white inside.

G. aquatica
Fremont gentian

2. Corolla blue.

G. prostrata
Moss gentian

1. Flowers usually more than 1, over 20mm long; plants usually more than 10cm tall; perennial.
    2. Corolla yellowish-white, spotted or streaked with purple, over 3cm long.

G. algida
Arctic gentian

    2. Corolla blue to purple, under 3cm long (except in *G. parryi*).
        3. Flowers 1–6 per stem, in terminal or subterminal clusters; floral bracts ovate to ovate-lanceolate.

G. parryi
Parry gentian

        3. Flowers several to many, some axillary well below apex of stem; floral bracts lanceolate or linear.
            4. Bracts rarely as long as the flower; upper leaves ovate to lanceolate.

G. affinis
Pleated gentian

            4. Bracts usually longer than the flowers; upper leaves linear or linear-lanceolate.

G. bigelovii
Prairie gentian

MOSS GENTIAN, *G. prostrata,* is a small, slender plant found in moss or in the grassy alpine tundra. Its blue, star-like flower is not over 10mm broad when open. It is so sensitive to light that it will close immediately if a cloud passes over, or even if a hand is held over it for an instant. FREMONT GENTIAN *G. aquatica* (fig. 202), is a closely related species, but its stem is erect with a whitish flower. Sometimes abundant in wet subalpine meadows but can be found lower. When it goes to seed, the capsule is pushed up above the calyx and splits, showing two spatula-shaped valves.

PARRY or MOUNTAIN GENTIAN, *G. parryi* (fig. 203), is the largest-flowered and

202. Fremont Gentian, ½X

203. Mountain Gentian, ½X

204. Prairie Gentian, ½X

205. Arctic Gentian, ½X

probably the most frequently seen of our gentians. The corollas are goblet-shaped, 4–5cm deep, bright, clear blue with dark greenish stripes on the outside; inconspicuous when closed. Several stems from a perennial crown; each stem may have 1 to several blossoms. Grows on open fields and meadows of the subalpine zone throughout our range.

PLEATED GENTIAN, *G. affinis,* is an erect plant with stems 15–38cm tall. Each one holds several to many bright blue, narrowly funnelform blossoms. The calyx is short and split at one side, and the corolla lobes are short and pointed. Grows in moist meadows and along streams of the montane zone. PRAIRIE GENTIAN, *G. bigelovii* (fig. 204), is a closely related plant that grows on dry fields of the mesas and foothills. It is a tufted plant with stems that are decumbent at base. It has blue flowers, which appear closed except in bright sunlight.

ARCTIC GENTIAN, *G. algida* (fig. 205), is a tundra plant and occurs in our region on grassy alpine meadows. It usually has several stems, each bearing a compact cluster of flowers. The deep, cup-shaped corollas are greenish-white with dark blue or purplish streaks on the outside of each lobe. The buds and closed flowers appear dark in color. Inside are 5 salmon-colored anthers. Occurs from Siberia and Alaska south through the high mountains into Utah and northern New Mexico. It is the latest of the alpine plants to bloom and is rarely seen in flower before August, by which time the fern-like leaves of its companion, alpine avens, have already turned maroon.

(5) *Gentianella*. Fringed gentian

1. Flowers mostly 4-merous; base of corolla lobes not fringed on inner surface.
                              G. propinqua ssp.
                              propinqua
                              Four-parted gentian
1. Flowers 4–5-merous; base of corolla lobes fringed on inner surface.
    2. Flowers solitary on a naked peduncle 2–10cm long; plants not over 10cm tall.
                              G. tenella
                              Slender or one-flowered
                              gentian

2. Flowers several to many in short-peduncled clusters; plants usually over 10cm tall.
   3. Calyx lobes conspicuously unequal and unlike.

                         G. heterosepala

   3. Calyx lobes somewhat unequal, but not conspicuously unlike.

                         G. amarella
                         Rose or northern gentian

AMARELLA or ROSE GENTIAN, *G. amarella,* is a variable plant of very wide distribution. Flowers 4–5-lobed, lobes pointed; a fringed crown at the throat of the corolla; 1 to many flowers. The plants are usually erect and often have several erect branches. Flowers vary from dull rose to lilac or pale blue. Usually found on marshy ground and occurs throughout our range from the montane zone to the alpine.

ONE-FLOWERED GENTIAN, *G. tenella,* is a tiny plant with a pale blue flower about 7–8mm long on the end of a thread-like stalk. Grows in grassy tundra from the Arctic south to Arizona.

(6) *Gentianopsis.* Fringed gentian

1. Plants annual; flowers on naked peduncles, not subtended by bract-like leaves.

                         G. detonsa var. *elegans*
                         Rocky Mountain fringed
                         gentian

1. Plants perennial; flowers short-peduncled in the axils of 2 bract-like leaves.

                         G. barbellata
                         Fragrant fringed gentian

ROCKY MOUNTAIN FRINGED GENTIAN, *G. dentosa* var. *elegans* (fig. 206). The flowers of this slender, erect, annual plant have 4 lobes with fringed margins. On a dull day, the lobes are erect and twisted together; but in bright sunlight they spread widely, displaying their brilliant purplish-blue color. Found in moist meadows and on margins of ponds and streams from the montane zone to the alpine zone throughout the Rockies. It was chosen the park flower of Yellowstone because of its abundance in that area.

FRAGRANT FRINGED GENTIAN, *G. barbellata,* is somewhat similar in general appearance. Its flowers vary from rather light grayish-blue to dark purplish but are never so bright as *G. detonsa,* and its corolla lobes are narrower and usually twisted. It is a rare plant, found at the edges of coniferous forests or on open slopes of the subalpine zone in the central and southern Rockies. It should be protected. Blooms late, often in flower in September.

206. Rocky Mountain Fringed
Gentian, ½ X

Only one species in North America, MARSH FELWORT, *Lomatogonium rotatum,* a small, white-flowered (sometimes blue) annual. Plants glabrous, 1–4dm high; leaves linear with midrib and 2 faint lateral nerves; corolla often surpassed by the sepals. Found in mountain bogs and wet meadows of northwestern America.

(8) *Swertia*. Star gentian

Only one species in this genus remains in North America: STAR GENTIAN, *Swertia perennis* (fig. 207), a perennial. Stems 15–30cm tall from a basal rosette of smooth, lanceolate or elliptic leaves, with a long narrow cluster of dull blue, star-shaped flowers. Corolla deeply 5- (or 4)-lobed with a basal pair of glands. Found in wet meadows and on stream banks in the subalpine and alpine zones of the Rocky Mountains and in boreal regions around the world.

DOGBANE FAMILY: APOCYNACEAE

Only one genus is likely to be found in our range. DOGBANE, *Apocynum androsaemifolium* var. *androsaemifolium* (fig. 208), is an erect, perennial herb having forked branching and smooth, ovate or oblong leaves; clusters of small,

207. Star Gentian, ½X   208. Dogbane, ⅖X

terminal, white or pink flowers; campanulate, 5-lobed corolla; the stamens alternate with the corolla lobes; and a milky juice that is said to be poisonous. The pods are in pairs, 7–15cm long and very slender; each seed has a tuft of silky hairs at one end. The opposite leaves usually hang down. Grows from the foothills to the subalpine zone on dry hillsides and along roads throughout our range. The stems contain strong fibers. A related species, INDIAN HEMP, *A. cannabinum,* which is taller, has a greenish-white corolla, and grows at lower altitudes, was used by the Indians for making string and rope.

MILKWEED FAMILY: ASCLEPIADACEAE

These plants have their flowers in umbels. They have milky juice and pods containing silky tufted seeds, as do the dogbanes, but milkweed flowers are elaborately constructed. In each individual flower, there is a *corona* of hooded appendages between the 5 stamens and the pistil. These flower parts are so arranged as to trap any insect that visits the flower, making it impossible for the insect to get away without carrying some pollen to the next flower it visits. Milkweed flowers are very attractive to butterflies, especially the handsome monarch butterfly, whose green and black banded caterpillars feed on the milkweed plants.

1. Corolla and crown orange or bright red.
    2. Corolla lobes orange; leaves mostly alternate; stems hirsute to rough-pubescent.
        Asclepius tuberosa ssp.
        interior
        Butterflyweed
    2. Corolla red; leaves opposite; glabrous to somewhat pubescent.
        A. incarnata
1. Corolla greenish, whitish, yellowish, or purplish.
    2. Hoods not crested or horned within.
        3. Mass of anthers and stigma nearly globose; leaves narrowly linear to linear-filiform.
            4. Hood truncate-emarginate at apex, a shorter tooth in the sinus; leaves usually over 3mm wide, usually opposite.
                A. stenophylla
        3. Mass of anthers and stigma definitely longer than wide; leaves lanceolate or broader.
            A. viridiflora
            Green-flowered milk-weed
    2. Hoods crested or horned within.
        3. Flowers large, corolla lobes 7–14mm long.
            4. Leaves oblong to lanceolate-ovate; corolla lobes pink to greenish-purple; fruit white-tomentose, often soft-spiny.
                A. speciosa
                Showy milkweed
            4. Leaves ovate-orbicular; corolla lobes greenish-white to greenish-yellow; fruit glabrous to puberulent, never soft-spiny.
                5. Plants small, to 20cm tall; corolla lobes 10–14mm long.
                    A. cryptoceras
                    Palid milkweed
                5. Plants over 20cm tall; corolla lobes 7–10mm long.
                    6. Stems and leaves canescent-tomentose; petioles usually over 8mm.
                        A. arenaria

6. Stems and leaves glabrous to puberulent; petioles rarely over 8mm.

A. latifolia

3. Flowers small, corolla lobes 3–6mm long.
    4. Leaves narrowly linear to linear-filiform.
        5. Leaves linear-filiform, irregularly alternate or obscurely whorled; stems tufted.

A. pumila
Low milkweed

        5. Leaves narrowly linear, whorled; stem 1 to several, but not tufted.

A. subverticillata
Whorled milkweed

    4. Leaves linear-lanceolate to ovate-lanceolate.
        5. Dwarf plants, not over 30cm tall.
            6. Leaf blades less than 5cm long, ciliate on margins and midrib; umbels sessile or subsessile.

A. uncialis
Dwarf milkweed

            6. Leaf blades over 5cm long; hairs not confined to margins and midrib; umbel short-peduncled.

A. brachystephana

        5. Taller plants, over 30cm tall.

A. hallii
Hall milkweed

209. Showy Milkweed, ²/₅ X

SHOWY MILKWEED, *Asclepias speciosa* (fig. 209), is our most common species. It has thick, short-petioled, oblong to ovate-lanceolate, opposite leaves 10–25cm long, which may be smooth or more or less woolly. The flowers are pink or whitish. The asymmetric, fat, usually rough pods may be single or in pairs. There are 2 ovaries, but sometimes only 1 matures. Grows from 45cm to 1.8m tall, forming clumps or extensive beds along roadsides and on ditch banks. Found throughout our area, usually at the lower altitudes, but seems to be extending its range into the montane zone and even higher.

BUTTERFLYWEED, *A. tuberosa* ssp. *interior,* with clusters of brilliant orange flowers and narrow, hairy leaves, is an exception in this group as it does not have the milky juice of its relatives. Occasionally seen along roadsides in the foothills.

POTATO OR NIGHTSHADE FAMILY: SOLANACEAE

Plants of this family have flowers with regular, united, usually 5-lobed corollas, which may be tubular to wheel-shaped; ovaries 2-celled, ripening into several-seeded berries of dry capsules. Often the fruits are enclosed in an enlarged calyx. Some of the species are poisonous, but the family also includes several important vegetables: potatoes, tomatoes, eggplant, and green and red peppers, along with tobacco and petunias.

1. Fruit a berry, sometimes enclosed in a spiny or papery husk.
   2. Fruiting calyx bladdery-inflated; anthers distinct.
      3. Corolla yellowish or greenish-yellow, rotate or rotate-campanulate.

                            Physalis, p. 250
                            Ground cherry

      3. Corolla purple, rotate.

                            Quincula p. 251
                            Ground cherry
   2. Fruiting calyx not enlarged or inflated; anthers connivent; corolla rotate.
                            Solanum p. 251
                            Nightshade or potato
1. Fruit a dry capsule; corolla funnelform to nearly salverform.
   2. Corolla 5cm long or longer, solitary in the axils of branching stems; capsules prickly.

                            Datura p. 249
                            Jimsonweed or thorn apple
   2. Corolla not over 4cm long, some in terminal racemes or panicles; capsules not prickly.
      3. Capsules circumscissile near apex, completely included in the calyx; stems viscid, short-villous.

                            Hyoscyamus p. 250
                            Henbane
      3. Capsules opening by longitudinal valves, not completely included in the calyx; clammy-pubescent herbs.

                            Nicotiana p. 250
                            Tobacco

(1) *Datura.* Jimsonweed or thorn apple (Narcotic-poisonous plants)

1. Stem purple; flowers violet or lavender.

                            D. stramonium var. *tatula*
1. Stem green; flowers white.
   2. Leaves irregularly sinuate-lobed.

                            D. stramonium var. *stramonium*

   2. Leaves entire, unequally ovate, sinuate.

                            D. meteloides

JIMSONWEED or THORN APPLE, *Datura stramonium*, is a stout, coarse plant with leaves 10–20cm long and a 5-angled calyx, the corolla funnel-form. The thorny, ovoid fruit is about 5cm long. This introduced plant has a worldwide distribution as a weed on waste or cultivated ground and is sometimes found in our foothill region.

(2) *Hyoscyamus*. Henbane

HENBANE, *Hyoscyamus niger,* the only species in our range, is another widely distributed, introduced weed. It is a coarse, strong-scented, sticky plant. Greenish-yellow flowers with purplish veins occur in the upper leaf axils and in terminal racemes. Leaves 6–20cm long, oblong, ovate to lanceolate, irregularly-lobed, cleft, or pinnatifid. The seed capsules are vase-shaped. Both seeds and foliage are poisonous. Found around buildings and along roadsides.

(3) *Nicotiana*. Tobacco

1. Stem leaves petiolate.

> *N. attenuata*
> Wild tobacco or coyote tobacco

1. Stem leaves sessile.

> *N. trigonophylla*
> Desert tobacco

WILD TOBACCO, *Nicotiana attenuata,* is an annual, sticky, usually-branched plant 30–60cm tall, with terminal clusters of greenish-white flowers. Corollas 2.5–5cm long, tubular with spreading lobes. The lower leaves are ovate, the middle ones lanceolate, the upper becoming linear. Found on sandy ground of the foothills from Montana to Utah and New Mexico.

(4) *Physalis*. Ground cherry

1. Annuals.
    2. Densely pubescent; corolla and berry yellow.

> *P. pubescens*
> Low hairy ground cherry

    2. Glabrous or nearly so; corolla yellow with purple center; berry purple.

> *P. ixocarpa*
> Tomatillo

1. Perennials; rhizomatous.
    2. Peduncles mostly 3–10mm long; plants 7–8dm tall; leaf blades rarely over 5cm long.
        3. Short stem-hairs glandular.

> *P. hederaefolia* var. *comata*
> Ivy-leaved ground cherry

        3. Stem-hairs forked or branched, nonglandular.

> *P. hederaefolia* var. *fendleri*
> Fendler ground cherry

    2. Peduncles mostly 10–20mm; leaf blades sometimes over 5cm long.
        3. Pubescence moderately dense, glandular; leaf blades mostly deltoid-ovate.

> *P. heterophylla* var. *heterophylla*
> Clammy ground cherry

        3. Pubescence sparse or absent, nonglandular; leaf blades lanceolate, elliptic, or rhombic-ovate.
            4. Hairs of middle stem largely directed downward.

> *P. virginiana*
> Virginia ground cherry

4. Hairs of middle stem largely directed upward, or spreading, or none.
    5. Flowering calyx and stems with spreading hairs 1mm or more
        long.

P. hispida

    5. Flowering calyx and stems with hairs 0.5mm or less long, or
        glabrous.

P. longifolia
Long-leaved ground
cherry

FENDLER GROUND CHERRY,
Physalis hederaefolia var.
fendleri (fig. 210), has ovate or
triangular leaves with somewhat
wavy margins and dull yellowish flowers with brownish
centers. The corolla is open
bell-shaped, 7–15mm wide. As
the fruit ripens, the calyx
enlarges and encloses the berry
in a papery, lantern-like husk.
These berries were gathered
for food by southwestern
Indians. The plants grow on
rocky ground of dry plains
and hills from Colorado and
Utah southward.

(5) *Quincula*. Ground cherry

One species occurs in North
America. PURPLE GROUND CHERRY,
Quincula lobata, is a low
perennial herb, diffusely-

210. Fendler Ground Cherry, ²/₅ X

branched. Corolla rotate, violet, the center with a 5–6-rayed white, woolly star;
globular-inflated fruiting calyx strongly 5-angled.

(6) *Solanum*. Nightshade or potato

1. Plants spiny; pubescence of leaves stellate.
    2. Taprooted annual; corolla yellow; inflorescence stellate-hairy, not glandular-
        villous.

S. rostratum
Buffalo bur

    2. Rhizomatous perennial; corolla blue or white; leaves entire to sinuate; underside
        of leaves heavily pubescent.

S. elaeagnifolium
Silver-leaved nightshade

1. Plants unarmed; pubescence not stellate, or lacking.
    2. Perennial; inflorescence few to 25-flowered.

3. Plant 2–3dm high from small tubers; fruit not in calyx; inflorescence few to several-flowered.

S. jamesii
Wild potato

3. Plant climbing or creeping; fruit a somewhat poisonous red berry; inflorescence 10–25-flowered.

S. dulcamara
Climbing nightshade

2. Annual; inflorescence 1–8-flowered; not climbing.
   3. Leaves pinnately-lobed or divided; fruit green.

S. triflorum

   3. Leaves entire, toothed, or sinuate.
      4. Stems with long-spreading hairs; fruit yellow or green.

S. sarrachoides
Hairy nightshade

      4. Stems glabrous or appressed-hairy; fruit black.

S. nigrum
Black nightshade

211. Buffalo-bur, ²⁄₅ X

BUFFALO BUR, *S. rostratum* (fig. 211), is a branched plant with saucer-shaped yellow flowers 2.5cm or more across. Irregularly pinnately-lobed leaves are covered with yellowish bristles. The leaf lobes are rounded. The fruit is enclosed in the bristly calyx. Seen on waste ground, roadsides, and overgrazed pastures of the high plains and lower foothills.

WILD POTATO, *S. jamesii,* is a true potato, closely related to our cultivated potato (native to South America), and bears small round tubers on its underground stems. The leaves are compound, made up of 5–9 lance-shaped to ovate-oblong leaflets. The corollas are 12–25mm across, usually white, deeply 5-lobed. Used by the Indians of the Southwest as food and grown by them in their gardens. Found in southern Colorado, Utah, Arizona, and New Mexico, growing mostly in the coniferous forests of the montane zone.

MORNING-GLORY FAMILY: CONVOLVULACEAE

This family includes some garden flowers and several very persistent weeds, as well as dodder, which is sometimes segregated. The stems climb by twining around any available supports, or they trail over the ground. Flowers 5-

merous, 5 stamens, ovary superior. In *Cuscuta* the plant separates from its roots once it is established on a host plant.

1. Plants lacking green cholorophyll, parasites on the stems of various hosts, no roots present when mature.

Cuscuta p. 253
Dodder

1. Plants with cholorophyll, not parasites.
   2. Styles 2, each 2-cleft.

Evolvulus p. 254
Evolvulus

    2. Style 1.
        3. Stigma 1, small, capitate or globose.

Ipomoea p. 254
Morning glory

        3. Stigmas 2, linear to oblong or ovate; style is 2-cleft at apex.

Convolvulus p. 253
Bindweed or morning glory

(1) *Convolvulus*. Bindweed or morning glory

Only one species is likely to be found in our range: BINDWEED, *Convolvulus arvensis* (fig. 212). The corolla is white to pink, funnel-form, and most conspicuous in the mornings. It has a deep, wide-spreading root, very difficult to eradicate, and it occurs as a weed in cultivated fields and along roadsides.

212. Bindweed, ²⁄₅ X

(2) *Cuscuta*. Dodder

1. Stigmas attenuate.
   2. Calyx membranous, the lobes acute and triangular.

C. epithymum
Common dodder

    2. Calyx more or less fleshy, the lobes broader than long.

C. approximata
Alfalfa dodder

1. Stigmas capitate.
   2. Corolla lobes acute or acuminate; appendages present inside corolla.
      3. Sepals broadly rounded at tip.

C. migalocarpa

        3. Sepals acute at tip.
            4. Flowers with small, rounded bumps (papillate).

C. indecora

            4. Flowers without papillae..

C. plattensis

    2. Corolla lobes not sharply acute; no appendages inside corolla.

C. occidentalis
Western dodder

(3) *Evolvulus.* Evolvulus

One species in our range: *Evolvulus nuttallianus.* Plant densely silky-villous or becoming tawny; solitary flowers axillary, rose to purple.

(4) *Ipomoea.* Morning-glory

Only one species is likely to be found in our range. *Ipomoea leptophylla* is an entirely glabrous, bushy plant, that does not twine. Flowers are rose to purple.

213. Buckbean, ½X

BUCKBEAN FAMILY: MENYANTHACEAE

Only one species is likely to be found in our range. BUCKBEAN, *Menyanthes trifoliata* (fig. 213), is a plant of cold bogs and ponds of the subalpine zone. The leaves are trifoliolate, the leaflets oval, thick, and on long petioles. Flowers, white, fringed, often pink-tinged, are in short spikes: 5-merous, stamens 5, style 1, stigma 2-lobed. Plants are sometimes seen standing in shallow water. Found in northern North America and occasionally in the Rockies. Often placed in the Gentian Family.

PHLOX FAMILY: POLEMONIACEAE

Plants in this family have regular, united corollas, either rotate, funnel-form, campanulate, or salverform. In the bud, the corolla lobes are always folded so that the edge of one overlaps the edge of the next, and they appear twisted. When buds are so arranged, they are said to be *convolute.* The 5 (4) stamens are attached inside the corolla tube; 3-celled, superior ovary.

1. Leaves lacking, the cotyledons persist; flower head surrounded by bracts; small annuals.
   Gymnosteris p. 256
   Gymnosteris
1. Leaves more or less well-developed; annuals and perennials.
   2. Calyx-tube of uniform texture.
      3. Leaves pinnately compound, the leaflets definite; calyx-tube herbaceous at anthesis (the opening of the flower).
         Polemonium p. 261
         Jacob's ladder
      3. Leaves entire to dissected, without definite leaflets; calyx-tube papery at anthesis.
         Collomia p. 255
         Collomia

2. Calyx-tube with green ribs separated by transparent intervals; if greenish uniformly, the leaves sessile and palmatifid into linear segments.
   3. Filaments unequally inserted; leaves entire, mostly opposite.
      4. Perennials; leaves mostly opposite.

                                     Phlox p. 260
                                     Phlox

      4. Annuals; upper leaves alternate.

                                     Microsteris p. 259
                                     Microsteris

   3. Filaments about equally inserted; leaves seldom at once opposite and entire.
      4. Leaves sessile, mostly palmatifid, the segments linear or awl-shaped.
         5. Perennials.
            6. Flowers nocturnal; plants shrubby; leaves firm and prickly; calyx intervals conspicuous.

                                     Leptodactylon p. 258
                                     Prickly gilia

            6. Flowers open day and night; plants woody only at base; leaves rather soft; calyx intervals narrow and inconspicuous.

                                     Linanthastrum p. 259
                                     Linanthastrum

         5. Annuals; leaves soft, mostly opposite.

                                     Linanthus p. 259
                                     Linanthus

      4. Leaves diverse, but not at once sessile and palmatified.
         5. Calyx-lobes rather obviously unequal; leaves mostly firm and prickly.

                                   Navarretia p. 259
                                     Navarretia

         5. Calyx-lobes roughly equal; leaf-segments or teeth sometimes inconspicuously spinulose-apiculate (tipped with an abrupt sharp point), but not bristle-tipped.
            6. Corolla more than 15mm long; if smaller, the flowers in dense heads.

                                   Ipomopsis p. 256
                                   Gilia

            6. Corolla about 10mm long, the flowers scattered in open inflorescence.

                                   Gilia p. 256

(1) *Collomia*. Collomia

1. Plants perennial; numerous stems, sprawling.
   2. Corolla mostly 15–25mm long.

                                     C. debilis var. *debilis*

   2. Corolla mostly 25–35mm long.

                                     C. debilis var. *ipomoea*
                                     Alpine collomia

1. Plants annual; single-stemmed generally, erect.
   2. Flowers large, corolla 20–30mm long.

                                     C. grandiflora
                                     Large-flowered collomia

   2. Flowers smaller, 4–15mm long.
      3. Stamens unequally inserted.

                                     C. linearis
                                     Narrow-leaved collomia

      3. Stamens equally inserted.

                                     C. tenella
                                     Diffuse collomia

NARROW-LEAVED COLLOMIA, *Collomia linearis,* is a weedy plant with narrow, tapering leaves and very small, trumpet-shaped, pinkish flowers 6–7mm across in a dense cluster, which is interspersed with leafy bracts. Found on disturbed ground and along roadsides. LARGE-FLOWERED COLLOMIA, *C. grandiflora,* has white to orange flowers that are 13–15mm across. Occurs in the western part of our range.

(2) *Gilia.* Gilia

This key separates *Gilia* and *Ipomopsis,* as is increasingly done. There are arguments for lumping them as *Gilia*; and the common name, gilia, is still applied to several different genera within the Phlox Family. All have salverform corollas with slender tubes, and most have divided leaves. In some cases, the leaves are so deeply divided that the segments appear to be narrow, simple leaves. Using the key to the genera will help avoid confusion.

1. Plant perennial or biennial; stamens exserted 1mm or more.
$$\text{G. pinnatifida}$$
Pinnate-leaf gilia
1. Plants annual; stamens exserted less than 1mm or not at all.
    2. Leaves mostly entire.
G. tenerrima
Delicate gilia
    2. Leaves pinnatifid or strongly toothed.
        3. Lower part of plant with cottony pubescence.
            4. Basal leaves often bipinnatifid; corollas 5–9mm long.
G. ophthalmoides
            4. Basal leaves pinnatifid; corollas 3–6mm long.
G. tweedyi
        3. Lower part of plant without cottony pubescence.
G. leptomaria
Great Basin gilia

PINNATE-LEAF GILIA, *Gilia pinnatifida,* is an inconspicuous plant with an erect branched stem, small bluish flowers, and sticky foliage. Grows on open sandy ground from Wyoming to New Mexico, at all altitudes up to timberline.

(3) *Gymnosteris.* Gymnosteris

1. Corolla showy, the tube 6–10mm long.
G. nudicaulis
Long-flowered gymnosteris
1. Corolla inconspicuous, the tube 2.5–5mm long.
G. parvula
Small-flowered gymnosteris

(4) *Ipomopsis.* Gilia

1. Plants annual.
    2. Corolla-tube less than 10mm long; flowers in dense heads.
        3. Style glabrous.
I. polycladon

3. Style pilose below.

I. pumila

2. Corolla 10–50mm long; flowers not in dense heads.
    3. Corolla-tube 10–15mm long.

I. laxiflora

3. Corolla-tube 30–50mm long.

I. longiflora

1. Plants biennial or perennial.
    2. Biennials.
        3. Leafy throughout; corolla-tube 14–45mm long.
            4. Corolla generally various shades of red, trumpet-shaped and flaring.
                5. Anthers all conspicuously exserted.

I. aggregata var.
formosissimus

5. All the anthers not conspicuously exserted.
    6. All but 1 anther included, the highest anther near orifice.

I. aggregata var. *collina*

6. Most anthers close to orifice.
    7. Corolla-tube about 1mm at base, flaring to orifice 2–3mm in diameter.

I. aggregata var.
attenuata

7. Corolla-tube at least 1.5mm at base, flaring to orifice 3–4mm in diameter.

I. aggregata var.
aggregata
Scarlet gilia

4. Corolla generally not red, salverform with non-flaring tube.
    5. Corolla-tube filiform, 10–22mm long.

I. aggregata var. *weberi*

5. Corolla-tube narrow but not filiform, 19–45mm long.
    6. Calyx-lobes 2mm long; inflorescence clustered.

I. aggregata var. *candida*
Fairy trumpet

6. Calyx-lobes 3–4mm long; flowers in spaced, open inflorescence.

I. tenuituba

3. Leaves mostly in a basal rosette; stems nearly naked; corolla-tube 15–20mm long.

I. subnuda
Canyonlands gilia

2. Perennials.
    3. Filaments shorter than the anthers.
        4. Inflorescence generally elongate.

I. spicata ssp. *spicata*
Spicate gilia

4. Inflorescence subcapitate.

I. spicata ssp. *capitata*
Alpine gilia

3. Filaments longer than the anthers.
    4. Leaves all entire; plants prostrate or nearly so.

I. congesta ssp. *crebifolia*
Ballhead gilia

4. Leaves (some of them) divided; plants upright or nearly so.
    5. Plants with a basal cluster of elongate, entire leaves.

I. congesta ssp.
pseudotypica

5. Plants without a basal cluster; all leaves somewhat similar.

I. congesta ssp. *congesta*
Ballhead gilia

214. Alpine
Gilla, ²/₅ X

ALPINE GILIA, *Ipomopsis spicata* ssp. *capitata* (fig. 214), has round, silky clusters of fragrant, cream-colored flowers, and cobwebby hairs on the stem and leaves. It is an alpine plant found on the high mountains of Colorado. The BALLHEAD GILIA, *I. congesta,* is a similar though taller plant, up to 30cm tall, with heads of small white flowers that have blue anthers. Found on dry and rocky or sandy hillsides and valleys of the foothills.

SPICATE GILIA, *I. spicata* ssp. *spicata,* is a woolly plant with tight balls of whitish or cream-colored flowers arranged in a spike-like inflorescence. Found on the high plains and foothills of Colorado, New Mexico, and Utah.

SCARLET GILIA, *I. aggregata* var. *aggregata* (fig. 215), is a biennial plant 3–9dm tall with many scarlet, trumpet-shaped

215. Scarlet Gilla, ²/₅ X

flowers, arranged along the stems in a narrow or open, elongated cluster. The pointed corolla-lobes are turned back, and often there are yellow markings on the inside of the corolla. Leaves are pinnately divided into narrow lobes. The rosettes of leaves become established in the first season, and the flowering stalk is developed the next summer. After it goes to seed, the plant usually dies; but if you look around carefully, you will discover some seedlings—rosettes preparing to furnish flowers for the next season. They bloom in July along roadsides and on banks in the foothills and montane zone throughout the Rocky Mountains. FAIRY TRUMPET, *I. aggregata* var. *candida,* is similar except that the corollas are sometimes longer and vary from a glistening pinkish-white to rose or salmon color. Found mainly in Colorado and New Mexico.

(5) *Leptodactylon.* Prickly gilia

1. Plant depressed and mat-forming; 4-merous; not over 7cm high.
　　　　　　　　　　　　　　　　　　　　　L. caespitosum
1. Plant not mat-forming; 5-merous; over 7cm tall.
　　2. All leaves opposite; plants woody only at base.
　　　　　　　　　　　　　　　　　　　　　L. watsonii
　　2. Some upper leaves alternate; plants woody well above base.
　　　　　　　　　　　　　　　　　　　　　L. pungens
　　　　　　　　　　　　　　　　　　　　　Spiny gilia

SPINY GILIA, *Leptodactylon pungens,* is a small, dry, spiny-looking shrub not over 30cm tall, with leaves deeply divided into slender, stiff, sharp segments. Its white, yellow, or pinkish flowers resemble those of a phlox. Grows in dry, rocky or sandy situations of the foothill and montane zones thoughout our range. The specific name *pungens* means *sharp-pointed.*

(6) *Linanthastrum.* Linanthastrum

1. Flowers numerous; leaves 7–3-lobed, rather soft.

> *L. nuttallii*
> Nuttall gilia

1. Flowers sparse; leaves 3–1-lobed, rather firm.

> *L. floribundum*

NUTTALL GILIA, *Linanthastrum nuttallii* (fig. 216), is a tufted plant with opposite leaves, deeply divided into 3–7 narrow linear segments, which appear as whorls or tiers on each stem. The flowers, on very short pedicels, are small, phlox-like, white with yellow eye and tube, clustered at the tips of the branches.

(7) *Linanthus.* Linanthus

Only one species is likely to be found in our range (although *Linanthastrum* is sometimes grouped with *Linanthus*): NORTHERN LINANTHUS, *L. septentrionalis.* Corolla 2.5–5mm long, white to light blue or lavender; seeds 2–8 in each locule.

216. Nuttall Gilia, ²/₅ X

(8) *Microsteris.* Microsteris

1. Main stem relatively tall, 8–25cm long, branched only above middle of stem.
> *M. gracilis* var. *gracilis*
> Collomia gilia
1. Main stem relatively short, 1–5 (8)cm tall, usually branched below.
> *M. gracilis* var. *humilior*

(9) *Navarretia.* Navarretia

1. Plants glandular-viscid.
> *N. breweri*
> Pincushion plant

1. Plants glabrous.
   2. Stems erect, simple or branched; corolla 4–7mm long.
> *N. entertexta* var.
> *propinqua*
> Needle-leaved navarretia
   2. Stems low, divaricately branched at the base; corolla less than 2mm long.
> *N. minima*
> Least navarretia

PINCUSHION PLANT, *Navarretia breweri,* is a low, stout, usually branched plant with spine-tipped leaves. Flowers white or yellow, crowded into dense heads surrounded by spine-tipped bracts. Grows in Yellowstone National Park and southward through Wyoming and western Colorado into Arizona.

(10) Phlox. Phlox

Our plants of this genus may be loosely spreading or very compact in dense mats, carpets, or cushions. Their leaves are opposite and often closely crowded. The corolla is always salverform, with slender tube and spreading limb. Usually there are 5 broad lobes. The buds show the characteristic convolute folding. Flowers on different species vary in size but are otherwise so similar as to be easily recognized once you learn one kind. There may be considerable color variation in any given species. Some species are more or less woody at base.

1. Plants densely caespitose, cushion-like; leaves crowded.
  2. Leaves with woolly, cobwebby hairs.
    3. Leaves densely arachnoid-lanate; corolla-tube longer than the calyx.
P. bryoides
Moss phlox

    3. Leaves sparsely arachnoid-lanate; corolla-tube little if any longer than the calyx.
P. hoodii
Carpet phlox

  2. Leaves not arachnoid-lanate.
P. pulvinata
Alpine phlox

1. Plants loosely branched, caespitose or not; leaves not crowded.
  2. Larger leaves usually over 25mm long; flowers 2 or more per stem.
P. longifolia
Long-leafed phlox

  2. Larger leaves usually less than 25mm long; flowers usually solitary, but sometimes more per stem.
    3. Leaves elliptic to lanceolate or oblanceolate, 2–5mm wide.
P. alyssifolia
Alyssum-leaved phlox

    3. Leaves mostly linear, lance-linear, or oblong, 2mm or less wide.
      4. Leaves 1mm or less wide at middle; base of leaves (and often the internodes) white, hyaline (nearly transparent); several-flowered.
P. andicola

      4. Leaves often wider, the leaf base and internodes often not hyaline.
        5. Leaf margins strongly ciliate; calyx often glandular-pubescent.
P. kelseyi
Marsh phlox

        5. Leaf margins not ciliate; calyx often pubescent but not glandular.
P. multiflora
Rocky Mountain phlox

LONG-LEAF PHLOX, *P. longifolia,* is a loosely-spreading plant with stems from 10 to 30cm long, which may be erect or decumbent. Flowers range from white to bright-pink or lilac, about 8mm across. Grows throughout our range, most commonly in the foothills of the Colorado Rockies, but also at most altitudes in Idaho.

ROCKY MOUNTAIN PHLOX, *P. multiflora* (fig. 217), has shorter stems and leaves than the preceding and forms rather loose mats. Flowers are white, bluish, or lavender. It is often found among shrubs, especially under sagebrush. This is our most widespread and commonly seen species, occurring throughout the Rockies.

MARSH PHLOX, *P. kelseyi*, is a low, half-shrubby plant of alkaline or salty marshes. It has bright-blue or lavender flowers and is found at low elevations in Montana, Wyoming, and Idaho.

✓ ALPINE PHLOX, *P. pulvinata* (fig. 218), is a compact, cushion plant with crowded leaves 6–12mm long, and short-stemmed white or bluish flowers. Grows on rocky alpine fields and infrequently at lower altitudes.

MOSS PHLOX, *P. bryoides*, is a very low-growing cushion plant with tiny, crowded, gray, 2-ranked leaves, which give the shoots a square appearance. Flowers are usually white and about 10mm across. Found on open, dry, rocky slopes and valleys of the foothills from Montana south to Colorado and Utah.

217. Rocky Mountain Phlox, ²/₅ X

218. Alpine Phlox, 1 X

CARPET PHLOX, *P. hoodii*, is a variable species as to size of flowers and leaves, but it is well described by its common name. It carpets large areas of gravelly soil on the foothills and high plains. In many of these areas it is the earliest flower to bloom. Its small white flowers appear in late April, and it is known as *Mayflower* by Wyoming children, which illustrates the fact that common plant names may not be meaningful outside the locality of their origin.

(11) *Polemonium*. Jacob's ladder

1. Corolla campanulate, as wide as long or wider; filaments pilose-appendaged at base; leaflets entire.
    2. Stems clustered from a branched caudex; taprooted; leaves mostly basal.
        P. pulcherrimum var. *pulcherrimum*
        Jacob's ladder
    2. Stems mostly solitary, from rhizomes; leaves mostly cauline.
        3. Inflorescence short and wide, flowers white to ochroleucous.
            P. foliosissimum var. *alpinum*
            Leafy polemonium
        3. Inflorescence long and narrow; flowers blue-purple (white).
            P. occidentale var. *occidentale*
            Western Jacob's ladder
1. Corolla funnelform, longer than wide; filaments not pilose-appendaged at base; leaflets often deeply 2–5-cleft and appearing whorled.

2. Plants alpine or subalpine, rarely lower; corolla blue or blue-purple, rarely white.
P. viscosum
Sky pilot
2. Plants mostly montane; corolla white or ochroleucous.
P. brandegei
Brandegee polemonium

JACOB'S LADDER, *P. pulcherrimum* (fig. 219). The ovate-lanceolate or elliptical leaflets are 2-ranked, suggesting a ladder. The plant's clustered stems are slender and decumbent, its leaflets are thin and pale green, its dainty, wide open flowers are sky blue. Usually found under "wind timber," the dwarf, bent and twisted trees at timberline, or in lightly shaded areas of the upper spruce forests. Grows in Wyoming, Colorado, New Mexico, Arizona, Utah, and Idaho.

SKY PILOT, *P. viscosum* (fig. 220), has its flowers crowded into compact clusters; the funnel-form corollas are violet-blue or purplish. Against this color the orange anthers of the 5 stamens show up beautifully. Leaves are mostly basal, of numerous tiny oval leaflets crowded along the rachis. Grows on the tundra, most frequently in places where the soil has been recently disturbed by road work or by pocket gopher diggings. Both white-flowered and cream-colored forms are seen.

219. Jacob's Ladder, ½X

220. Sky Pilot, ½X

LEAFY POLEMONIUM, *P. foliosissimum* (fig. 221), is a stout, branched, sticky plant 3–9dm tall, with a disagreeable, skunk-like odor. It has many short, open, funnel form, blue flowers. Its leaves have 12–25 2-ranked, lanceolate to oblong leaflets, the upper ones often confluent. Found along ditches and roadsides at middle elevations throughout our area. WESTERN JACOB'S LADDER, *P. occidentale,* is a more slender, unbranched plant with similar flowers. Grows in moist meadows and open woods of the montane and subalpine zones from the Arctic south to Colorado, Utah, and California.

221. Leafy Polemonium, ½ X

WATERLEAF FAMILY: HYDROPHYLLACEAE

The flowers in this family have 5-lobed, bell-shaped corollas with 5 stamens, which are attached to the tube of the corolla. The filaments and styles are longer than the corollas, giving most of these flowers a fringed appearance. Their leaves are usually more or less lobed or divided.

1. Leaves all basal; peduncles 1-flowered.

> *Hesperochiron* p. 264
> Hesperochiron

1. Leaves not all basal.
  2. Corolla convolute in bud.
    3. Perennials; stamens exserted; many-flowered clusters.

> *Hydrophyllum* p. 264
> Waterleaf

    3. Annuals; stamens included; 1-flowered, axillary or terminal pedicels.
      4. Leaves with 2 pairs of lateral lobes; calyx with a reflexed appendage at each sinus.

> *Nemophila* p. 265
> Nemophila

      4. Leaves with 4 or more pairs of lateral lobes; calyx naked at the sinuses.
> *Ellisia* p. 264
> Ellisia

  2. Corolla imbricated in bud.
    3. Flowers in scorpioid (curved or rolled inward) cymes or spicate racemes; leaves ovate-elliptic, obtuse, and petiolate.

> *Phacelia* p. 265
> Phacelia

3. Flowers solitary in the leafy forks of the stem; leaves linear, acute, not distinctly petiolate.

Nama p.
Nama

(1) *Ellisia*. Ellisia

One species in North America: *Ellisia nyctelea*. Annuals; lower leaves opposite, the others alternate, pinnately-divided; flowers solitary in upper leaf axils; corolla white to lavender.

(2) *Hesperochiron*. Hesperochiron

1. Corolla rotate or saucer-shaped; flowers usually 1–5; leaves glabrous on lower (or both) surfaces.

H. pumilus
Dwarf hesperochiron

1. Corolla campanulate or funnelform; flowers usually more than 5; leaves often short-hairy on both sides.

H. californicus
California hesperochiron

(3) *Hydrophyllum*. Waterleaf

1. Leaflets mostly entire; peduncle shorter than the petiole.

H. capitatum
Ball-head waterleaf

1. Leaflets coarsely toothed; peduncle longer than the petiole.

H. fendleri
Fendler waterleaf

FENDLER WATERLEAF, *Hydrophyllum fendleri* (fig. 222). A plant 3–9dm tall; large leaves pinnately divided into 9–13 long-pointed, main divisions, which are toothed or again divided. Flowers pale, in loose clusters on stalks taller than the leaves. Found in moist, shaded situations of the foothill and montane zones from Wyoming to New Mexico and Utah.

BALL-HEAD WATERLEAF, *H. capitum,* is a smaller plant with round clusters of lavender or light purple flowers on stalks shorter than the leaf stalks. Found in moist locations of the montane zone from the Canadian Rockies south to western Colorado.

222. Fendler Waterleaf, ²⁄₅X

(4) *Nama*. Nama

1. Plant very hispid-hairy.
    2. Branched from the base, the branches prostrate.

        N. densum var.
        parviflorum
        Matted nama

    2. Branched well above the base, the branches ascending.

        N. hispidum

1. Plant smooth or pubescent, hot hispid-hairy.

        N. dichotomum

(5) *Nemophila*. Nemophila

1. Leaves all alternate; corolla about 2mm wide, lavender.

        N. breviflora

1. Leaves all opposite (some of the upper may be alternate); corolla 2–6mm wide, white or whitish.

        N. parviflora

(6) *Phacelia*. Phacelia

1. Leaves simple and entire, or nearly so.
    2. Annuals; stamens included (shorter than corolla).
        3. Corolla 1–2mm long, usually 4-lobed.

        P. tetramera

        3. Corolla 2.5–13mm long, usually 5-lobed.
            4. Leaf blades mostly linear.

        P. linearis
        Threadleaf phacelia

            4. Leaf blades broader, ovate, obovate, orbicular, or elliptic-oblong.
                5. Plants glandular, hairy or pubescent.
                    6. Corolla 5–8mm long.

        P. demissa

                    6. Corolla 3–4.5mm long.

        P. incana
        Hoary phacelia

                5. Plants pilose-hirsute, but not glandular.

        P. submutica

    2. Biennials or perennials; stamens exserted.
        3. Stems clustered, ascending; stems and leaves white-canescent.

        P. hastata
        Silver scorpionweed

        3. Stems mostly single and erect; stems and leaves hispid or hirsute, but not canescent.

        P. heterophylla
        Scorpionweed

1. Leaves coarsely toothed or pinnately lobed or divided.
    2. Annuals; stamens included.
        3. Corolla lobes jagged or ragged on margins.

        P. denticulata

        3. Corolla lobes mostly entire.
            4. Corolla 4mm or more long, the blades of the petals lavender to blue.
                P. glandulifera
                Sticky phacelia

            4. Corolla 1–4mm long; petal blades white to yellowish.

5. Leaves mostly lobed over halfway to midrib.
>> *P. ivesiana*

5. Leaves mostly toothed, or lobed less than halfway to midrib.
6. Style 1–2mm long in fruit, conspicuously exserted from the calyx.
>> *P. scopulina*
>> Yellow phacelia

6. Style 0.5–1mm long, hardly exserted from the calyx.
>> *P. salina*
>> Yellow phacelia

2. Annuals or perennials; stamens exserted.
3. Annuals.
4. Corolla tubular.
>> *P. integrifolia*

4. Corolla campanulate.
5. Corolla lobes jagged or ragged on margins; flowers mostly white (or bluish).
>> *P. neomexicana* var. *alba*

5. Corolla lobes entire; flowers blue, violet, or purple.
6. Stamens 2–3mm longer than corolla.
>> *P. formosula*

6. Stamens 4–7mm longer than corolla.
7. Stems glabrous to glandular-puberulent.
>> *P. splendens*

7. Stems densely villous or minutely hispid and glandular.
>> *P. glandulosa*
>> Glandular phacelia

3. Perennials.
4. Plants scarcely caespitose; pubescent or finely hirsute; filaments short-exserted.
>> *P. idahoensis*

4. Plants caespitose; silky; filaments long-exserted.
5. Densely hairy, sericeous; petioles not strongly ciliate.
>> *P. sericea* var. *sericea*

5. Less densely hairy, more strigose; petioles clearly ciliate.
>> *P. sericea* var. *ciliosa*
>> Purple fringe

PURPLE FRINGE, *Phacelia sericea* (fig. 223), has silky-silvery foliage and stems 10–60cm tall, which are often clustered. The crowded, spike-like inflorescences are composed of purple flowers from which the stamens and pistils protrude conspicuously. Grows on roadsides, banks, and sandy or gravelly hillsides, especially in disturbed soil, from the montane through the alpine zones throughout our range.

SILVER SCORPIONWEED, *P. hastata,* is a branching, spreading plant with paler flowers in coiled clusters and silvery, irregularly-lobed foliage. *P. heterophylla* is closely related, and both are common weeds on disturbed soil throughout our range.

GLANDULAR PHACELIA, *P. glandulosa,* is an erect, glandular-hairy species that always looks dark-colored. Its insignificant flowers are on erect, 1-sided racemes of which there are usually several. Widely distributed as a weed.

THREADLEAF PHACELIA, *P. linearis,* is an annual, erect, rough-hairy plant with a simple or branched stem 10–38cm tall. Its flowers are saucer-shaped, bright blue or white, about 8mm broad. Grows in Yellowstone National Park and on plains and hills in Idaho and Utah.

BORAGE FAMILY: BORAGINACEAE

Our plants of this family have united, 5-lobed, symmetrical corollas with 5 stamens attached to the corolla-tube and alternating with its lobes. The flowers are usually white, blue, or yellow. With the exception of the mertensias and lithospermums, which have long corolla-tubes, they are on the same pattern as forget-me-not flowers: a short, incon-spicuous tube with 5 spreading, rounded lobes and little crests at the throat. Ovary superior. The fruit consists of 4 separate seeds or nutlets, though sometimes not all four mature. The branches of the inflorescence are often scorpioid (coiled to one side). In many species the plants are bristly-hairy.

223. Purple Fringe, ½X

1. Ovary undivided or only 4-grooved; style terminal, entire or absent.
    Heliotropium p. 271
    Heliotrope
1. Ovary 4-lobed or 4-sided; style central.
    2. Nutlets armed with barbed prickles; fruit bur-like.
        3. Prickles completely covering nutlets; corolla reddish-purple.
            Cynoglossum p. 270
            Hound's tongue
        3. Prickles mostly along the margins of nutlets; corolla blue or white.
            4. Pedicels ascending or erect in fruit; inflorescence bracteate.
                Lappula p. 271
                Stickseed
            4. Pedicels recurved or deflexed in fruit; inflorescence naked or nearly so.
                Hackelia p. 271
                Stickseed
    2. Nutlets unarmed, or the prickles (if any) not barbed.
        3. Nutlets attached laterally to a pyramid-like projection of the receptacle.
            4. Annuals.
                5. Leaves all alternate.
                    Cryptantha p. 268
                    Cryptantha
                5. Leaves opposite below.
                    6. Flowers orange or yellow.
                        Amsinckia p. 268
                        Tarweed

6. Flowers white.

Plagiobothrys p. 275
Popcorn flower

4. Biennials or perennials.
5. Nutlets prickly on the margin; flowers blue.

Eritrichium p. 271
Alpine forget-me-not

5. Nutlets not prickly on the margin.
6. Flowers white or yellow.

Cryptantha p. 268
Cryptantha

6. Flowers blue (rarely white).

Mertensia p. 273
Bluebell

3. Nutlets attached by their bases.
4. Nutlets small, smooth, and membranous; flowers blue, in bractless racemes.

Myosotis p. 275
Mountain forget-me-not

4. Nutlets bony and polished; flowers yellow, in leafy or bracteate racemes.

Lithospermum p. 272
Puccoon

(1) *Amsinckia*. Tarweed

1. Corolla throat obstructed by well-developed hairy scales.

A. lycopsoides
Tarweed fiddleneck

1. Corolla throat open, glabrous.
2. Stem spreading-hipsid, also puberulent or strigose with shorter and softer, retrorse (backward or downward) hairs; orange or orange-yellow.

A. retrorsa
Rigid fiddleneck

2. Stem spreading-hispid, shorter and softer hairs nearly absent below inflorescence; light yellow.

A. menziesii
Small-flowered fiddleneck

RIGID FIDDLENECK, *Amsinckia retrorsa,* is a rough plant covered with bristly hairs. The orange flowers are arranged in a one-sided, coiled spike that resembles a fiddle neck. Found in Idaho, Utah, and western Colorado on fields and waste ground.

(2) *Cryptantha*. Cryptantha

1. Annual plants; slender stems; flowers inconspicuous, white.
2. Calyx circumscissile at maturity (only the basal part persisting).

C. circumscissa
Winged cryptantha

2. Calyx not circumscissile.
3. Margins of the nutlets winged.

C. pterocarya
Winged cryptantha

3. Margins of nutlets rounded or angled, not winged.
4. Nutlets of 2 kinds, definitely differing in size, color, shape, or surface markings.
5. Spikes with foliaceous bracts subtending some of the flowers on the spike.

C. minima

   5. Spikes naked, or with a few bracts near base.
             C. kelseyana
             Kelsey's cryptantha
  4. Nutlets all alike or nearly so.
    5. Nutlets spiculate or tuberculate, at least toward apex.
      6. Nutlets and calyx bent and recurved, granulate.
             C. recurvata
      6. Nutlets and calyx straight, not recurved.
        7. Nutlets ovate, tuberculate.
             C. ambigua
        7. Nutlets lanceolate, spiculate.
             C. scoparia
    5. Nutlets smooth and more or less shiny.
      6. Ventral groove of nutlet off center.
             C. affinis
             Slender cryptantha
      6. Ventral groove of nutlet central.
        7. Nutlets broadly ovate.
             C. torreyana
             Torrey's cryptantha
        7. Nutlets oblong-ovate to lanceolate.
          8. Usually 1 nutlet (rarely 2-3) developing.
             C. gracilis
             Slender cryptantha
          8. Usually 4 nutlets developing.
           9. Margins of nutlets acute-angled, at least above.
             C. watsonii
             Watson's cryptantha
           9. Margins of nutlets rounded or obtuse.
             C. fendleri
             Fendler's cryptantha
1. Biennial or perennial plants.
  2. Corolla-tube longer than the calyx (6mm or more long).
    3. Nutlets smooth; corolla yellow.
             C. flava
             Golden cryptantha
    3. Nutlets wrinkled on both sides; corolla white with yellow throat.
             C. flavoculata

  2. Corolla-tube about equal to or shorter than the calyx (5mm long or less).
    3. Dorsal surface of nutlets wrinkled, spiny, or tuberculate.
      4. Caudex branched; leaves mostly near base; plants 15cm or less tall.
        5. Dorsal surface of nutlets muricate (with short, hard points); leaf
        hairs closely appressed, silvery-canescent.
             C. cana
             Hoary cryptantha
        5. Dorsal surface of nutlets roughened or tuberculate; leaf hairs
        spreading beyond margins of leaf.
             C. caespitosa
             Matted cryptantha
      4. Stems solitary or few; leaves conspicuous on stems; plants often well
      over 15cm tall.
        5. Bracts of the inflorescence greatly exceeding the flower clusters;
        inner surface of nutlets smooth.
             C. virgata
             Miner's candle
        5. Bracts of the inflorescence may exceed flower clusters, but not as
        above.
          6. Basal leaves with silvery, uniform appressed hairs on upper
          surface, mixed with coarse, pustulate hairs on lower surface.
             C. sericea
             Silky cryptantha

6. Basal leaves with coarse, spreading, pustulate hairs on surfaces.
   7. Ventral surface of nutlets smooth.
                 C. stricta
   7. Ventral surface of nutlets wrinkled, tuberculate, or muricate.
      8. Inflorescence broad and rounded in outline.
                 C. thyrsiflora
                 Bushy cryptantha
      8. Inflorescence narrow, spicate.
                 C. celosioides
                 Northern miner's candle.
3. Dorsal surface of nutlets smooth or nearly so.

                 C. cinerea var. *jamesii*

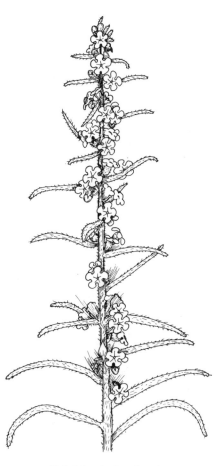

224. Miner's Candle, ½ X

MINER'S CANDLE, *Cryptantha virgata* (fig. 224). Stems 2.5–6dm tall, unbranched, covered with bristly hairs; flowers white, like forget-me-nots, in short, coiled clusters along the erect stem, interspersed with linear, leaf-like bracts longer than the flower clusters. Found on open dry fields and hillsides of Wyoming and Colorado from 5,000 to 9,000 feet. The common MINER'S CANDLE in Montana and Wyoming is *C. celosioides,* found on dry plains and hills. It is even more prickly than *C. virgata.* (There are several other bristly species with similar white flowers in the semi-arid regions of our area.)

BUSHY CRYPTANTHA, *C. thyrsiflora,* with 1 to several stout stems and a broad round-topped inflorescence, is common on dry foothill slopes from the eastern side of the Continental Divide in Wyoming south to New Mexico. These plants die after blooming, but the dry, bristly stalks often stand through the winter.

GOLDEN CRYPTANTHA, *C. flava,* is a yellow-flowered species having several stems with narrow crowded inflorescences. Occurs on dry foothill fields and slopes from Wyoming and Utah through western Colorado to New Mexico and Arizona.

(3) *Cynoglossum.* Hound's tongue

One species in our range, an introduced weed: HOUND'S TONGUE, *Cynoglossum officinale.* Tall, stout, erect, with oblong to lance-shaped, white-velvety leaves; scorpioid racemes of small, reddish-purple flowers, the corollas rotate-salverform. Found along roads and around buildings.

(4) *Eritrichium*. Alpine forget-me-not

1. Mature leaves densely appressed-hairy; hairs not forming tuft at apex of leaf.
   E. howardii
   Howard eritrichium
1. Mature leaves loosely long-hairy; hairs sometimes forming tuft at apex of leaf.
   E. nanum var. *elongatum*
   Alpine forget-me-not

ALPINE FORGET-ME-NOT, *Eritrichium nanum* var. *elongatum* (fig. 225), is a cushion plant of the alpine meadows, with fragrant, brilliantly blue (sometimes white), yellow-eyed flowers. Its stems, 2.5–7.5cm tall, rise from compact rosettes of tiny, silver-hairy leaves. HOWARD ERITRICHIUM, *E. howardii*, is similar but may be a little taller with darker blue flowers. Found on dry hilltops in Montana.

225. Alpine Forget-me-not, ½X

(5) *Hackelia*. Stickseed

1. Corolla limb 1.5–3mm wide; annual or biennial.
   H. deflexa var. *americana*
1. Corolla limb over 3mm wide; biennial or perennial.
   2. Corolla white (or tinged with blue).
   H. patens var. *patens*
   Spreading stickseed
   2. Corolla blue.
      3. Usually single stem from taproot; corolla limb 3–6mm wide.
      H. floribunda
      Many-flowered stickseed
      3. Usually several stems from taproot and branched caudex; corolla limb 6–11mm wide.
      H. micrantha
      Blue stickseed

STICKSEED FORGET-ME-NOT, *Hackelia floribunda* (fig. 226), is a coarse plant up to 9dm in height with blue flowers 4–10mm across. The 4 (3) nutlets are armed with hooked prickles, which assures their distribution, as they catch in animal fur or human clothing. Often grows among shrubs.

(6) *Heliotropium*. Heliotrope

1. Perennial; plant glabrous or glaucous; leaves oblanceolate or narrower; fruit 4-lobed.
   H. curassavicum var. *obovatum*
1. Annual; plant strigose-canescent; leaves ovate to linear; fruit 2-lobed.
   H. convolvulaceum

(7) *Lappula*. Stickseed

1. Nutlets with marginal prickles in at least 2 rows.
   2. Nutlets including prickles about 3mm long.
   L. echinata

2. Nutlets including prickles 4–5mm long.

L. cenchrusoides

1. Nutlets with marginal prickles in 1 row only.
   2. Bristles free.

L. redowskii
Common stickseed

   2. Bristles united in a cup around the dorsal face.

L. texana
Western stickseed

COMMON STICKSEED, *Lappula redowskii,* is similar to the stickseed forget-me-not but has much smaller flowers and seeds. Its flowers are usually white, less than 7mm in diameter. Both plants grow around buildings and along trails and roadsides throughout our area.

(8) *Lithospermum*. Puccoon

Our plants are yellow-flowered with narrow, pubescent leaves. Their nutlets, which are very hard, are whitish, shiny, and smooth, very well described by the generic name *Lithospermum*, which means *stone-seed*.

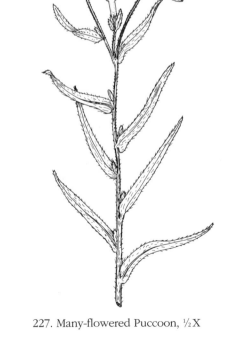

226. Stickseed Forget-me-not, ½X   227. Many-flowered Puccoon, ½X

1. Corolla greenish-white to pale-yellow; leafy stems branched from base; flowers inconspicuous.

L. ruderale
Wayside gromwell

1. Corolla bright yellow.
   2. Corolla-tube 3–4 times the length of the calyx, the lobes fringed.

L. incisum
Narrow-leaf puccoon

   2. Corolla-tube 2 times the length of the calyx, lobes rounded.

L. multiflorum
Many-flowered puccoon

NARROW-LEAF PUCCOON, *L. incisum,* has stems 15–45cm tall, usually several in a clump. The yellow corolla is 14–40mm long, salverform, with a slender tube and toothed or finely slashed lobes. Found on dry valleys and slopes of the mesas, foothills, and lower montane zone throughout the Rockies.

MANY-FLOWERED PUCCOON, *L. multiflorum* (fig. 227), has smaller, orange-yellow flowers and corolla-lobes with entire margins. Occurs from the foothills to the subalpine zone from Wyoming to Mexico.

WAYSIDE GROMWELL, *L. ruderale,* has many unbranched stems and very small, pale-yellow flowers almost hidden by long upper leaves. Its big, woody root was cooked and eaten by Indians.

(9) *Mertensia.* Bluebell

BLUEBELL, MERTENSIA, or CHIMING BELLS. *Mertensia* has many species in the mountains of western North America, recognized by the combination of blue, tubular, often pendant, 5-lobed corollas and the 4-parted ovary, which develops into 4 nutlets. Buds are often lavender or pinkish, and the leaves always have entire margins.

1. Stems 4dm or more tall; moist sites.
   2. Leaves glabrous, sometimes minute warts above.
      3. Calyx-lobes 1.5–3mm long; corolla-tube 6–8mm long.

M. ciliata var. *ciliata*
Chiming bells

      3. Calyx-lobes 3–4mm long; corolla-tube 4–6mm long.

M. arizonica var.
leonardii

   2. Leaves with wart-like based hairs on upper surface.

M. franciscana

1. Stems mostly less than 4dm high; open, dry habitats, or moist ground.
   2. Filaments attached well into the corolla-tube; anthers not projecting beyond the throat.
      3. Leaves glabrous except on margins.

M. humilis

      3. Leaves hairy above, glabrous below.
         4. Calyx-lobes glabrous outside.

M. alpina
Alpine mertensia

         4. Calyx-lobes hairy outside.

M. brevistyla

   2. Filaments attached near throat of corolla; anthers projecting beyond the throat.
      3. Limb of the corolla shorter than the tube.

4. Limb only slightly shorter than the tube.

M. viridis
Green mertensia

4. Limb conspicuously shorter than the tube.

M. obolongifolia

3. Limb of the corolla equal to or longer than the tube.
    4. Leaves hairy above, glabrous below; calyx-lobes hairy outside.

M. fusiformis
Utah mertensia

    4. Leaves hairy on one or both sides; calyx-lobes usually glabrous outside
        (or sometimes hairy if leaves are hairy on both surfaces).
        5. Leaves mostly linear-oblong, elliptic, or lanceolate; mostly plains
           and montane.

M. lanceolata
Foothill mertensia

        5. Leaves broader, elliptic, lanceolate to ovate; mostly subalpine to
           alpine.

M. viridis
Green mertensia

228. Chiming Bells, ½X

CHIMING BELLS, *M. ciliata* (fig. 228), has smooth, entire, bluish-green leaves with a few strong lateral veins; dusty blue flowers that hang in clusters; buds pinkish. This is one of the lushest plants of the high mountains. It grows from Montana to New Mexico, often forming large clumps or beds along streams and in wet places. *M. franciscana* is similar but has short, stiff, appressed hairs on the upper leaf surface. Gradually replaces the former in the southwest part of our range.

FOOTHILL MERTENSIA, *M. lanceolata,* is 25–38cm tall with stems often slanting upward; leaves usually smooth beneath and more or less pubescent above, dull green or distinctly bluish. The pink-tinged buds are in tight clusters that later become panicles. Blooms in early spring. Frequently seen on mesas, foothills, and in the montane zone on dry, sunny fields and slopes, from Saskatchewan to New Mexico. The smaller *M. oblongifolia* is found in the western part of our area from Montana through Wyoming and Utah.

UTAH MERTENSIA, *M. fusiformis,* has erect stems from deep, spindle-shaped rootstocks. May be distinguished from other similar species by the character of its leaf hairs, which, on the upper leaf surface, always point toward the margin of the leaf. Montane and subalpine zones, western Colorado and Utah.

(10) *Myosotis.* Mountain forget-me-not

1. Calyx strigose; plants of moist soil or shallow water.

                                                        M. scorpioides

1. Calyx with spreading hairs, uncinate (hooked at top).
   2. Corolla relatively large, the limb 4–8mm wide; blue.

                                         M. alpestris
                                         Wood forget-me-not

   2. Corolla relatively small, the limb 1–4mm wide.
      3. Calyx asymmetrical, 3 lobes smaller than other 2; white.

                                         M. verna
                                         Early forget-me-not

      3. Calyx symmetrical, all lobes similar; blue or white.
         4. Fruiting pedicels equal to or longer than calyx.

                                         M. arvensis
                                         Field forget-me-not

         4. Fruiting pedicels shorter than calyx.

                                         M. micrantha
                                         Blue forget-me-not

MOUNTAIN or WOOD FORGET-ME-NOT, *Myosotis alpestris* (fig. 229), is closely related to the cultivated varieties and resembles them, but its flowers are usually a deeper blue. Stems from 10 to 30cm tall, usually tufted. It grows in meadows and other moist places in the montane and subalpine zones from northwestern Colorado northward to Alaska.

(11) *Plagiobothrys.* Popcorn-flower

Only one species is likely to be found in our range: POPCORN FLOWER, *Plagiobothrys scouleri* var. *penicillatus,* a much-branched annual of sandy or muddy places at the lower altitudes in our area. Small, inconspicuous, white-flowered weed; leaves opposite.

229. Mountain Forget-me-not, ½X

VERBENA FAMILY: VERBENACEAE

The plants of this family have opposite or whorled leaves, which, in our species, are toothed, more or less lobed, or divided. The blue, purple, or rose-colored flowers have 4 stamens; the united corollas, which may be slightly irregular, have 4–5 lobes. Most species of this family grow at altitudes below our range.

1. Corolla 4-lobed; nutlets 2; spikes lateral; stamens 4, didynamous.
　　　　　　　　　　　　　　　　　　　　　Phyla p. 276
　　　　　　　　　　　　　　　　　　　　　Fogfruit
1. Corolla 5-lobed; nutlets 4; spikes terminal on stems and branches; stamens 4.
　　2. Flowers small, spicate; anthers without a gland-like appendage.
　　　　　　　　　　　　　　　　　　　　　Verbena p. 276
　　　　　　　　　　　　　　　　　　　　　Verbena or vervain
　　2. Flowers large, capitate, becoming spicate; anthers of the longer stamens with a
　　　　gland-like appendage.
　　　　　　　　　　　　　　　　　　　　　Glandularia p. 276
　　　　　　　　　　　　　　　　　　　　　Vervain

(1) *Glandularia*. Vervain

One species in our range: SHOWY VERVAIN or WILD VERBENA, *Glandularia bipinnatifida* var. *bipinnatifida*. Found on the high plains and lower foothills along the eastern and southern fringes of the Rockies. A low, branching plant; leaves deeply pinnatifid; corolla-tube much exceeding calyx; clusters of showy rose-purple flowers.

(2) *Phyla*. Fogfruit

One species in our range: *Phyla cuneifolia*. Perennial; plants trailing and rooting at the nodes; flowers white, pink, to purple.

(3) *Verbena*. Verbena or vervain

1. Spike slender; bracts inconspicuous.
　　2. Plants glabrate or sparsely rough-hirsute; 9–15dm tall.
　　　　　　　　　　　　　　　　　　　　　V. hastata
　　　　　　　　　　　　　　　　　　　　　Blue vervain
　　2. Plants densely soft-pubescent; 3–7dm tall.
　　　　　　　　　　　　　　　　　　　　　V. stricta
　　　　　　　　　　　　　　　　　　　　　Hoary vervain
1. Spike thick; bracts surpassing the flowers; plant branched from the base.
　　　　　　　　　　　　　　　　　　　　　V. bracteata
　　　　　　　　　　　　　　　　　　　　　Vervain

VERVAIN, *Verbena bracteata,* is similar in branching habit to *Glandularia bipinnatifida,* but it has small, light blue or purplish flowers in conspicuously bracted spikes. Occurs as a weed in gardens, fields, and along roadsides over most of North America. Found up to 7,500 feet in Colorado and Wyoming.

MINT FAMILY: LABIATAE

The plants of this family have opposite leaves, square stems, irregular flowers with 2-lipped corollas, and some have aromatic foliage. Leaves are simple, but usually their margins are more or less toothed or finely scalloped. The stamens are 4 or 2, and the 4-lobed ovary ripens into 4 small nutlets, each containing one seed.

1. Anther-bearing stamens 2.
　　2. Corolla 2-lipped, irregular.

3. Connective of anther laterally elongated, one end bearing a perfect cell, the other a modified abortive one or none.

Salvia p. 281
Sage

3. Connective short.
    4. Flowers large, in one or few dense capitate clusters.

Monarda p. 279
Horsemint or bergamot

    4. Flowers small, in many axillary clusters.

Hedeoma p. 278
Pennyroyal

2. Corolla nearly regular.

Lycopus p. 279
Water hoarhound

1. Anther-bearing stamens 4.
    2. Calyx teeth 10; stems white-woolly.

Marrubium p. 279
Hoarhound

    2. Calyx teeth 5 or less; stems usually not white-woolly.
        3. Calyx 2-lipped, the lips entire, calyx with a crest-like gibbosity (swelling) on the upper side.

Scutellaria p. 281
Skullcap

        3. Calyx not 2-lipped or, if so, the lips lobed or toothed; no crest present.
            4. Ovary only 4-lobed, the style not basal.

Teucrium p. 282
Germander

            4. Ovary deeply 4-parted, the style basal.
                5. Upper (posterior) pair of stamens as long as the lower (anterior) pair.

Mentha p. 279
Mint

                5. Upper pair of stamens not the same length as the lower pair.
                    6. Upper pair of stamens longer than the lower pair.
                        7. Anther sacs parallel or nearly so.

Agastache p. 278
Giant hyssop

                        7. Anther sacs divergent.
                            8. Coarse, erect herbs.
                                9. Perennial; corolla well exserted from calyx.

Nepeta p. 280
Catnip

                                9. Annual or biennial; corolla barely exserted.

Dracocephalum p. 278
Dragonhead

                            8. Trailing herbs.

Glecoma p. 278
Ground ivy

                    6. Upper pair of stamens shorter than the lower pair.
                        7. Calyx 2-lipped, closed in fruit.

Prunella p. 280
Self-heal

                        7. Calyx subequally 5-toothed.
                            8. Membranous and inflated in fruit.

Physostegia p. 280
False dragonhead

                            8. Not membranous, unchanged in fruit.
                                9. Corolla-tube hairy within.

Leonurus p. 278
Motherwort

                                9. Corolla-tube glabrous within.
                                    10. Leaves small (about 1cm long).

Monardella p. 280
Monardella

10. Leaves large (4–8cm long).
Stachys p. 282
Woundwort or hedge-nettle

(1) *Agastache*. Giant hyssop

1. Leaves pale beneath, with felt-like hairs; flowers blue.

A. foeniculum
Horsemint

1. Leaves green on both sides; flowers violet to purple.

A. urticifolia
Giant hyssop

GIANT HYSSOP, *Agastache urticifolia,* has stems from 6–18dm tall; terminal, compact, spike-like clusters of whitish, rose, or purplish flowers 5–15cm long. Its ovate to triangular-ovate leaves are usually smooth, without felt-like hairs beneath. Occurs on moist soil of the montane and lower subalpine zones throughout our range.

(2) *Dracocephalum*. Dragonhead

1. Bracts of infloresence strongly toothed or lobed.

D. parviflorum

1. Bracts of inflorescence entire or nearly so.

D. thymiflorum

DRAGONHEAD, *Dracocephalum parviflorum,* has dense spikes of blue or pinkish flowers interspersed with spine-tipped, pectinate (comb-like) bracts. Frequently seen on burned-over areas.

(3) *Glecoma*. Ground ivy

One species in North America: *G. hederacea*. Pubescent perennial with creeping stems; leaves petiolate, crenate; flowers blue-violet.

(4) *Hedeoma*. Pennyroyal

1. Hispid-pubescent annual; 8–15cm tall.

H. hispida

1. Puberulent, ash-grayish perennial; 10–20cm tall.

H. drummondii
Pennyroyal

PENNYROYAL, *Hedeoma drummondii,* is a bushy little gray plant, 10–20cm tall, with slender pink blossoms in the leaf axils. Grows on dry ground of the foothill ridges throughout the Rockies.

(5) *Leonurus*. Motherwort

One species in our range. *L. cardiaca* is a puberulent perennial 5–10dm high; leaves long-petioled; the lower suborbicular and dentate; the upper oblong-lanceolate, 3-lobed or deeply cleft; corolla purple to white.

(6) *Lycopus*. Water horehound

1. Some or all of the leaves cut deeply (incised); nonalkaline sites.

L. americanus
Cut-leaved water
hoarhound

1. Leaves only serrate or dentate; tolerates alkali.

L. asper
Rough bugleweed

(7) *Marrubium*. Horehound

Only one species in North America. *M. vulgare* is a taprooted perennial; leaves petiolate, usually crenate; flowers whitish, in axillary clusters.

(8) *Mentha*. Mint

1. Flowers in terminal spikes.

M. spicata
Spearmint

1. Flowers in axillary whorls.

M. arvensis
Wild or field mint

WILD MINT, *Mentha arvensis,* is an aromatic plant with pale lavender or pinkish flowers in dense clusters in the leaf axil; grows from 10 to 35cm tall. Leaves oblong, rounded at base, pointed at tip, the margins with rounded or sharp teeth. A circumpolar species found on moist meadows and along streams in the Rockies up to 9,500 feet. Its leaves are useful for flavoring beverages.

(9) *Monarda*. Horsemint or bergamot

1. Heads solitary; upper lip of corolla erect and straight, the stamens exserted beyond it.

M. fistulosa var.
menthifolia
Wild bergamot

1. Heads 2 or more; upper lip of corolla falcate (scythe-shaped), the stamens usually not exserted.

M. pectinata
Whorled monarda

HORSEMINT or BERGAMOT, *Monarda fistulosa* var. *menthifolia* (fig. 230). A showy perennial plant with rounded, rose-colored heads. Each corolla is slashed into 2 lips; the stamens protrude from the upper one; the lower is 3-lobed and spreading. Stems 3–6dm tall with fragrant, opposite, ovate-lanceolate, serrate leaves on short petioles. Grows in meadows and ravines, along roadsides, often in clumps or patches, in the foothill and montane zones throughout our range. Frequently visited by bees and hummingbirds

WHORLED MONARDA, *M. pectinata*, is an annual plant branched from the base, 16–38cm tall, with smaller pinkish flowers arranged in tiers around the stems. Occurs in the foothill zone from Colorado and Utah south to Mexico.

(10) *Monardella*. Monardella

One species in our range. CLOVERHEAD HORSEMINT, *Monardella odoratissima* (meaning most fragrant), is similar in appearance to *Monarda* but with 4 stamens and strongly aromatic foliage. It has a base of woody twisted stems with branches not over 3dm tall, which are softly hairy. Flowers pale purple, in terminal, globose clusters, surrounded by purplish bracts. Occurs in the montane zone from Montana through western Colorado to Utah and New Mexico.

(11) *Nepeta*. Catnip

1. Corolla whitish; plant widely distributed.
   N. cataria
1. Corolla blue or purple; central Wyoming.
   N. mussinii

(12) *Physostegia*. False dragonhead

One species in our range: *P. parviflora*. Stems slender, leafy, 3–6dm tall; leaves serrate to subentire; flowers lavender, rose, or purple.

(13) *Prunella*. Self-heal

1. Middle cauline leaves about half as wide as long, with broadly rounded base; plant often dwarf and prostrate; adventive.
   P. vulgaris var. *vulgaris*
   Self-heal
1. Middle cauline leaves about a third as wide as long, tapering toward the base; plant ascending or erect; native.
   P. vulgaris var. *lanceolata*

230. Monarda, ½X

231. Prunella, ½X

PRUNELLA or SELF-HEAL, *Prunella vulgaris* var. *vulgaris* (fig. 231), has short, blunt spikes of purple flowers interspersed with broad, clasping bracts; leaves opposite and entire. Grows on more or less shaded, moist soil of meadows, streambanks, and roadsides of the upper foothill and montane zones. Widely distributed in the cooler part of North America, Europe, and Asia.

(14) *Salvia*. Sage

1. Perennial; stems 30–100cm tall.
    2. Upper lip of calyx entire; corolla 15–30mm long.

    S. azurea ssp. *grandiflora*

    2. Upper lip of calyx 3-toothed; corolla 10–14mm long.

    S. nemorosa

1. Annual; stems not over 30cm tall.

    S. reflexa
    Lanceleaf sage

LANCELEAF SAGE, *Salvia reflexa,* is an inconspicuous plant with pale blue or whitish flowers in the leaf axils. Usually seen as a weed in gardens, on vacant lots, or on disturbed soil.

(15) *Scutellaria*. Skullcap

1. Flowers in racemes, axillary or terminal.

    S. lateriflora
    Blue skullcap

1. Flowers solitary in the axils.
    2. Plants more or less glandular.

    S. brittonii
    Skullcap

    2. Plants not glandular.
        3. Lower lip of corolla papillate.

    S. galericulata
    Marsh skullcap

        3. Lower lip of corolla long-hairy; deep blue-violet.

    S. angustifolia
    Narrow-leaved skullcap

SKULLCAP, *Scutellaria brittonii* (fig. 232), is a plant with erect stems 10–25mm tall, which grows from creeping underground rootstocks. The rich, purplish-blue flowers are in pairs, standing erect in the leaf axils. This plant takes its common name from the cap-shaped calyx. Found on slopes of the foothill and montane zones from Wyoming to New Mexico. MARSH SKULLCAP, S. *galericulata*, is a similar but larger plant that occurs on wet ground around lakes and ponds and on ditch banks of the foothills and lower montane zone. Widely distributed in northern North America, Europe, and Asia.

NARROW-LEAVED SKULLCAP, S. *angustifloia*, has long-petioled lower leaves, much narrower, almost sessile, upper leaves. The axillary flowers are from 2.5 to 4cm long. Occurs on moist soil from Montana and Idaho to Utah.

232. Skullcap, ½X   233. Woundwort, ½X

(16) Stachys. Woundwort or hedge-nettle

One species in our range: WOUNDWORT, *Stachys palustris* var. *pilosa* (fig. 233), is a rhizomatous perennial resembling *Agastache urticifolia* but smaller and distinctly hairy. Rarely exceeds 7.5dm in height, and its infloresence is less dense. The flowers, rose, lavender, or purple, splotched with white, are usually in threes, in the axils of leafy bracts. Grows on moist soil; widely distributed in North America.

(17) *Teucrium*. Germander

1. Leaves undivided, serrate; flowers rose to purple.
$\qquad$ *T. canadensis* var. *occidentale*
1. Leaves pinnately parted or incised; flowers white or whitish.
$\qquad$ *T. laciniatum*

GERMANDER, *Teucrium canadensis,* a plant 3–6dm tall, has unbranched stems, except sometimes branching in the inflorescence. Flowers small, purple, in slender compact spikes 10–20cm long; calyx covered with long, soft, sticky hairs; stem 4-angled; leaves narrow, pointed, opposite, 5–10cm long. Moist soil in foothills of the Rockies; widespread throughout North America.

MARE'S TAIL FAMILY: HIPPURIDACEAE

A family with a single genus and species, sometimes found lumped with the Haloragaceae. MARE'S TAIL, *Hippuris vulgaris,* is a water plant with leaves simple, entire, linear, in whorls of 8–12. The tiny flowers are axillary, solitary, perfect, apetalous; stamen 1. Usually the entire plant is submersed, but part of the plant may extend above the water. Found from Greenland throughout North and South America, and in Europe and Asia. Occurs in ponds up to 10,000 feet.

WATER-STARWORT FAMILY: CALLITRICHACEAE

Only one genus in this family. Leaves opposite and entire; flowers axillary, without perianth, polygamous (having both perfect and imperfect flowers on the same plant), sometimes with 2 bracts.

1. Bracts present; emersed leaves obovate.

Callitriche verna
Water-starwort

1. Bracts absent; all leaves linear.

C. hermaphroditica
Autumn water-starwort

WATER STARWORT, *C. verna,* is a water plant with narrow, one-nerved, submersed leaves about 2.5cm long, and shorter, broader floating leaves arranged in little rosettes. Tiny 4-seeded fruits in the leaf axils. Found in shallow ponds and in rock pools in Colorado up to 11,500 feet and is widely distributed throughout the northern hemisphere. AUTUMN WATER STARWORT, *C. hermaphroditica,* is a completely submersed plant having only narrow, 1-nerved leaves. It has the same distribution.

PLANTAIN FAMILY: PLANTAGINACEAE

Only one genus in our range. The flowers are very small, without bright color and crowded into compact spikes. Leaves are basal and usually prominently ribbed. The flowers are regular, 4-merous; stamens 4 or 2; style 1. The genus includes some common lawn weeds that have been introduced from Europe and naturalized here, but there are also several native species.

1. Stamens 4; flowers perfect; corolla not closing over the capsule.
    2. Leaves 3–8-nerved, glabrous or pubescent but not silky-lanate; spikes not woolly.
        3. Capsule circumscissile near the middle.
            4. Leaves broad; mostly oval.

Plantago major
Common plantain

            4. Leaves lanceolate.

P. lanceolate
English plantain

        3. Capsule circumscissile near the base.
            4. Crown (an appendage on the claw of petals) not woolly; plant subalpine.

P. tweedyi
Tweedy's plantain

            4. Crown with brown or red wool, involving the base of the petioles; plant found in saline or alkaline meadows.

P. eriopoda
Redwool plantain

2. Leaves linear, silky-lanate; spikes woolly.

P. patagonica
Indian wheat

1. Stamens 2; plants subdioecious or polygamo-dioecious; corolla-tubes closing over the capsule.

P. elongata
Slender plantain

REDWOOL PLANTAIN, *Plantago eriopoda* (fig. 234), has a basal rosette of oblong or oblanceolate leaves 8–20cm long and 3–9-ribbed, which have reddish woolly hairs at their bases. Flower spikes 5–13cm long on stalks 13–38cm tall. Occurs on plains, hills, and mountain slopes (up to 10,000 feet in Colorado).

OLIVE FAMILY: OLEACEAE
1. Trees with pinnate leaves; fruit a samara.
    2. Leaves lanceolate.

Fraxinus pennsylvanica
Green ash

    2. Leaves broadly ovate or cordate.

Fraxinus anomala

1. Shrubs with simple leaves; fruit a drupe.

Forestiera pubescens
Mountain privet

234. Redwool Plantain, ½X

MOUNTAIN PRIVET, *Forestiera pubescens* (fig. 235), is a large shrub with pale or greenish bark and opposite branching. Found in valleys and on canyon sides of the foothills of southwestern Colorado, New Mexico, and Arizona. Its small, clustered yellow, dioecious flowers appear before the leaves; the fruits are oval or oblong, bluish-black berries. GREEN ASH, *Fraxinus pennsylvanica*, is found from eastern Wyoming northwestward. *F. anomala* is found from southern Colorado through Utah to Nevada.

FIGWORT FAMILY:
SCROPHULARIACEAE
A large family of beautiful wildflowers. Their corollas are irregular, mostly 2-lipped, and the lips are usually lobed. The superior ovary ripens into a 2-celled seed capsule. Leaves frequently, but not always, opposite. Stamens are 2, 4, or 5; but, except in *Verbascum*, never more than 4 are anther-bearing. A more or less reduced,

sterile stamen, called a *staminode*, is sometimes present. Some plants of this family, which have opposite leaves, also have 4-angled stems and might be confused with members of the mint family. But plants of the figwort family may always be distinguished by the 2-celled ovary, which ripens into a dry, many-seeded capsule, whereas members of the mint family have fruits consisting of 4 nutlets. The cultivated snapdragons and several other garden flowers belong here.

235. Forestiera, ½ X

1. Anther-bearing stamens 5.

> *Verbascum*, p. 306
> Mullein

1. Anther-bearing stamens 4 or 2.
  2. Corolla spurred or saccate at base on lower side.

> *Linaria*, p. 291
> Toadflax or
> butter-and-eggs

  2. Corolla not spurred, but sometimes somewhat inflated or swollen on one side.
    3. Stamens 5, four anther-bearing, the fifth often rudimentary and sterile (a staminode).
      4. Staminode reduced to a gland or scale on the upper side of the corolla-tube.
        5. Small annuals; peduncles 1-flowered.

> *Collinsia*, p. 289
> Blue-eyed Mary

        5. Large, coarse perennials; peduncles several-flowered.

> *Scrophularia*, p. 305
> Figwort

      4. Staminode conspicuous, elongated.
        5. Calyx deeply 5-parted or divided.

> *Penstemon*, p. 296
> Penstemon or
> beardtongue

        5. Calyx obtusely 5-lobed.

> *Chionophila*, p. 290
> Snowlover

    3. Stamens all anther-bearing.
      4. Stamens 2.
        5. Calyx of 5 nearly distinct lobes.

> *Gratiola*, p. 290
> Hedge hyssop

        5. Calyx 4-parted.
          6. Leaves alternate, mostly basal.

> *Besseya*, p. 286
> Kittentails

          6. Leaves opposite, at least the lower.

> *Veronica*, p. 306
> Veronica or speedwell

      4. Stamens 4.
        5. Corolla regular or nearly so.

6. Small prostrate plants, rooting at the nodes, mud-loving or
aquatic.

Limosella, p. 290
Mudwort

6. Stems erect, usually branched.

Agalinis, p. 286
Agalinis

5. Corolla distinctly bilabiate (2-lipped).
6. Stamens not enclosed in the upper lip.

Mimulus, p. 291
Monkey flower

6. Stamens ascending under (or enclosed in) the upper lip.
7. Anther cells equal, parallel.
8. Calyx inflated and veiny in fruit; leaves opposite.

Rhinanthus, p. 305
Yellow rattle

8. Calyx not inflated; leaves alternate or whorled.

Pedicularis, p. 293
Lousewort

7. Anther cells unequal, one pendulous by its apex.
8. Lips of the corolla very unequal.

Castilleja, p. 287
Paintbrush

8. Lips of the corolla equal or nearly so.
9. Calyx nearly equally 4-lobed (the lips deeply
cleft).

Orthocarpus, p. 292
Owl-clover

9. Calyx-lips not cleft, but the lower sometimes
absent.

Cordylanthus, p. 290
Bird beak

(1) *Agalinis.* Agalinis

1. Corolla pubescent on the outside; capsule oblong.

A. aspera

1. Corolla glabrous; capsule globose.

A. besseyana

(2) *Besseya.* Kittentail

1. Corolla wanting; filaments conspicuously colored.

B. wyomingensis

1. Corolla present; filaments not conspicuously colored.
2. Corolla violet-purple

B. alpina
Alpine kittentails

2. Corolla white to yellow (may be purple-tinged).
3. Corolla yellow.

B. ritteriana

3. Corolla white to pinkish or purple-tinged.

B. plantaginea
Kittentails

KITTENTAILS, *Besseya plantaginea* (fig. 236), has a basal rosette of rounded or
oval, petiolate leaves and erect spikes from 20 to 38cm tall. The white to pinkish

individual flowers are very small and densely crowded. Occurs on moist wooded slopes of the foothill and montane zones from Wyoming to New Mexico. In Colorado, it is found only on the eastern side of the Continental Divide, but a similar plant with yellowish flowers, *B. ritteriana,* occurs in the mountains of southwestern Colorado.

ALPINE KITTENTAILS, *B. alpina,* with somewhat woolly foliage and fuzzy purplish spikes from 5 to 15cm tall, occurs among rocks of the alpine zone from Wyoming and Utah to New Mexico. Easily confused with purple fringe, *Phacelia sericea,* but the latter has fern-like leaves. This genus was named in honor of Charles Edwin Bessey, an eminent botanist at the University of Nebraska.

236. Kittentails, ²/₅ X

(3) *Castilleja.* Paintbrush

Many plants with bright colors belong to this group, but the flowers themselves are inconspicuous. The showy effect is produced by red, orange, pink or white calyxes and leafy bracts, which surround the flowers in brushlike inflorescences. Some of the species have a wide range of color variation, and several are difficult to identify accurately. The corolla is 2-lipped; the upper lip is extended into a long beak-like *galea* that includes the 4 stamens; the much shorter lower lip is tucked up under it.

1. Plants annual; stems mostly solitary, erect, 30–80cm tall; bracts red or red-tipped; calyx green, cleft on both sides to about the middle; corolla yellow; in wet places.
C. exilis

1. Plants perennial; stems often clustered.
    2. Galea mostly less than 1/2 the length of the corolla-tube.
        3. Lower lip of corolla relatively small, usually less than 1/3 the length of the galea; bracts yellow, calyx yellowish.
C. flava
Yellow paintbrush
        3. Lower lip of corolla relatively conspicuous, usually 1/3 to 1/2 the length of the galea.
            4. Corolla-tube 3–5cm long; bracts green; calyx greenish or yellowish; plants well below 8,000 feet.
C. sesssiliflora
Downy painted-cup
            4. Corolla 2cm or less long; altitude various.

5. Calyx about equally cut above and below; upper stems and bracts puberulent to villous-hirsute.

C. longispica
White paintbrush

5. Calyx cut (or cleft) very unequally.
  6. Plants puberulent.
    7. Calyx cut less than half as deeply laterally as above and below.

C. pallescens
Palish paintbrush

  ✔7. Calyx cut more deeply below than above.

C. puberula

  6. Plants villous or viscid-villous.
    7. Bracts equaling or exceeding flowers, usually yellow.

C. cusickii

    7. Bracts shorter than the flowers, purplish or reddish.

C. pulchella
Showy paintbrush

2. Galea mostly more than 1/2 the length of the corolla-tube.
  ✔3. Calyx much more deeply cut below than above; bracts and calyx red or scarlet.

C. linariaefolia
Wyoming paintbrush

3. Calyx about equally cut above and below.
  4. Plants more or less densely tomentose.
    5. Bracts crimson.

C. integra
Foothills paintbrush

    5. Bracts dull yellowish.

C. lineata

  4. Plants either glabrous or hairy, but not tomentose.
    5. Plants alpine.
      6. Bracts greenish-yellow, sometimes streaked with red.

C. occidentalis
Western yellow paintbrush

      6. Bracts rose, crimson, or purplish.

C. haydenii

    5. Plants below alpine (montane to subalpine).
      6. Upper leaves deeply cleft into linear, divaricate lobes; plants with both short and long hairs.

C. chromosa
Nelson or desert paintbrush

      6. Leaves entire or shallowly lobed near apex; one kind of hair present (if any).
        7. Inflorescence yellow or yellowish.

C. sulphurea
Northern paintbrush

        ✔7. Inflorescence rose to red or orange.
          8. Stems 1–3dm tall; subalpine mostly, rarely lower.

C. rhexifolia
Rosy or subalpine paintbrush

          8. Stems usually over 3dm tall; montane mostly, rarely higher; moist areas.

C. miniata
Scarlet paintbrush

FOOTHILLS or ORANGE PAINTBRUSH, *C. integra,* is a spring-blooming plant of mesas and lower foothills, growing in clumps, with stems from 10 to 30cm tall.

The showy bracts are mostly entire. Found from Wyoming to Oregon, south to New Mexico. NELSON PAINTBRUSH, *C. chromosa*, is rather similar in color, but the upper leaves and bracts are divided into narrow lobes. Occurs in Wyoming on dry, high plains, mesas, and foothills; southward to New Mexico.

✓ WYOMING PAINTBRUSH, *C. linariaefolia* (fig. 237), is a branched plant 6dm or more in height. Green, red-edged corollas protrude conspicuously from its "brushes" of scarlet bracts and calyxes. Often grows among shrubs, frequently with sagebrush or antelope brush. Found mostly in the montane zone and occurs from Wyoming to Oregon, south to New Mexico and California. It is the state flower of Wyoming. SCARLET PAINTBRUSH, *C. miniata*, is another species usually found in the montane zone. It has erect, unbranched stems 3–6dm tall, several of which may come from the same root. Its corollas do not project so conspicuously as in the Wyoming paintbrush. Grows from the Canadian Rockies southward throughout our range.

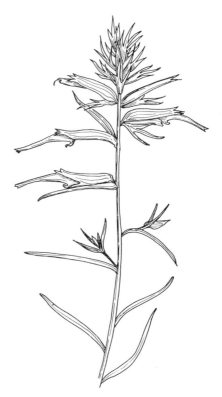

237. Wyoming Paintbrush, ½X

ROSY or SUBALPINE PAINTBRUSH, *C. rhexifolia*, is sometimes hard to distinguish from the last, but in general its bracts are not scarlet. Instead, they may be crimson, rose, purple, pink, or even 2-toned. These varying color forms probably indicate intergrading, not uncommon in this genus. A subalpine species.

✓ NORTHERN PAINTBRUSH, *C. sulphurea*, has whitish bracts; its stems are solitary or few together, sometimes branched at least once. Grows in moist aspen groves and shaded meadows of the montane and subalpine zones in the Rockies, and also in moist meadows of eastern Canada and New England.

WESTERN YELLOW PAINTBRUSH, *C. occidentalis* (fig. 238), grows in clumps with stems 8–20cm tall. The bracts are greenish-yellow or sometimes purplish. An alpine and subalpine species of Colorado, Utah, and New Mexico.

(4) *Collinsia*. Blue-eyed Mary

One species in our range: BLUE-EYED MARY, *Collinsia parviflora*, is a delicate winter annual with pale-blue or bicolored corollas not over 6mm long. The pointed, sometimes purplish, leaves are opposite on the lower stem and whorled above. Usually 2–5 flowers in the whorls. Grows on shaded ground where other vegetation is sparse, in the foothill and montane zones throughout the Rockies.

(5) *Chionophila.* Snowlover

One species in our range: *Chionophila jamesii,* a small subalpine to mostly alpine perennial called SNOWLOVER, is closely related, and similar, to the penstemons. A dwarf, 8–15cm tall, with short, 1-sided spikes of white, cream, or greenish-white flowers. The staminode is shorter than the stamens and smooth. Grows on moist, gravel slopes of high mountains in southern Wyoming and Colorado.

(6) *Cordylanthus.* Birdbeak

1. Calyx diphyllous (with 2 parts), subtended by 2–4 small bractlets, as well as by larger bracts.
   2. Corolla purplish; plant glabrous to lightly puberulent.
          C. wrightii
   2. Corolla dull-yellow; plant ash-grayish-puberulent.
          C. ramosus
          Branched birdbeak
1. Calyx monophyllous, subtended only by larger bracts; corolla purplish; plant glandular puberulent or glandular-villous.
          C. kingii

BRANCHED BIRDBEAK, *Cordylanthus ramosus,* has grayish, finely-divided leaves and head-like clusters of yellowish flowers. Individual flowers are 15–20mm long with tubular, 2-lipped corollas that resemble birds' beaks. Occurs on dry foothills from Montana through Colorado.

(7) *Gratiola.* Hedge hyssop

One species in our range. *G. neglecta* var. *neglecta* is an annual, with solitary, pedicelate flowers in the leaf axils; corolla-tube white to yellowish, limb purplish.

(8) *Limosella.* Mudwort

One species in our range. *L. aquatica* is a glabrous plant with

238. Western Yellow Paintbrush, ½X

239. Butter-and-Eggs, ½ X

fibrous roots; tufts 3–6cm high, often stoloniferous; corolla usually white or pink; muddy flats and shores.

(9) *Linaria*. Toadflax or butter-and-eggs

1. Flowers blue to whitish; annuals.

L. canadensis var. *texana*
Toadflax

1. Flowers yellow; perennial.
   2. Stem leaves cordate-clasping, upper leaves ovate to ovate-lanceolate.

L. dalmatica

   2. Stem leaves not clasping, linear to linear-lanceolate.

L. vulgaris
Butter-and-eggs

BUTTER-AND-EGGS, *Linaria vulgaris* (fig. 239), is an introduced plant seen around settlements and along roadsides. It has racemes of pale-yellow and bright-orange, long-spurred flowers.

TOADFLAX, *L. canadensis*, is a native species with blue flowers 2.5cm or less long. About 1/3 of this length consists of the spur. Grows in the eastern part of our range below 7,500 feet.

(10) *Mimulus*. Monkey-flower

Flowers of these plants have 5-lobed and more or less 2-lipped corollas, 2 ridges extending back from the lower lip into the throat. The leaves are opposite and the stems square. Grows in moist places.

1. Perennials; plants from rhizomes or stolons.
   2. Corolla crimson or red (or pink, rose, violet).

M. lewisii
Lewis mimulus

   2. Corolla yellow (sometimes marked with red or maroon, or washed with pale-purple.
      3. Upper calyx tooth markedly larger than others.
         4. Corolla 1–2cm long with throat open.
            5. Fruiting pedicels 10–20mm long.

M. glabratus var.
fremontii

            5. Fruiting pedicels 25–50mm long.

M. glabratus var.
utahensis

         4. Corolla 2–4cm long, with throat nearly closed.
            5. Plants with rhizomes (may also have stolons); usually under 25cm high.

M. tilingii var. *tilingii*
Large mountain monkey-
flower

            5. Plants with stolons, rhizomes usually lacking; often over 25cm high.

M. guttatus ssp. *guttatus*
Yellow monkey-flower

      3. Upper calyx tooth not evidently larger than others; herbage viscid-villous; corolla 1–3cm long, only slightly irregular.

M. moschatus var.
moschatus
Musk-flower

1. Annuals; without rhizomes or stolons.
    2. Mature calyx strongly inflated; corolla yellow, 7–14mm long.

> *M. floribundus*
> Many-flowered monkey-
> flower

    2. Mature calyx little or not inflated.
        3. Corolla yellow, 5–6mm long; calyx teeth glabrous.

> *M. suksdorfii*
> Suksdorf monkey-flower

        3. Corolla red or rose, 6–10mm long; calyx teeth usually ciliate.

> *M. rubellus*
> Little red mimulus

240. Monkey-flower, ½X

COMMON or YELLOW MONKEY-FLOWER, *M. guttatus* (fig. 240), has stems that may be erect or decumbent and range from being quite low up to 45cm long. It has yellow flowers spotted with red; corollas 2.5cm or more long with closed throats. Found around springs and on mossy stream banks, sometimes in wet gravel, usually in the subalpine zone, throughout our range. MANY-FLOWERED MONKEY-FLOWER, *M. floribundus*, is a smaller, branched plant of the montane zone, with similar but smaller flowers with open throats.

LITTLE RED MIMULUS, *M. rubellus*, is a small, annual plant with flowers 6–12mm long on stalks longer than the flowers. The funnelform corollas are usually red, sometimes white or yellow. A rare plant, growing on mossy banks of the montane and subalpine zones.

LEWIS MIMULUS, *M. lewisii*, has unbranched, finely-hairy, and rather sticky stems 3–6dm tall; rose-red corollas 2.5–5cm long. The anthers are hairy, and the 2 ridges on the lower part of the throat are bearded with orange hairs. It is a showy plant, abundant along mountain streams, in moist ravines, and around seepage areas of the montane and subalpine regions of Grand Teton, Yellowstone, Glacier, and Mt. Rainier National Parks and the surrounding areas. Named in honor of Captain Meriwether Lewis.

(11) *Orthocarpus*. Owl-clover

1. Corolla yellow; spike dense.
    2. Pubescent and hirsute; stem straight and narrow.

> *O. luteus*
> Yellow owl-clover

    2. Puberulent only; stem usually branched above.

> *O. tolmiei*

1. Corolla white, becoming rose-purple.

> *O. purpureo-albus*
> Purple owl-clover.

Yellow Owl-clover, *Orthocarpus luteus* (fig. 241), is an erect, annual plant 10–30cm tall; narrow leaves; flowers in a terminal, clubshaped inflorescence. Grows abundantly on dry fields and slopes of the foothill and montane zones, blooming in early summer. Often seen in overgrazed pastures. Purple Owl-clover, *O. purpureo-albus,* is similar in habit, but has white, pink or purplish flowers. Occurs in the montane zone of southwestern Colorado, New Mexico, and Arizona. These plants are not related to the true clovers of the pea family.

241. Yellow Owl-clover, ½X

(12) *Pedicularis.* Lousewort

The louseworts have curiously shaped flowers. The corolla is strongly 2-lipped, the upper lip (galea) arched and sometimes elongated into a beak. Stamens 4 in 2 pairs, one pair longer than the other, all included in the galea. In most species the leaves are pinnately divided, sometimes into comb-like divisions, or, in other species, into fern-like patterns. They usually grow in moist places. The name *lousewort* is an old English plant name, and *Pedicularis* is merely the Latin form of it. It was once believed that cows who ate this plant became infested with lice.

1. Galea prolonged into a slender beak, curved outward and upward, the beak over 8mm long.
          P. groenlandica
          Elephanthead
1. Galea beakless or with short or incurved beak not over 6mm long.
  2. Calyx usually 2-lobed or toothed; leaves toothed or pinnately lobed.
    3. Beak of galea 3–5mm long; corolla white.
          P. racemosa var. *alba*
          Sickletop lousewort
    3. Galea beakless; corolla usually not white.
      4. Leaves pinnately lobed, the sinuses extending 1/2 to 2/3 the distance toward the midrib; corolla yellow or reddish.
          P. canadensis
          Canada lousewort or wood betony
      4. Leaves double crenate-toothed, but not lobed.
          P. crenulata
          Purple lousewort
  2. Calyx usually 5-lobed or toothed; many of the leaves deeply pinnately lobed (to or nearly to the midrib) or bipinnatifid.
    3. Galea with a beak 4–6mm long, curved downward and partially hidden by the lower lip.
      4. Corolla white to ochroleucous; calyx glabrous or glabrate.
          P. contorta var. *contorta*
      4. Corolla purplish; calyx usually villous at base.
          P. contorta var. *ctenophora*

3. Galea beakless or with straight beak 3mm or less long.
    4. Stems leafy, leaves often over 3cm wide; basal leaves may be wanting.
        5. Galea 13–25mm long, entire or with 2 rounded teeth at tip.

> *P. bracteosa* var.
> *paysoniana*
> Fernleaf lousewort

        5. Galea of some flowers over 30mm long, with 2 acute teeth near the tip.

> *P. procera*
> Indian warrior

    4. Stems not leafy or with few leaves, reduced in size, rarely exceeding 2cm wide; leaves mostly basal.
        5. Galea with beak 1–3mm long.
            6. Inflorescence conspicuously long-hairy; corolla pink to purplish.

> *P. parryi* ssp. *purpurea*

            6. Inflorescence glabrous or sparsely hairy; corolla white to yellow.

> *P. parryi* ssp. *parryi*
> Parry lousewort

        5. Galea beakless or with 2 projections less than 1mm in length.
            6. Corolla whitish to yellow.

> *P. oederi*

            6. Corolla pink, red, or purple.
                7. Bracts similar to the leaves.

> *P. pulchella*
> Dwarf lousewort

                7. Bracts very different from the leaves.

> *P. cystopteridifolia*
> Fern-leaved lousewort

ELEPHANTHEAD or ELEPHANTELLA, *P. groenlandica* (fig. 242), can always be recognized by its long, twisted beak that resembles an elephant's trunk. Grows from 1.5 to 6dm tall, and its slender spike of purple or rose-colored flowers is frequently seen in wet meadows, on pond margins, and along stream banks, most commonly in the subalpine zone, but sometimes at either lower or higher altitudes. Ranges across Arctic America and southward in the mountains to New Mexico and California. PURPLE LOUSEWORT, *P. crenulata*, also has rose-colored flowers, but its long, slender leaves are finely scalloped, and there are long hairs on the stem. Grows in wet montane meadows, sometimes becoming a weed in hay fields.

SICKLETOP LOUSEWORT or MOUNTAIN FIGWORT, *P. racemosa* (fig. 243), usually has several stems in a clump. Stems and leaves are reddish; flowers white or creamy with a curved and somewhat twisted galea. Grows in dry coniferous forests of the upper montane and subalpine zones from Canada south to New Mexico and California.

PARRY LOUSEWORT, *P. parryi* (fig. 244), is an alpine plant with creamy or yellowish flowers on erect stalks 10–30cm tall; leaves divided nearly to the midrib into regular, narrow, toothed segments. Found from Montana south to New Mexico and Arizona. CANADA LOUSEWORT or WOOD BETONY, *P. canadensis*, has yellow (occasionally reddish) flowers; leaves are divided not more than 2/3 of the way to midrib. Occurs in meadows and open woods; widely distributed throughout North America south of the Arctic.

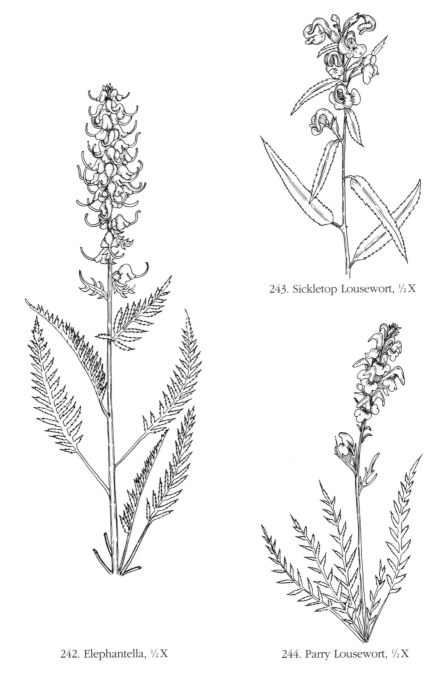

243. Sickletop Lousewort, ½X

242. Elephantella, ½X

244. Parry Lousewort, ½X

FERNLEAF LOUSEWORT, *P. bracteosa* var. *paysoniana*, has stems 3–9dm tall and fern-like leaves to 15cm long. The yellowish flowers are in a dense spike; the erect galea is curved at the top. A plant of moist woods and shaded meadows of the subalpine zone from Montana south to Colorado and Utah. INDIAN WARRIOR,

P. procera, is even larger, with stems to nearly 12dm tall, leaves sometimes 3dm or more in length. Flowers are dingy-yellow, often streaked with red. Grows in montane woods from Wyoming and Utah to New Mexico and Arizona.

(13) *Penstemon*. Penstemon or beardtongue

The generic name means *5 stamens* and refers to the fact that, although most members of Scrophulariaceae have 4 stamens, this genus regularly has 5. But the fifth has no anther and is, therefore, not functional (sterile). It is called a *staminode* and differs from the 4 fertile stamens in shape. In many species, the staminode has a tuft or brush of hairs at its tip or along one side, accounting for the common name *beardtongue*. The united corolla is irregular and 2-lipped; the upper lip has 2 lobes, the lower 3. The corolla may be tubular-funnelform, or it may be bulged or inflated above a short, slender tube. Leaves opposite; inflorescence often one-sided.

As there are so many species in this genus, it is customary to divide the key into groups according to color, size, or some other more dependable character. The blue and purple forms vary a good deal in color, sometimes having pink or white forms. Normally tall species may begin blooming when their stems are very short, especially if the season or location is unfavorable. One should seek additional plants of the same species to get an estimate of average size or height. Several penstemon species are restricted in their distribution to small areas in the Colorado mountains, and in a different locality they will be replaced by a similar, but different, species.

1. Inflorescence bright red or light yellow.
  2. Inflorescence bright red.
    3. Corolla glabrous outside.
      4. Corolla conspicuously bilabiate.
        5. Anthers glabrous.

> *P. barbatus* var. *torreyi*
> Scarlet bugler

        5. Anthers hairy.

> *P. barbatus*. var. *trichander*

      4. Corolla hardly bilabiate; anthers puberulent.
        5. Stems puberulent.

> *P. eatonii* var. *undosus*
> Firecracker penstemon

        5. Stems glabrous.

> *P. eatonii* var. *eatonii*

    3. Corolla glandular-pubescent outside; anthers usually glabrous.

> *P. rostiflorus*
> Bridges penstemon

  2. Inflorescence light yellow, sometimes whitish or slightly brownish; anthers glabrous; plant glabrous throughout.

> *P. confertus*
> Yellow penstemon

1. Inflorescence blue, pink to purple, or white (sometimes with a purple tinge).
  2. Plants with anthers densely long-woolly.
    3. Plant woody, a shrub; calyx glandular-pubescent.

> *P. fruticosus*
> Bush penstemon

3. Plant herbaceous (if woody, only at base); corolla glabrous outside.
    4. Calyx glabrous.
                              P. comarrhenus
    4. Calyx glandular-pubescent.
        5. Leaves glabrous, rounded at tip; plant caespitose, low.
                              P. ellipticus
                              Rockvine penstemon
        5. Leaves pubescent, pointed at tip; plant over 10cm tall.
                              P. montanus var.
                              montanus
                              Mountain beardtongue or
                              cordroot penstemon
2. Plants with anthers either glabrous or pubescent, but not densely long-woolly.
    3. Low plants, either creeping or caespitose, under 10cm tall.
                              Group I, p. 298
    3. Taller plants with erect stems, over 10cm tall.
        4. Inflorescence glandular-pubescent outside.
                              Group II, p. 299
        4. Inflorescence mostly glabrous outside, not glandular.
            5. Anthers short- or long-hairy only on the side opposite dehiscence.
                              Group III, p. 301
            5. Anthers glabrous on the side opposite dehiscence.
                              Group IV, p. 303

SCARLET BUGLER, *P. barbatus* var. *torreyi* (fig. 245), has stems 4–9dm tall, solitary or in clumps, with a spire-like panicle of brilliant scarlet tubular flowers. The 3 lobes of the lower corolla lip are reflexed. The leaves and smooth stem are often dark and tinged reddish. Found from the Pikes Peak region south and west to New Mexico and Arizona in the montane zone. EATON or FIRE-CRACKER PENSTEMON, *P. eatonii*, is similar. Stems are somewhat shorter; the corollas scarcely 2-lipped, their short lobes not reflexed. Occurs from Mesa Verde National Park southwestward.

BRIDGES PENSTEMON, *P. rosti-florus*, has stems to 6dm tall, the upper part often branched, from a somewhat woody base. The scarlet corollas are strongly 2-lipped, the upper lip erect, the lower reflexed. Grows in the oak-piñon-juniper areas of the southwest with a range similar to that of *P. eatonii*.

245. Scarlet Bugler or Torrey Penstemon, ⅖X

YELLOW PENSTEMON, *P. confertus*. The slender stems are usually clustered, 18–60dm tall, with interrupted inflorescences made up of clusters of light yellow or whitish, tubular corollas. Brown hairs inside the lower lip. The range is northern; common in Glacier National Park.

BUSH PENSTEMON, *P. fruticosus*, forms a wide, dense clump 10–50cm high from a woody base. The leaves are thick and shiny, more or less toothed, 7–25mm long. The inflorescence is somewhat 1-sided with bright lavender-blue flowers. It has hairy anthers and a yellow-bearded staminode. Found on rocky slopes in openings of the coniferous forests from the mountains of Montana and Wyoming westward. CORDROOT PENSTEMON, *P. montanus*, is a similar but smaller plant, with sharply toothed leaves and a smooth staminode. Occurs on stony montane and subalpine slopes in Wyoming, Idaho, and Montana.

ROCKVINE PENSTEMON, *P. ellipticus*, forms loose mats. It is somewhat woody at base with smooth leaves; long and showy, light violet flowers; the staminode heavily bearded with long, yellow hairs. A low plant found on rock slides and cliff faces above and below timberline, from the Canadian Rockies into western Montana and northern Idaho. Especially conspicuous in Glacier National Park

Group I
Low plants, under 10cm tall, either creeping or caespitose; inflorescence glandular-puberulent or glandular-pubescent; anthers glabrous; staminode more or less bearded.

1. Flowers sessile, flowering stems under 1cm long.
      P. acaulis
1. Flowering stems longer, 2–10 (15)cm long.
  2. Leaves over 5mm wide.
      P. barbourii
      Harbour alpine penstemon
  2. Leaves under 5mm wide.
    3. Calyx lobes usually with prominent scarious margins.
      P. crandallii
      Crandall penstemon
    3. Calyx lobes without scarious margins.
      4. Leaves linear, not over 1.5mm wide.
      P. teucrioides
      Grayleaf creeping penstemon
      4. Leaves narrowly oblanceolate or spatulate, over 1.5mm wide.
      P. caespitosus
      Mat penstemon

CRANDALL PENSTEMON, *P. crandallii* (fig. 246), forms a mat 10–15cm thick of small green leaves, mostly under 2.5cm long and under 5mm wide. May become covered, in May or early June, with blue or purple flowers 2.5cm long. Grows on loamy soils of the montane and lower subalpine zones from central Colorado south and west into Utah. C. S. Crandall was professor of botany at Colorado Agricultural College (now Colorado State University) from 1889 to 1899, and made extensive botanical collections in the state.

GRAYLEAF CREEPING PENSTEMON, *P. teucrioides*, has slender stems and small narrow leaves covered with fine gray hairs; the flowers are pale-purple, nearly 2.5cm long. A low, creeping plant usually found among sagebrush on dry slopes of the montane zone in central and southwestern Colorado.

HARBOUR ALPINE PENSTEMON, *P. harbourii*, is a creeping plant, the stems not longer than 15cm, the leaves dull-green or grayish to 2.5cm long. The corollas are almost 2.5cm long, light-purple, and hairy inside and outside. Strictly an alpine species, found only on scree or gravel slopes of the high Colorado peaks.

246. Crandall Penstemon, ½X

MAT PENSTEMON, *P. caespitosus*, has stems not over 8cm long but spreads into mats as large as 12dm in diameter. The leaves, usually under 2.5cm long, may be grayish or green; flowers up to 2cm long, light-blue or purplish. Occurs in western Wyoming, northwestern Colorado, and northeastern Utah on clay or rocky soil in sagebrush country between 6,000 and 9,000 feet.

Group II

Taller plants with erect stems, over 10cm tall; inflorescence glandular-pubescent outside.

1. Anthers glabrous on the side opposite dehiscence.
   2. Corolla white or whitish.
      3. Corolla glandular-hairy inside; leaves toothed.

         P. deustus
         Hot-rock penstemon

      3. Corolla glabrous inside; leaves not toothed.
         4. Staminode glabrous or bearded only at tip.

         P. whippleanus
         Whipple penstemon

         4. Staminode bearded for 1/3 of its length.

         P. attenuatus
         Sulfur penstemon

   2. Corolla blue, violet, or dark purple.
      3. Leaves of lower stem wholly glabrous, bright or dark-green.
         4. Ovary and capsule glandular-puberulent at top.

         P. whippleanus
         Whipple penstemon

         4. Ovary and capsule glabrous.
            5. Upper leaves somewhat toothed; flowers clustered.
               6. Corolla 10–17mm long.

         P. virens
         Bluemist or greenleaf penstemon

               6. Corolla 15–23mm long.

         P. gracilis
         Slender penstemon

            5. Leaves entire; corolla 12–20mm long.

6. Leaves linear or lance-linear, mostly basal in rosettes; staminode bearded at tip.

P. aridus
Stiff-leaved penstemon

6. Leaves mostly oblanceolate, not mostly basal; staminode short-bearded from tip downward.

P. hallii
Hall's alpine penstemon

3. Leaves of lower stem either puberulent or pubescent throughout, or at least on veins and margins; dull or grayish-green

4. Corolla somewhat flattened beneath, with 2 strongly marked longitudinal grooves on underside.

5. Basal rosette well-developed.

P. oligantus

5. Basal rosette absent.

P. radicosus
Matroot penstemon

4. Corolla-tube rounded, without marked grooves.

5. Plant a subshrub, woody below.

P. linarioides var. *sileri*
Colorado penstemon

5. Plants herbaceous.

6. Corolla-tube 24–36mm long; staminode conspicuously exserted.

P. eriantherus
Crested tongue penstemon

6. Corolla-tube 8–14mm long; staminode included.

P. humilis
Lowly penstemon

1. Anthers pubescent on the side opposite dehiscence.

2. Corolla 16–22mm long; staminode bearded at tip.

P. saxosorum

2. Corolla 20–35mm long; staminode sparsely bearded at tip.

P. subglaber

✔ Whipple Penstemon, *P. whippleanus*, is easily recognized in its most common form by the very dark wine-red or purple, almost black, bell-shaped corollas. The flowers are in whorls more or less separated, or sometimes in single, head-like clusters. The stems vary from 10 to 45cm in height. Found on rocky or openly wooded slopes or along trails, from the montane to the alpine zones, most abundantly in the subalpine region. Idaho, Wyoming, Colorado, Utah to Arizona and New Mexico. Forms of this species have whitish or dingy-blue flowers.

Hall's Alpine Penstemon, *P. hallii,* usually has several erect stems 10–20cm tall from tufts of smooth, oblanceolate leaves. The flowers are in rather few-flowered clusters on the shorter plants, or in an elongated, 1-sided inflorescence on the taller ones. The corollas are about 2.5cm long, bright reddish-purple or violet with white markings at the throat, and are abruptly inflated. Grows in rocky or gravelly situations on the higher peaks of central and southern Colorado, from 10,000 to 13,000 feet.

Slender Penstemon, *P. gracilis*, has stems 25–50cm tall, which usually grow alone, occasionally a few together. Lower leaves usually entire, the upper finely toothed and long-pointed, the lower blunt. Flowers are pale blue or lavender, somewhat sticky outside, with hairs inside. The staminode is densely bearded but not exserted. This is an eastern species, growing from Ontario and Wisconsin to the eastern slope of the Rockies; found along the base of the mountains among scrub oak or pines and in foothill meadows, from Alberta to New Mexico.

MATROOT PENSTEMON, *P. radicosus*, forms clumps of slender stems up to 35cm tall; leaves are grayish; flowers, in rather compact clusters, vary from pale to dark-blue. Found in Yellowstone National Park and from Montana southward through western Wyoming and Idaho to north central Colorado. Grows in the montane zone, often in rocky or sagebrush areas.

COLORADO PENSTEMON, *P. linarioides* var. *sileri*, is a plant with wiry stems 10–25cm tall from a heavy woody base. Its leaves are narrow, crowded, and grayish. The inflorescence is 1-sided, of lavender or lilac-colored flowers, the corollas strongly inflated and lightly bearded inside. Found on rocky ground and among sagebrush on the upper foothill and montane zones of southwestern Colorado and northwestern New Mexico.

BLUEMIST OR GREENLEAF PENSTEMON, *P. virens*, has numerous erect stems, (10) 15–35cm tall, from mats of bright-green, shiny leaves. Blooms in such masses on some foothill slopes as to give the hillside a misty-blue appearance from a distance. Range extends up to 10,500 feet along the eastern slope of the Continental Divide from southern Wyoming through Colorado. Blooms in late May in the scrub-oak country, in June or early July at higher altitudes.

LOWLY PENSTEMON, *P. humilis*, is rather similar to *P. virens,* with duller leaves. Found in Yellowstone National Park and through western Wyoming into northwestern Colorado and Idaho.

CRESTED TONGUE PENSTEMON, *P. eriantherus*, has stems 12–36cm tall. Stems and leaves are sticky-hairy; flowers usually pinkish or pale-purple; the staminode, bearded with long yellow hairs, protrudes conspicuously. A plant of the high plains and foothills; North Dakota to northern Colorado and northwestward to Canada.

Group III
    Taller plants with erect stems, over 10cm tall; inflorescence mostly glabrous outside; anthers lanate or short-pubescent on the side opposite dehiscence.

1. Anthers lanate-pubescent on side opposite dehiscence.
    2. Sepals with long-attenuate tips; staminode glabrous or bearded.
                                                                P. caryi
    2. Sepals rounded, with short-acute tips.
        3. Corolla 15–21mm long; staminode bearded at tip.
                                                                P. cyanocaulis
        3. Corolla 22–30mm long; staminode usually bearded but may be glabrous.
                                                                P. strictus
                                                                Rocky Mountain
                                                                penstemon
1. Anthers short-pubescent on side opposite dehiscence.
    2. Corolla 15–22mm long.
        3. Stems and leaves pubescent.
                                                                P. fremontii
                                                                Fremont penstemon

3. Stems and leaves glabrous or nearly so.
   4. Plants to 2dm high; all leaves linear to lanceolate.

   > *P. paysoniorum*
   > Payson penstemon

   4. Plants to 8dm high; upper leaves ovate to subcordate.

   > *P. cyananthus*
   > Wasatch penstemon

2. Corollas usually over 22mm long.
   3. Sepals ovate to suborbicular, rounded and ragged at top.
      4. Upper leaves somewhat clasping.

      > *P. cyaneus*

      4. Upper leaves usually not clasping.

      > *P. glaber*

   3. Sepals lanceolate to ovate, acute to acuminate at top.
      4. Corolla glabrous inside.

      > *P. subglaber*

      4. Corolla hairy inside, if sparsely.

      > *P. alpinus*
      > Mountain beardtongue

247. Mountain Beardtongue, ½X

ROCKY MOUNTAIN PENSTEMON, *P. strictus*, has large, dark-blue flowers and is somewhat 1-sided. Stems 20–70cm tall; staminode usually bearded. Widely distributed from southern Wyoming to Arizona but most common on the western side of the Continental Divide.

MOUNTAIN BEARDTONGUE, *P. alpinus* (fig. 247), is another showy species. Its stems are stouter and usually not so tall as *P. strictus*; the stalks often grow in clumps, and the 1-sided inflorescence is rather stubby at apex. The flowers are usually a clear, intense blue. Occurs on gravelly soils, being especially luxuriant on the red granitic gravel of the Pikes Peak region. This is one of the species collected by Edwin James on the first ascent by white men of this famous mountain in July of 1820. Grows from southeastern Wyoming to northern New Mexico, extending up to 11,000 feet along the eastern slope.

WASATCH PENSTEMON, *P. cyananthus*, is a tall, handsome, blue-flowered species found from southwestern Montana through western Wyoming and eastern Idaho into the Wasatch Mountains of Utah. Its anthers have short, stiff hairs.

Group IV

Taller plants with erect stems, over 10cm tall; inflorescence mostly glabrous outside, not glandular; anthers glabrous on the side opposite dehiscence.

1. Staminode usually glabrous.
    2. Plant a woody shrub; corolla 20–25mm long, pink.

                    P. ambiguus
                    Bush or sand penstemon

    2. Plant herbaceous.
        3. Corolla 17–33mm long, blue to purplish.

                    P. virgatus ssp. *asa-grayi*

        3. Corolla 17–20mm long, white to bluish-white.

                    P. lentus

1. Staminode more or less bearded.
    2. Corolla 3.5–5cm long; upper leaves ovate to cordate.

                    P. bradburii

    2. Corolla under 3cm long; upper leaves various.
        3. Stamens conspicuously exserted from corolla; leaves glaucous, the upper ones ovate and somewhat clasping.

                    P. cyanthophorus
                    North Park penstemon

        3. Stamens included or nearly so.
            4. Corollas mostly 6–10mm long; flowers clustered.

                    P. procerus var. *procerus*
                    Clustered or tiny-bloom penstemon

            4. Corollas 10–30mm long.
                5. Leaf blades linear or filiform, mostly a basal tuft, glabrous or nearly so.

                    P. laricifolius ssp. *exilifolius*
                    Nelson larchleaf penstemon

                5. Leaf blades mostly broader than linear, entire.
                    6. Leaf blades narrowly lanceolate to oblanceolate, 1–12cm long, strongly glaucous.

                    P. angustifolius
                    Skyblue penstemon

                  6. Leaves broadly lanceolate to ovate on the stem; corollas various.
                    7. Sepals 2–3.5mm long; corolla 11–17mm long.

                    P. watsonii
                    Watson penstemon

                  7. Sepals 4mm or more long.
                    8. Sepals with ragged-scarious margins.
                      9. Stem leaves broadly ovate to suborbicular, glaucous; calyx 4–6mm long; corolla 15–22mm long.

                    P. pachyphyllus

                    9. Stem leaves lanceolate to ovate, glabrous to occasionally glaucous; flowers clustered.
                    10. Calyx 3–7mm long; corolla 11–16mm long, hairy within; leaf blades 2–12cm long.

                    P. rydbergii
                    Rydberg penstemon

                    10. Calyx 5–7mm long; corolla 14–20mm long, mostly glabrous within; leaf blades acuminate, 4–9cm long.

                    P. osterhoutii
                    Osterhout penstemon

                  8. Sepals with margins entire, often purplish.

9. Corollas 17–27mm long; note racemes 1-sided.
>P. secundiflorus
>One-sided penstemon or sidebells
9. Corollas 10–18mm long, inflorescence not 1-sided.
>10. Lower bracts of inflorescence broadly ovate, about twice as long as wide; calyx 4-7mm long.
>>P. nitidus
>10. Lower bracts of inflorescence lanceolate to narrowly ovate, at least 3 times as long as wide; calyx 3.5–6mm long.
>>P. arenicola
>>Red Desert penstemon

SAND PENSTEMON, *P. ambiguus*, with phlox-like pink and white flowers, occurs on the eastern plains and foothills of our range. CLUSTERED OR TINY-BLOOM PENSTEMON, *P. procerus* (fig. 248), has erect stems 10–30cm tall. Its flowers are small, tubular, dark-blue, slightly 2-lipped, in whorls or separated clusters. Grows in moist meadows of the montane and subalpine zones, even up to 12,000 feet. Occurs from Alaska to south-central Colorado, especially in the eastern part of the range. RYDBERG PENSTEMON, *P. rydbergii*, a similar but larger species, is found farther west.

WATSON PENSTEMON, *P. watsonii*, usually grows in clumps, having several stems 3–6dm tall. There is no basal rosette, and the upper leaves are up to 6.5cm long and larger than the lower ones. It has many blue or purplish flowers; staminode bearded with long yellow hairs. Occurs from southwestern Wyoming through western Colorado and Utah and northwestern Arizona, and in Idaho and Nevada. Grows among sagebrush or other brushy vegetation.

ONE-SIDED PENSTEMON or SIDEBELLS, *P. secundiflorus* (fig. 249), with stems 10–50cm tall, is one of our most charming wildflowers. The leaves clasp the stem and are smooth, thick, and glaucous. The flowers are rose, lilac, or purple, and the slender, definitely 1-sided inflorescence often has its tip slightly curved down,

248. Clustered Penstemon, ½X

249. Sidebells Penstemon, ½X

especially when still partly in bud. The corolla flares into an open throat, showing the staminode strongly-bearded with dark-yellow hairs. Grows on hillsides of the foothill and montane zones from southern Wyoming through central Colorado into New Mexico.

SKYBLUE PENSTEMON, *P. angustifolius*, has rather stout stems 10–25cm tall with glaucous leaves, which have prominent midribs, and cylindrical inflorescences of pale or bright blue flowers. The buds and some of the corollas are often tinged with pink. A plant of the mesas and lower foothills along the eastern slope. Occurs from North Dakota and eastern Montana to New Mexico.

NELSON LARCHLEAF PENSTEMON, *P. laricifolius* ssp. *exilifolius*, is a plant with several slender stems 13–23cm tall, from a cushion of very narrow bright green leaves. The dainty corolla may be white, greenish, or pinkish, the tube abruptly inflated and with spreading lips. Occurs on the dry, rocky plains and hills of north-central Colorado and southeastern Wyoming between 7,000 and 9,000 feet.

(14) *Rhinanthus*. Yellow rattle

One species in our range: YELLOW RATTLE, *Rhinanthus minor*, is 20–60cm tall with narrow leaves; flowers small, yellow, in a leafy-bracted spike. The 4-toothed, papery calyx, which becomes inflated after flowering, makes the plant conspicuous. Grows in meadows of the montane and subalpine zones from Alberta to Arizona.

(15) *Scrophularia*. Figwort

One species in our range: LANCE-LEAF FIGWORT, *Scrophularia lanceolata* (fig. 250), is a weedy plant found on disturbed soil of roadbanks and around buildings. Grows 45–120cm tall, is coarse, and has a narrow inflorescence of small green or greenish-brown flowers. The 2 lobes of its upper lip stand erect. Occurs occasionally throughout our region in the foothill and montane zones.

250. Lance-leaf Figwort, ²/₅X

(16) *Verbascum*. Mullein

1. Plant stout, densely woolly.

V. thapsus
Common mullein

1. Plant slender, glabrate.

V. blattaria
Moth mullein

COMMON MULLEIN, *Verbascum thapsus,* is a tall weed of roadsides, banks, and waste ground. A biennial, during the first season it develops pale-green rosettes, which may become as large as plates. The leaves are covered with tangled, yellowish, wool-like hairs. During the second season, an erect stem develops which may grow 1.8m high, bearing a long spike of yellow saucer-shaped flowers which are only slightly irregular. The corolla is 5-lobed, always folded in bud, so that one lobe is entirely outside the others. Five fertile stamens, the three upper ones having longish hairs on their filaments. MOTH MULLEIN, *V. blattaria,* is smaller with almost smooth leaves and is less frequently seen. Its stamens all have knob-tipped, purple hairs along the filaments. Both species have been introduced from Europe.

(17) *Veronica*. Veronica or speedwell

Our species of this genus are small plants that grow in wet or moist places. Flowers blue or rarely whitish, in terminal or axillary racemes. The corollas are rotate, 4-lobed; and there are only 2 stamens.

1. Small annual.

V. peregrina var. *xalapensis*
Purslane speedwell

1. Plants perennial.
　　2. Flowers in terminal spikes.
　　　　3. Capsule oblong, notched at the summit.

V. wormskjoldii
Alpine veronica

　　　　3. Capsule orbicular, obcordate.

V. serpyllifolia var. *humifusa*
Thyme-leaved speedwell

　　2. Flowers racemose, axillary.
　　　　3. Leaves oblong to oval.
　　　　　　4. Stem leaves sessile.

V. anagallis-aquatica
Water pimpernel

　　　　　　4. All the leaves petiolate.

V. americana
American brooklime

　　　　3. Leaves linear or nearly so.

V. scutellata
Marsh speedwell

ALPINE VERONICA, *V. wormskjoldii* (fig. 251), is a slender, erect plant, usually about 10–20cm tall; terminal raceme of dark-blue flowers; leaves opposite, sessile, ovate to oblong. The flattened capsule is slightly notched at the top. Occurs in moist tundra among grasses and sedges, and in wet mountain meadows of northern North America from the Arctic south to New Mexico and California. Found in the Rockies up to 12,500 feet.

AMERICAN SPEEDWELL, *V. americana*, is a trailing plant of wet locations with opposite, ovate or lanceolate leaves and axillary racemes of small blue flowers. It grows in shallow water or mud around springs and along streams. Widely distributed in North America. THYME-LEAVED SPEEDWELL, *V. serpyllifolia* var. *humifusa*, is a creeping plant with stems to 3dm long, which root at the lower axils and end in racemes of blue or white flowers. Widely distributed in moist soil.

251. Alpine Veronica, $^2/_5$ X

BROOMRAPE FAMILY: OROBANCHACEAE

Only one genus in our range. Parasitic herbs without green coloring, they have flowers similar to those in the figwort family. The stems are fleshy, finely hairy, and pinkish-brown.

1. Flowers spicate, sessile or subsessile, bracteate.
    2. Corolla 2cm or more long; anthers woolly.

    Orobanche multiflora

    2. Corolla less than 2cm long; anthers glabrous.

    O. ludoviciana

1. Flowers pendunculate and bractless.
    2. Flowers solitary or few.

    O. uniflora

    2. Flowers several to many.

    O. fasciculata
    Tufted broomrape

TUFTED BROOMRAPE, *Orobanche fasciculata* (fig. 252), is the species most frequently seen. Flowers dull yellow or purplish on thick straight stalks standing erect from the fleshy stem. Occurs throughout our range. Parasitic especially on species of *Eriogonum* and *Artemisia*, but not exclusively.

BLADDERWORT FAMILY: LENTIBULARIACEAE

Members of this family grow in water or in very wet situations. Most of them are insectivorous. Corollas irregular and usually spurred.

1. Calyx 5-lobed, forming an upper lip of 3 and a lower lip of 2; flowers solitary, bractless; leaves entire, basal; terrestrial herbs in wet places.

    Pinguicula vulgaris
    Butterwort

1. Calyx 2-lobed; flowers solitary or racemose, each subtended by a bract; leaves dissected; submerged aquatic plants with emergent peduncles.
    2. Leaves pinnately divided; corolla broad (12mm), with permanent spur; bladders 3–5mm long.

Utricularia vulgaris
Common bladderwort

    2. Leaves dichotomously divided; corolla half as broad; bladders few, often none, not over 2mm long.

U. minor
Lesser bladderwort

basal leaf

252. Tufted
Broomrape, ½X

253. Harebell, ½X

COMMON BLADDER-WORT, *Utricularia vulgaris*, is a floating plant with finely divided leaves. There are little inflated bladders on the leaf segments, which act as floats and also trap insects. Yellow flowers about 12mm long. Seen occasionally in ponds of our region.

BUTTERWORT, *Pinguicula vulgaris,* grows on mud or wet rocks. It has a basal rosette of yellowish-green leaves; leaves have a grease-like, sticky substance on their surfaces that traps and digests insects. Violet-like flowers on leafless stalks 5–10cm tall. A northern plant, seen in Glacier National Park.

BELLFLOWER FAMILY: CAMPANULACEAE
    Members of this family have united 5-lobed corollas and inferior ovaries. The native species have blue flowers. *Lobelia* and *Porterella*, sometimes segregated and put in the Lobeliaceae, are included here.

1. Filaments and anthers distinct; corolla regular.
    2. Perennials; corolla campanulate; flowers pedicellate or pedunculate.
Campanula, p. 309
Bellflower

2. Annuals; corollas various; flowers sessile or subsessile.
 3. Corolla campanulate, shallowly lobed.

Heterocodon, p. 310
Heterocodon

 3. Corolla rotate, deeply lobed (to below the middle).

Triodanis, p. 310
Venus's looking-glass

1. Filaments and anthers united in a tube, 2 of the anthers smaller than the others; corolla irregular; flowers pedicellate.
 2. Perennial; corolla-tube cleft at least halfway to base on dorsal side; flowers mostly in bracteate racemes.

Lobelia, p. 310
Lobelia

 2. Annual; corolla-tube not deeply cleft dorsally; flowers axillary.

Porterella, p. 310
Porterella

(1) *Campanula.* Bellflower

1. Tall, coarse plant; basal and stem leaves coarsely serrate; rhizomatous; flowers in long raceme, purple; an escaped cultivar.

C. rapunculoides
Creeping bellflower

1. Stems smooth, not very leafy; flowers 1 to several.
 2. Anthers 1–2.5mm long; corolla deeply and narrowly lobed; flowers mostly solitary; usually alpine.

C. uniflora
Alpine harebell

 2. Anthers 4–6.5mm long; corolla shallowly and broadly lobed; leaves linear, the basal leaves sometimes broader; foothills to alpine.
  3. Plants usually several-flowered; lower leaves with scabrous or entire margins at base.

C. rotundifolia
Harebell

  3. Plants often 1-flowered; lower leaves ciliate at base.

C. parryi var. *parryi*
Parry harebell

HAREBELL, *Campanula rotundifolia* (fig. 253), with its truly bell-like, lavender-blue flowers hanging from slender stems, is frequently seen from the foothills to the alpine zone throughout the Rockies. Despite its specific name, which means *round leaf,* its leaves are seldom round. The basal leaves are usually somewhat cordate, and the stem leaves are linear or linear-lanceolate. From 30cm at lower altitudes to 5–8cm at timberline or above. This plant is widely distributed in the northern countries and is the bluebell of Scotland.

254. Alpine Harebell, ½X

ALPINE HAREBELL, *C. uniflora* (fig. 254), is a diminutive relative with slender stems 10cm tall and a corolla under 12mm long, the lobes slender and pointed. Grows in the dense grass and sedge sod of the tundra. Widely distributed in arctic and alpine regions.

PARRY HAREBEL, *C. parryi* (fig. 255), has slender, usually 1-flowered stems; corollas erect, violet-colored or purple, and open funnel-form. A plant of moist meadows and edges of aspen groves in the montane and subalpine zones from Wyoming and Utah to New Mexico and Arizona.

(2) *Heterocodon.* Heterocodon

One species occurs in our range: *Heterocodon rariflorum.* Annual to 3dm tall; leaves sessile, ovate to cordate, toothed or lobed; corolla blue.

(3) *Lobelia.* Lobelia

1. Slender spike of scarlet flowers; corolla irregular, 2.5–4cm long.
   L. cardinalis
   Cardinal flower
1. Large spike of blue flowers, striped with white; plant 30–100cm tall; corolla irregular.
   L. siphilitica
   Great lobelia

CARDINAL FLOWER, *Lobelia cardinalis,* is found in our range only in New Mexico and Arizona, in moist situations up to 7,000 feet. Occurs occasionally in southeastern Colorado, but below 5,000 feet. Easily recognized by its brilliant red, irregular corolla, 2.5–4cm long.

(4) *Porterella.* Porterella

One species occurs in our range: *Porterella carnosula.* Annual; glabrous; fibrous-rooted; flowers pedicellate, solitary in axils, blue; corolla irregular, 6–16mm long; in wet areas, up to 3dm tall.

(5) *Triodanis.* Venus's looking-glass

1. Leaves linear to lanceolate, not clasping the stem; blue-violet.
   T. leptocarpa
1. Leaves broader, ovate or orbicular to cordate.
   2. Leaves and flower bracts ovate-cordate, clasping; blue or purple.
      T. perfoliata var. *perfoliata*
      Venus's looking-glass
   2. Leaves and flower bracts mostly ovate or deltoid, usually not clasping; blue or purple.
      T. holzingeri

VENUS'S LOOKING-GLASS, *Triodanis perfoliata* (fig. 256), is an annual plant with stems that are usually erect and unbranched, and with alternate, cordate-clasping leaves and bracts, the leaves merging into the bracts, the blue or purplish flowers in their axils. Occurs occasionally on dry slopes of the foothills and lower montane zone throughout our range.

MADDER FAMILY: RUBIACEAE

Plants of this family have 4-angled stems with opposite or whorled leaves; small flowers with regular, 4-lobed corollas; and a 2-celled inferior ovary that becomes a 2-lobed fruit. Only one genus of the family is likely to be seen in our area, and most of the species are inconspicuous.

1. Annuals; fruit hispid or hirsute.
    2. Stems rough on the angles, reclining; leaves 6–8 in each whorl.
            Galium aparine
    2. stems smooth, erect; leaves mostly in fours.
            G. bifolium
1. Perennials; fruit various
    2. Stems entirely herbaceous; flowers perfect.
        3. Leaves 3-nerved; fruit canescent, becoming glabrous.
            G. boreale
            Northern bedstraw
        3. Leaves 1-nerved; fruit smooth or hispid.

255. Parry Harebell, ½X

256. Venus' Looking-
glass, ½X

4. Leaves cuspidate.

G. triflorum
Fragrant bedstraw

4. Leaves not cuspidate.
    5. Fruit hispid with hooked hairs.
        6. Leaves 4 in a whorl.

G. trifidum var. *trifidum*

        6. Leaves 4–6 in a whorl, at least some with 5.

G. trifidum var. *pacificum*

    5. Fruit only granulate-scabrous.

G. asperrimum

2. Stems somewhat woody at base; flowers dioecious.

G. coloradoense

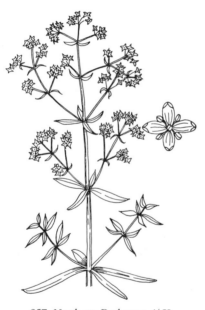

257. Northern Bedstraw, ½X

NORTHERN BEDSTRAW, *Galium boreale* (fig. 257), is an erect plant 20–60cm tall with 4 narrow, 3-nerved leaves at each node and masses of tiny, white, fragrant flowers. Grows throughout the area from foothills to timberline. Expect a considerable variation in appearance between plants growing in sunny, dry situations and those in moist, shaded places. Shaded plants are larger, greener, and less compact. FRAGRANT BEDSTRAW, *G. triflorum*, has longer stems, which are inclined to lean on the surrounding vegetation. Its 1-nerved leaves are in whorls of 5 or 6, and its few flowers are greenish or purplish. In this species, it is the foliage that is fragrant. Grows in moist, shaded locations of the montane zone throughout the Rockies.

HONEYSUCKLE FAMILY: CAPRIFOLIACEAE

Most of the plants of this family are woody. Our species are shrubs or vines. All have opposite leaves, and some have paired flowers. Corollas are united, with 5 stamens inserted in the corolla-tube; ovaries inferior. The fruits are fleshy and berry-like in all but one of our genera.

1. Corolla short, rotate or urn-shaped; stigma subsessile; fruit drupaceous.
    2. Leaves pinnately compound.

Sambucus, p. 314
Elderberry

    2. Leaves simple.

Viburnum, p. 316
Viburnum

1. Corolla tubular or campanulate; style elongated; fruit dry or berry-like.
   2. Herb, slightly woody; twin flowers.

   Linnaea, p. 313
   Twinflower

   2. Shrubs.
      3. Corolla regular; fruit 2-seeded.

      Symphoricarpos, p. 315
      Snowberry
      or buckbrush

      3. Corolla irregular; fruit few- or several-seeded.

      Lonicera, p. 313
      Honeysuckle

(1) *Linnaea.* Twinflower

One species in our range: TWINFLOWER, *Linnaea borealis* (fig. 258), a low, trailing vine, slightly woody, with paired evergreen leaves. The flower stalks are very slender, 5–10cm tall, from the leaf axils. Flowers in pairs, pinkish, funnelform. Grows in moist, shaded situations, usually under coniferous trees. Occurs in the upper montane and subalpine zones throughout our range, and in favorable locations across the northern United States and Europe. This was the favorite flower of the Swedish botanist Carolus Linnaeus, and is named in his honor.

258. Twinflower, ½ X

(2) *Lonicera.* Honeysuckle

1. Vines or climbing shrubs; upper leaves mostly connate-perfoliate; flowers in capitate clusters.
   2. Corolla orange-yellow to orange-red, slightly bilabiate.

   L. ciliosa
   Orange honeysuckle

   2. Corolla seldom orange, strongly bilabiate.

   L. dioica var.
   glaucencens
1. Erect shrubs; flowers in axillary pairs; terminal leaves not perfoliate.
   2. Bracts broad, foliaceous.

   L. involucrata var.
   involucrata
   Involucred honeysuckle
   or twinberry

   2. Bracts narrow or absent.
      3. Leaves pale; fruit blue-black.

      L. caerulea

      3. Leaves green; fruit red

      L. utahensis
      Utah honeysuckle

TWINBERRY or INVOLUCRED HONEYSUCKLE, *Lonicera involucrata* (fig. 259), is an erect, leafy shrub up to 2.7m tall. Leaves opposite, 5–15cm long, obovate to

259. Twinberry Honeysuckle, ²/₅X

260. Red Elderberry, ²/₅X

oval or ovate-oblong, pointed at tip. Flowers in pairs on a common stalk in the upper leaf axils; tubular, light yellow, enclosed in an involucre of leafy bracts, which enlarges and turns red as the twin berries ripen and become glossy black. Commonly seen along mountain streams and on moist slopes of the sub-alpine zone throughout our range. UTAH HONEYSUCKLE, *L. uta-hensis*, is a smaller shrub up to 2m in height. Leaves pale, glaucous beneath, 2.5–8cm long. Lacks the conspicuous involucres of the last, but has a handsome, bright red, 2-lobed berry.

ORANGE or ARIZONA HONEY-SUCKLE, *L. ciliosa*, is a clambering plant with clustered, red or orange flowers. The upper pair of leaves is joined around the stem like a collar. Found in the montane zone of Utah, Arizona, and New Mexico.

(3) *Sambucus*. Elderberry

In our region, these are stout shrubs with large, pinnately compound, opposite leaves. The clusters of small white flowers are followed by many small berries.

1. Cyme flat-topped, broadly rounded.
    2. Stem tree-like, long enduring, not stoloniferous; fruit glaucous.
            S. cerulea
            Blue elderberry
    2. Stem short-lived, only slightly woody, stoloniferous; fruit not glaucous.
            S. canadensis
            Elderberry
1. Cyme not flat-topped, pyramidal or oval.
    2. Fruit black or purple-black; leaflets somewhat hairy below.
            S. racemosa var.
            melanocarpa
            Black elderberry

2. Fruit red to yellowish.
   3. Leaflets glabrous or nearly so.

S. racemosa var.
microbotrys
Red elderberry

   3. Leaflets very hairy below.

S. racemosa var. *pubens*
Red elderberry

BLUE ELDERBERRY, *S. cerulea*, has large, flat-topped clusters of small, white flowers and bluish berries. At lower altitudes it becomes a small tree. Grows in the foothill and montane zones. BLACK ELDERBERRY, *S. racemosa* var. *melanocarpa*, grows along stream banks and on moist slopes of the montane and lower subalpine areas. RED ELDERBERRY, *S. racemosa* var. *pubens* (fig. 260), is found on subalpine slopes, widely distributed throughout our region.

(4) *Symphoricarpos*. Snowberry or buckbrush

1. Corolla short-campanulate, 3–4mm long.
   2. Stamens and style exserted.

S. occidentalis
Wolfberry

   2. Stamens and style included.
      3. Leaves and new twigs pubescent.

S. albus var. *albus*
Snowberry or buckbrush

      3. Leaves and new twigs mostly glabrous.

S. albus var. *laevigatus*

1. Corolla long-campanulate to tubular-funnelform, 6–12mm long.
   2. Leaves finely-pubescent or tomentulose.

S. rotundifolius
Snowberry

   2. Leaves glabrous or nearly so.

S. oreophilus var.
utahensis
Mountain snowberry

MOUNTAIN SNOWBERRY, *Symphoricarpos oreophilus* (fig. 261), is a much-branched shrub 3–9dm tall, with shredding bark and roundish to ovate leaves. Small, tubular or funnel-shaped, pink flowers are followed by white berries. Grows on slopes and in ravines of the montane zone from the Canadian Rockies south through Colorado. BUCK-BRUSH or SNOWBERRY, *S. albus* (fig. 262), is similar, but its flowers are bell-shaped. Found in valleys and on hillsides of the montane and lower subalpine zones from Montana through Colorado.

261. Mountain Snowberry, ½X

262. Buckbrush or Snowberry, ½ X

WOLFBERRY, *S. occidentalis*, is an erect shrub from 45dm to 1.5m tall, with oval leaves, sometimes wavy-margined or lobed, 2.5–10cm long. Short spikes of pinkish flowers in the upper leaf axils followed by clusters of pale-greenish berries. Grows in ravines and along sandy stream banks in the foothills and lower montane zone throughout our range.

(5) *Viburnum*. Viburnum

1. Leaves 3-lobed, palmately-veined; fruit red.
   2. Cyme with the exterior flowers ray-like.

      V. opulus
      Snowball tree

   2. Cyme without radiant flowers.

      V. edule
      High-bush cranberry

1. Leaves not lobed, pinnately-veined; fruit blue or black.

      V. lentago
      Nannyberry

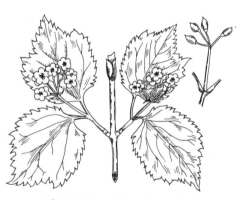

263. Highbrush Cranberry, ⅖ X

HIGH-BUSH CRANBERRY, *Viburnum edule* (fig. 263), is a shrub 9–18dm tall; leaves opposite, toothed, some lobed like maple leaves. Small clusters of white flowers, followed by red, juicy, edible, berry-like drupes containing flat stones. Widely distributed in Canada and the northern United States, where the fruits are used for jelly. Extends south in the montane and subalpine zones of the Rockies as far as central Colorado.

MOSCHATEL FAMILY: ADOXACEAE

A family containing only one genus and one species: MUSK-ROOT, *Adoxa moschatellina* (fig. 264), which is a delicate, smooth herb. Long-petioled basal leaves divided into 3-lobed segments; a pair of 3-parted leaves on the flower stalk. Flowers few, small, greenish-white or yellowish. The top flower has a 4–5-lobed corolla, and the others are 5–6-lobed; stamens twice as many as the lobes of the corolla, inserted in pairs on its tube. Grows in moist soil of the subalpine zone of the Rocky Mountains, but seldom seen.

VALERIAN FAMILY: VALERIANACEAE

Only one genus is represented in our region. Its flowers are small, with calyx lobes modified into plumose bristles, which unroll as the seeds ripen and act as parachutes to aid their dispersal.

1. Plant with stout taproot, short-branched caudex; leaves tapering to a petiolar base, veins nearly parallel to the midrib, mostly entire.

Valeriana edulis
Tall valerian

1. Plants with a stout rhizome or caudex, numerous fibrous roots; leaves with definite petiole, veins pinnate, mostly compound.
    2. Corolla small, 2–4mm; some flowers perfect, others chiefly staminate.
        3. Small plants, 1–4dm tall; not very leafy, the lateral lobes of the cauline leaves less than 1cm wide.

V. dioica

        3. Larger plants, 3–9dm tall; leafy, the lateral lobes of the cauline leaves often more than 1cm wide.

V. occidentalis
Western valerian

    2. Corolla larger, 4–9mm; flowers mostly perfect; stamens generally longer than corolla lobes; basal leaves generally without lobes.

V. acutiloba
Subalpine valerian

√ TALL OR EDIBLE VALERIAN, *V. edulis*, is usually 3–6dm tall. Inflorescence very open and branched. The thick, smooth, pale-green, basal leaves have several veins parallel to the midrib, and usually some of them have a few lobes. Grows from a stout taproot, which has a distinctive odor. These roots were cooked and used for food by the Indians of the Northwest. They are said to be poisonous unless cooked. The plant grows on moist meadows, ravines, and open hillsides of the montane and subalpine zones throughout the Rockies.

SUBALPINE VALERIAN, *V. acutiloba*, is a smooth plant with stems 15–45cm tall and clusters of tiny, pink or whitish flowers. The clusters are at first compact but become more open as they develop.

264. Musk-root, ½ X

Leaves in 2–4 pairs on the stem, and they usually have 2 or more thumb-like lobes. A plant of moist, shaded ground, often found in aspen groves and edges of spruce forest, sometimes above timberline, from Wyoming to New Mexico and Arizona. WESTERN VALERIAN, *V. occidentalis* (fig. 265), is similar, but the flowers are even smaller and the leaves more lobed. Found in the foothill and montane zones from Montana to Arizona. It is sometimes seen in early spring on moist foothill slopes beginning to bloom with a short-stemmed cluster of tiny, pink buds centered in the basal rosette of pinnately-lobed leaves.

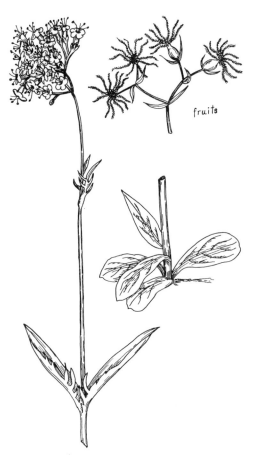

fruits

265. Western Valerian, ½X

COMPOSITE FAMILY: COMPOSITAE

The composites make up one of the largest of the seed-plant families. Members vary greatly, but all have an inflorescence arranged according to a similar pattern. Their structure provides for great efficiency in reproduction and distribution. Species of this family are considered difficult to identify, but they are not impossible if one is willing to devote some time to learning the fundamental characteristics of the family and its genera. A good 10x hand lens is necessary, as many of the individual flowers are very small. They are called *composites* because their inflorescences have become crowded into compact heads that simulate individual blossoms. Sunflowers, thistles, and dandelions are well-known examples.

If you cut open a dandelion or sunflower head, you will find that it is made up of numerous separate flowers. In the dandelion all the flowers are alike. Each has a strap-shaped corolla formed into a short tube at the base, which surrounds a ring of 5 stamens. The ovary is inferior, 1-seeded; very slender style with 2 stigmas. Attached to the top of the ovary (and surrounding the base of the corolla) is a tuft of white hairs. This represents the calyx and is called the *pappus*. In the case of the sunflower, you will find that the head is made up of two kinds of flowers. Around the outside are flowers similar to those of the dandelion: long, flat corollas called *ray* flowers. In the center of the head are numerous flowers with tubular, 5-toothed corollas, called *disk* flowers. In the sunflower, the pappus consists of scales or awns.

In both cases, the flowers are all attached to the enlarged top of a stalk, and this structure is called the *receptacle*. Originating at the edge of the receptacle, and surrounding the cluster of flowers, is a series of green bracts, the *phyllaries*. The receptacle and the phyllaries constitute the *involucre*, and this involucre simulates the calyx. In the dandelion head, all the flowers are perfect; each produces a seed-like fruit called an *achene*. In the sunflower, the ray flowers are pistillate but sterile, and the disk flowers are perfect, each producing an achene.

Many variations occur in different members of the family. In some species the ray flowers produce achenes, while the disk flowers are sterile. In some species, only the disk flowers are present. The pappus may consist of soft hairs, of stiff bristles, or of scales. In many species, the pappus serves to insure wide distribution of the seeds. Those achenes that have a pappus of soft hairs or bristles are carried long distances by the wind; those with scales are often blown along the ground; those with barbed bristles hook into the hair of animals for transportation.

The receptacle may be flat, convex, or concave; sometimes it is elevated into a cone or a column; sometimes it is pitted; and often it bears scales or bristles between the flowers. Often the anthers cohere in a ring around the styles, which are usually 2-branched. At one stage, the stigmas are elevated to protrude above the ring of anthers; at another stage, the anther ring is pushed up. The phyllaries are usually separate, but they may be joined together; they may be in one or more rows; often they overlap symmetrically, like shingles on a roof, and are said to be *imbricated*; sometimes they come in colors other than green; sometimes they are thin and papery. Such numerous possible variations account for the difficulties in identification.

Some of the composites bloom in early spring and summer, but a majority of the composite species are late-blooming. From mid-August until frost, the wildflower show is dominated by plants of this family.

The family is represented by over 500 species in the Rocky Mountain region. Because of the efficiency of its distributional devices, the family includes a large number of weeds. Many of the species are abundant in our flora yet little noticed because their flowers are inconspicuous. Not all the species can be included in a volume of this scope. In many cases, the Latin generic names have been applied as common names, because the common names in this group are particularly indefinite and confused. Consequently, only generic names are used in the key. For convenience, the family may be divided into 3 groups. Group 1: genera having both ray and disk flowers. Group 2: genera usually having only disk flowers. Group 3: genera having only strap-shaped (ligulate) corollas.

Group 1.
    Heads usually with both disk and ray flowers.
1. Ray flowers white, pinkish, lavender, or purple (never yellow); disk flowers yellow (except white in *Achillea*).
    2. Leaves pinnately- or palmately-lobed, or finely divided.
        3. Leaves pinnately-lobed or 1–4 times ternately or palmately-lobed.
            4. Plants tall, 3–6dm high; rays white, about 2.5cm long.
                                                    Chrysanthemum, p. 338
            4. Plants 1–2.5dm high; rays white or purple, 8–15mm long.
                                                    Erigeron, p. 344
        3. Leaves deeply and finely divided.
            4. Heads about 1cm broad.
                                                    Achillea, p. 322

4. Heads about 2.5cm broad.
Haplopappus, p. 353
2. Leaves entire or wavy-margined.
3. Ray flowers conspicuous, several or many.
4. Plants usually branched, with few to many heads.
5. Phyllaries (bracts) flat, tips not recurved; plants perennial.
Aster, p. 332
5. Phyllaries with recurved tips; plants usually annual or biennial.
Machaeranthera, p. 364
4. Plants caespitose or taller; stems little-branched, with single or few heads.
5. Phyllaries many, imbricated.
Townsendia, p. 378
5. Phyllaries fewer, usually in 1 series.
Erigeron, p. 344
3. Ray flowers small, inconspicuous.
4. Basal leaves large, triangular, white beneath.
Petasites, p. 366
4. Lower cauline leaves spatulate or linear-spatulate, not white beneath.
Erigeron, p. 344
1. Ray flowers yellow or orange.
2. Leaves (at least the lower ones) opposite.
3. Plants smooth; leaves divided into thread-like divisions.
Thelesperma, p. 378
3. Plants hairy; leaves not divided.
4. Pappus conspicuous, of white or tawny bristles.
Arnica, p. 326
4. Pappus none, or of scales or awns.
5. Pappus none; foliage with very fine hairs.
Viguiera, p. 381
5. Pappus of scales or awns; foliage often very rough.
6. Achenes flattened with thin edges.
Helianthella, p. 356
6. Achenes not flattened.
Helianthus, p. 357
2. Leaves either all basal or alternate on stem (not opposite).
3. Plants with leaves all, or mostly, basal.
4. Leaves 10–30cm long.
5. Leaves green (ovate, oblong, or lanceolate).
Wyethia, p. 381
5. Leaves gray (arrowhead-shaped).
Balsamorhiza, p. 336
4. Leaves 2.5–10cm long; silvery or white-woolly.
Hymenoxys, p. 361
3. Plants with some leaves along the stems (alternate).
4. Disk flowers dark-colored.
5. Disk flowers dark red.
Gaillardia, p. 350
5. Disk flowers black, brown, or greenish
6. Receptacle elongate (cylindrical or columnar).
Ratibida, p. 366
6. Receptacle conical.
Rudbeckia, p. 367
4. Disk flowers yellow or orange.
5. Heads 12mm or more broad.
6. Involucres rotate, or having 2–3 series of narrow phyllaries.
Helenium, p. 356
6. Involucres campanulate, hemispheric, or cylindrical.
7. Phyllaries tightly appressed.
8. Involucre hemispheric.
Bahia, p. 336
8. Involucre cylindrical.
9. Plants more or less rough-hairy.
Heterotheca, p. 359

9. Plants smooth or more or less cottony or woolly.
Senecio, p. 367
7. Phyllaries many, loose, and leafy; or narrow and bristle-tipped or with tips strongly recurved.
8. Phyllaries with strongly recurved tips.
Grindelia, p. 352
8. Phyllaries loose and leafy or bristle-tipped.
Haplopappus, p. 353
5. Heads less than 12mm broad , numerous.
6. Plants bushy with very narrow leaves.
Gutierrezia, p. 353
6. Stems usually unbranched; inflorescence often 1-sided and arched.
Solidago, p. 373

Group 2. Heads with disk flowers only (or rarely with a few inconspicuous ray flowers).
1. Phyllaries scarious or satiny; variously-colored heads.
2. Phyllaries satiny; heads white, pink, or brownish; with fibrous roots.
Gnaphalium, p. 351
2. Phyllaries papery; heads white, pink, or brownish; with fibrous roots.
3. Basal leaves usually forming conspicuous tufts; stems seldom leafy; plant dioecious; 5–25cm tall, or if taller, of montane woods.
Antennaria, p. 324
3. Basal leaves soon deciduous; stem leaves well-developed and numerous; pistillate heads with a few staminate flowers; largely subalpine; 25cm or taller.
Anaphalis, p. 323
1. Phyllaries green, gray, yellowish, or purplish, but not scarious.
2. Plants with spiny foliage; stems often spiny.
3. Pappus bristles barbellate (with short barbs).
Carduus, p. 338
3. Pappus bristles plumose; but pappus of outer flowers sometimes barbellate.
Cirsium, p. 341
2. Plants without spines on foliage.
3. Flowers bright yellow.
4. Plants woody, at least at base.
5. Phyllaries and flowers only 4 in each head.
Tetradymia, p. 377
5. Phyllaries and flowers more than 4 in each head.
Chrysothamnus, p. 339
4. Plants not woody.
5. Leaves opposite.
6. Leaves triangular or cordate, smooth, the tips very attenuate.
Pericome, p. 366
6. Leaves lanceolate, glandular-hairy.
Arnica, p. 326
5. Leaves alternate.
6. Phyllaries in 1 even row (often a few shorter ones at base of involucre).
Senecio, p. 367
6. Phyllaries in 2 or more overlapping rows.
Erigeron, p. 344
3. Flowers not bright yellow.
4. Leaves opposite or in whorls.
5. Leaves opposite; heads cream-colored, nodding.
6. Pappus plumose.
Kuhnia, p. 362
6. Pappus smooth or barbellate.
Brickellia, p. 337
5. Leaves whorled; heads whitish or purple.
Eupatorium, p. 350
4. Leaves alternate or all basal.
5. Leaves all basal.
Petasites, p. 366

5. Leaves alternate.
   6. Numerous, small, grayish or yellowish heads; foliage aromatic.
                Artemisia, p. 328
   6. Heads more than 12mm high.
      7. Leaves entire; heads bright rose-purple.
                  Liatris, p. 363
      7. Leaves pinnately divided; heads whitish or pinkish.
         8. Phyllaries green; pappus scales 4–10.
                  Chaenactis, p. 338
         8. Phyllaries with papery margins; pappus scales 10–20.
                  Hymenopappus p. 360

Group 3. Flowers are ligulate (with strap-shaped corollas); sap milky.
1. Leaves strictly basal; stalks unbranched, with 1 head to each stalk.
  2. Leaves with white tomentose margins.
                Nothocalais, p. 365
  2. Leaves without white tomentose margins.
    3. Phyllaries imbricated in several rows.
                Agoseris, p. 323
    3. Phyllaries in 1 main row (often a few short ones at base of involucre); stalks
    hollow.
                Taraxacum, p. 376
1. At least a few leaves or bracts on the stalks.
  2. Leaves mostly at or near base, or upper leaves much smaller; stalks branched.
    3. Flowers rose, pinkish, or blue.
      4. Flowers bright blue.
                Cichorium, p. 340
      4. Flowers rose or pinkish.
        5. Pappus bristles plumose.
                Stephanomeria, p. 376
        5. Pappus bristles simple.
                Lycodesmia, p. 364
    3. Flowers yellow.
      4. Pappus of small scales, each tipped with a feathery bristle.
                Microseris, p. 365
      4. Pappus not of bristle-tipped scales.
        5. Pappus double: outer ring of scales, the inner ring of long bristles.
                Krigia, p. 362
        5. Pappus single: long, soft bristles.
         6. Pappus bristles usually white.
                Crepis, p. 343
         6. Pappus bristles usually tan.
                Hieracium, p. 359
  2. Stalks equally leafy to the inflorescence.
    3. Leaves long, grass-like; fruiting heads spherical, 5–8cm in diameter.
                Tragopogon, p. 380
    3. Leaves not grass-like; fruiting heads smaller.
      4. Flowers rose; inflorescence a raceme.
                Prenanthes, p. 366
      4. Flowers blue or dark purplish; inflorescence an open panicle.
                Lactuca, p. 363

(1) *Achillea.* Yarrow

1. Plant 3–10dm; margins of involucral bracts pale to brownish; lower to middle elevations.
                A. millefolium var.
                lanulosa.
1. Plant 1–3dm tall; margins of involucral bracts dark brown to blackish; higher elevations.
                A. millefolium var.
                alpicola.

YARROW, *Achillea millefolium* var. *lanulosa* (fig. 266), is a strongly aromatic plant with grayish leaves finely divided into numerous, extremely narrow segments; and flat-topped clusters of numerous small, white heads. Widely distributed throughout our area.

(2) *Agoseris*. False dandelion

1. Plant an annual.
   A. heterophylla
1. Plant a taprooted perennial.
   2 Flowers yellow.
      3. Plants glabrous or sparsely ciliate on the lower part of leaves and petioles.
         A. glauca var.*glauca*
      3. Plants somewhat pubescent, especially on and below the involucre.
         4. Leaves oblanceolate or broader, usually obtuse, entire to weakly
            laciniate below.
               A. glauca var.
               dasycephala
         4. Leaves more lanceolate, acute or acuminate, mostly laciniate.
            5. Plants usually over 25cm tall; lower part of leaves and petioles
               glabrous or nearly so.
                  A. glauca var.. *agrestis*
            5. Plants usually under 25cm tall; low part of leaves and petioles
               arachnoid.
                  A. glauca var. *laciniata*
   2. Flowers burnt orange.
      3. Involucral bracts narrow and pointed, scarcely imbricated or not at all.
         A. aurantiaca var.
         aurantiaca
      3. Involucral bracts broader and blunt, noticeably imbricated, mottled with purple.
         A. aurantiaca var.
         purpurea

PALE AGOSERIS or FALSE DANDELION, *Agoseris glauca* (fig. 267). This dandelion-like plant may be recognized by its pointed, dark-colored phyllaries, which have conspicuous light borders, and its pale-green leaves. Usually grows on drier ground than the common dandelion; often seen among grasses on open slopes from the foothills to the alpine zone throughout the range. BURNT-ORANGE AGOSERIS, *A. aurantiaca*, is similar but often smaller, and its flowers are a deep orange. It has the same range and extends into the alpine zone.

266. Yarrow, ½ X

(3) *Anaphalis*. Pearly everlasting.

One species recognized in North America; PEARLY EVERLASTING, *Anaphalis margaritacea* (fig. 268), a rhizomatous plant. The little round heads appear entirely white because of their white, papery, imbricated phyllaries. Foliage is whitish or gray from a coating of cobwebby hairs. Frequently seen in masses along trails and roads, and on rocky slopes of the upper montane and subalpine zones throughout the Rockies.

268. Pearly Everlasting, ½X

267. Pale Agoseris, ½X   269. Mountain Pussytoes, ½X

(4) *Antennaria*. Pussytoes

1. Plants mat-forming, with numerous stolons, usually leafy.
   2. Plants low, mostly under 5cm tall; heads among the leaves, usually solitary.
      3. Leaves 5–10cm long; stems 1cm long or less.

A. rosulata

      3. Leaves over 10cm long, stems often over 1cm long.
         4. Heads 3–8; stolons leafy.

A. parvifolia
Mountain or
Nuttall pussytoes

         4. Heads solitary; stolons naked.

A. flagellaris

   2. Plants over 5cm tall, heads well above the basal leaves.

3. Basal leaves clearly less pubescent above than below, becoming green
   above and white tomentose beneath.
   4. Heads in an open, elongate inflorescence, with long peduncles.
      A. racemosa
   4. Heads in a crowded cyme.
     5. Involucral bracts whitish near apex, brown in lower half; upper
       stem leaves with a strap-like scarious tip.
      A. neglecta
     5. Involucral bracts generally greenish in lower half; upper stem
       leaves with a hair-like scarious tip, or not scarious.
      A. neodioica
3. Basal leaves similar on both sides, pubescent, silver or gray.
   4. Terminal, scarious part of outer involucral bracts brownish to
     blackish-green.
     5. Terminal part of involucral bract entirely blackish-green and often
      sharply pointed.
      A. media
      Alpine pussytoes
     5. Terminal part of involucral bract becoming white, or the whole
      scarious part remaining brownish; blunt at tip.
      A. umbrinella
      Umber pussytoes
   4. Terminal, scarious part of involucral bracts white or pink, sometimes
     with a dark spot at base.
     5. Involucral bracts with conspicuous dark spot at base of scarious part.
      A. corymbosa
      Meadow or flat-topped
      pussytoes
     5. Involucral bracts without dark spot at base of scarious part.
      6. Involucres mostly 7–11mm long; corollas mostly 5–8mm long.
       A. parvifolia
       Mountain or
       Nuttall pussytoes
      6. Involucres mostly 3–7mm long; corollas mostly 2.5–4.5mm long.
       A. microphylla
       Pink or meadow
       pussytoes
1. Plants usually not mat-forming; without stolons, but sometimes caespitose.
  2. Plants low, under 5cm tall; heads solitary, barely exceeding basal leaves, caespitose.
    A. dimorpha
    Low pussytoes
  2. Plants over 5cm tall; heads several to many, clearly exceeding the basal leaves.
   3. Involucre scarious throughout, glabrous or nearly so.
    A. luzuloides
    Woodrush pussytoes
   3. Involucre hairy, less scarious on lower parts of bracts.
    4. Plants under 2dm high; involucre blackish, dark-brown, or green, but
     may be white at tip.
     A. lanata
     Woolly pussytoes
    4. Plants 2–5dm tall; involucre brownish to whitish.
     5. Involucre mostly brownish-green.
     A. pulcherrima
     Showy pussytoes
     5. Involucre mostly whitish.
     A. anaphaloides
     Tall pussytoes

MOUNTAIN PUSSYTOES, *Antennaria parvifolia* (fig. 269), is perhaps the most common species, forming silvery mats on dry meadows and open wooded slopes from the foothills to timberline throughout our range. The stems vary greatly in height, from under 5cm to 15cm or more. Each has a cluster of 3–8 heads. Basal leaves are obovate or rounded; phyllaries are usually more or less brownish at base and papery-white or pale-pink at the tips. UMBER or ALPINE PUSSYTOES, *A. umbrinella*, is similar, but its phyllaries are definitely brown, and it may have even shorter stems.

PINK or MEADOW PUSSYTOES, *A. microphylla*, is a plant with stems usually 20–25cm high, but they may reach 40cm. Its basal leaves are spatulate, with distinct petioles; its phyllaries are usually bright-pink, the heads less than 5mm long. Occurs in moist meadows and along stream banks from foothills to the alpine zone from Alaska to Montana and south to Colorado and California.

TALL PUSSYTOES, *A. anaphaloides*, is a gray or whitish woodland plant that does not have the mat-forming habit of the previous species. Its stems are solitary or few together, 25–40cm tall. It resembles pearly everlasting, as indicated by the name *anaphaloides*, which means *like Anaphalis*. Basal leaves are 10–15cm long, narrow and strongly 3-nerved. This leaf character (and its different habit of growth) distinguish the tall pussytoes from the everlasting. Never forms big beds; occurs scatteringly in aspen or open spruce or moist pine forests from Montana south through Colorado.

(5) *Arnica*. Arnica

This genus contains about 30 species, all occurring in the northern hemisphere, and about half of them in the Rocky Mountains. Most of them have bright yellow rays. Stem leaves are opposite; the phyllaries are equal in length (see Plate G, fig 68); the pappus is made up of pure white, cream, or brownish bristles, which may be either minutely barbed or plumose. One usually needs a hand lens to see this character. Arnica has long been used in medicine.

1. Stem leaves usually 5–10 pairs.
2. Involucral bracts obtuse to acute, with tuft of long hairs near or at the tip.
     A. chamissonis
     ssp. *foliosa*
     Leafy arnica
2. Involucral bracts definitely acute, the bracts either pubescent or not.
     A. longifolia
     Long-leaf arnica
1. Stem leaves usually 2–4 pairs.
2. Ray flowers lacking; young heads nodding.
     A. parryi
     Nodding or Parry arnica
2. Ray flowers present; young heads erect.
  3. Pappus subplumose to plumose, tawny in color; usually without basal leaves.
     A. mollis
     Subalpine or hairy arnica
  3. Pappus barbellate, white to whitish; basal leaves often present; leaf blades relatively broad.
   4. Involucral bracts acute to acuminate, not hairy inside at the apex.
    5. Involucre and peduncle below head sparsely to densely long-hairy; both basal and lower stem leaves strongly cordate.
      A. cordifolia
      Heart-leaf arnica
    5. Involucre and peduncle below head without long hairs or nearly so; basal and lower stem leaves seldom cordate.

6. Involucral bracts 10–17mm long, 1–3mm wide; heads mostly 1–3; moist habitats.

A. latifolia
Broad-leaf arnica

6. Involucral bracts 7–9mm long, 1–1.5mm wide; heads mostly 5–9; rocky habitats.

A. gracilis
Slender arnica

4. Involucral bracts obtuse to acutish, hairy inside at the apex; leaf blade relatively narrow.

5. Heads with 7–10 rays; subalpine and alpine.

A. rydbergii
Alpine arnica

5. Heads with 10–23 rays.

6. Old leaf bases with tufts of brown wool.

A. fulgens
Meadow or orange arnica

6. Old leaf bases without tufts of brown wool.

A. sororia
Twin arnica

HEART-LEAF ARNICA, *A. cordifolia* (fig. 270), is the most showy of all species. Its large, bright yellow, usually solitary heads are conspicuous in moist coniferous forests of the montane and subalpine zones in late spring and early summer. The seed heads attract attention later, with tufts of white pappus. The base of the involucre and the stalk just below the head are clothed with long white hairs. Occurs in the Keewennaw Peninsula of Michigan and from Alaska down the length of the Rockies to New Mexico and southern California, and on the North Rim of the Grand Canyon. The BROAD-LEAF ARNICA, *A. latifolia*, is very similar and occurs throughout the Rockies. It blooms later, lacks the white hairs at the base of the involucre, and it may have from 1–5 heads to a stalk.

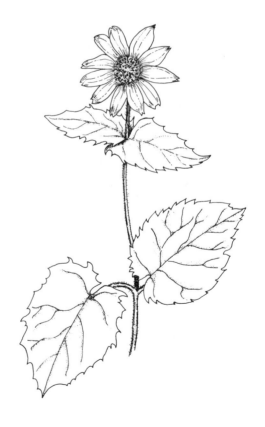

270. Heartleaf Arnica, ²∕₅X

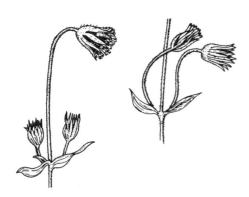

271. Parry Arnica, ½X

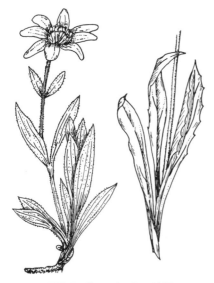

272. Rydberg Arnica, ½X

PARRY ARNICA, *A. parryi* (fig. 271), is a tall, hairy, slightly sticky plant of moist slopes and open woods throughout the range. The rayless heads, which hang down, are brownish-yellow.

Several of the arnicas are most likely to be found in meadows. MEADOW or ORANGE ARNICA, *Arnica fulgens*, is a strongly-scented, sticky-hairy plant with 1–3 orange heads. Abundant in montane and subalpine meadows from western Canada south to Colorado and Nevada. SUBALPINE ARNICA, *A. mollis*, is found in moist meadows in the subalpine and alpine zones of our region. Widely distributed from Quebec to the mountains of Maine and New Hampshire, and from Alberta and British Columbia to Colorado and California. LEAFY ARNICA, *A. chamissonis* ssp. *foliosa*, may be recognized by its several comparatively small heads, 2.5–4cm broad, and its more numerous leaves, which taper to both ends. In meadows of the montane and lower subalpine zones from western Canada to New Mexico. LONG-LEAF ARNICA, *A. longifolia*, is somewhat glandular-sticky, with several pairs of leaves whose bases sheathe the stem. Grows in wet meadows of the upper montane and lower subalpine zones from Montana and Idaho to western Colorado and Nevada.

ALPINE or RYDBERG ARNICA, *A. rydbergii* (fig. 272), has 3-nerved leaves and a spreading rootstock. Forms patches on rocky slopes and ridges at timberline and above from Alberta and British Columbia to Colorado.

SLENDER ARNICA, *A. gracilis*, is found at higher altitudes in our range, from Wyoming north, and it is abundant above timberline in Glacier National Park.

(6) *Artemisia*. Sagewort or Sagebrush

This group has many representatives in western North America, including both woody and herbaceous plants. Most of them are strongly aromatic. Their foliage is usually silvery; their small flower heads, arranged in elongated clusters, have no colorful characters. They bloom in late summer, and their wind-blown pollen is a cause of hay fever.

1. Flowers all alike, perfect, fertile; the plants shrubs, not spiny.
   2. Leaves 3-toothed or 3–6-parted at apex.
      3. Leaf segments long-linear or filiform.
         4. Plants well over 15cm tall.
            A. tripartita ssp. *tripartita*
         4. Plants seldom 15cm tall.
            A. tripartita ssp. *rapicola*
      3. Leaves toothed or lobed at apex, or entire.
         4. Heads few, rather large, in a narrow panicle; leaves lobed.
            5. Involucres mostly twice as long as wide, or more.
               A. arbuscula var.
               arbuscula
               Dwarf sagebrush
            5. Involucres usually less than twice as long as wide.
               A. longiloba
         4. Heads many, small; leaves 3-toothed.
            5. Shrubs 4–20dm tall; involucres tomentose.
               6. Plants often over 1m tall; flower stalks rising to different levels above the foliage, making an uneven top.
                  A. tridentata var.
                  tridentata
               6. Plants mostly under 1m tall; flower stalks rising more or less evenly above the foliage giving a flatter top.
                  7. Plants 5–10dm high.
                     A. tridentata var.
                     vaseyana
                  7. Plants under 5dm high.
                     A. tridentata var.
                     wyomingensis
            5. Shrubs under 4dm tall; involucres glabrate.
               A. nova
               Black sagebrush
   2. Leaves all linear, entire or nearly so.
      3. Leaves filiform, less than 1mm wide.
         A. filifolia
         Sand sage
      3. Leaves broader, 1mm or more wide.
         4. Leaves greenish, usually less than 4mm wide and 5cm long.
            A. cana var. *viscidula*

         4. Leaves silver-gray, often over 4mm wide and 5cm long.
            A. cana var. *cana*
            Silver sage
1. Flowers of 2 kinds: marginal flowers pistillate, the inner flowers perfect but sometimes sterile, the ovaries aborting; herbs or subshrubs.
   2. Plants spiny; subshrub; white tomentose.
         A. spinescens
   2. Plants not spiny.
      3. Inner flowers sterile, the style entire, the ovary aborting; plants herbaceous (or woody at base only).
         4. Herbage green, glabrous; leaves mostly entire.
            A. dracunculus
            Tarragon
         4. Herbage grayish-pubescent or canescent; leaves pinnatifid or dissected.

5. Heads 2–3mm wide; flower yellow.
  6. Plants tall, 3–7dm high.

A. campestris ssp. *caudata*

  6. Plants very small, 5–15cm high.

A. pedatifida
Birdsfoot sage

5. Heads 4–5mm wide; flowers brown.

A. campestris ssp. *borealis*
Northern wormwood

3. Inner flowers also fertile, the style 2-cleft.
  4. Receptacle beset with long wooly hairs.
    5. Plants mostly 4–12dm tall; cauline leaf blades at least 3cm long.

A. absinthium
Absinthe

    5. Plants mostly 1–4dm tall; cauline leaf blades mostly less than 3cm long.
      6. Heads small, 4–5mm wide, and numerous.

A. frigida
Fringed mountain sage

      6. Heads large, 6–12mm wide, few or solitary.
        7. Heads several in a spike-like raceme.

A. scopulorum
Rocky Mountain sage

        7. Heads solitary or few in a close terminal cluster.

A. pattersonii
Alpine mountain-sage

  4. Receptacle without woolly hairs among the flowers.
    5. Plants silky-pubescent or glabrate, never tomentose; leaves twice 3–7-parted, primarily basal.

A. arctica ssp. *saxicola*
Boreal mountain sage

    5. Plants tomentose, at least on the lower surface of the leaves; leaves mostly cauline.
      6. Involucres densely tomentose.
        7. Leaves permanently tomentose on both surfaces.
          8. Leaves with revolute margins.
            9. Leaves entire, long linear.

A. longifolia
Long-leaved sage

            9. Leaves pinnatifid, narrowly linear or filiform lobes.

A. carruthii

          8. Leaf margins not revolute.

A. ludoviciana
ssp. *ludoviciana*
Prairie or Louisiana sage

        7. Leaves usually glabrate on upper surface in age.

A. ludoviciana ssp. *mexicana*

      6. Involucres glabrous or glabrate, at least in age; leaves glabrous above.
        7. Divisions in the leaves broadly linear to lanceolate.

A. francerioides

        7. Divisions in the leaves narrowly linear; leaves 1–2-pinnately parted.

A. ludoviciana
ssp. *incompta*

BIG SAGEBRUSH, *A. tridentata* (fig. 273), is a rigid, much-branched shrub, quite variable in height (see key). Leaves bluish-silvery, wedge-shaped, 3-toothed at apex, strongly aromatic; heads numerous, in narrow panicles. Sagebrush usually grows in colonies, often covering many acres or square miles on dry, open valleys, foothills, and lower montane slopes, especially where there is deep but dry soil.

ALPINE or ROCKY MOUNTAIN SAGE, *A. scopulorum*, is a small, silvery plant with few or several tufted stems 10–25cm tall; leaves mostly basal, divided into narrow segments. Occurs from Montana to New Mexico and Utah. ALPINE MOUNTAIN SAGE, *A pattersonii* (fig. 274), found in Colorado and New Mexico, is similar but somewhat shorter. Both occur on rocky or grassy alpine slopes. BOREAL MOUNTAIN SAGE, *A. arctica* ssp. *saxicola*, is another alpine species, occurring from Alaska south along the western mountains to Colorado and California. Clustered, reddish stems 10–30cm tall; nodding heads with dark-margined phyllaries.

FRINGED or PASTURE MOUNTAIN SAGE, *A. frigida* (fig. 275), is an attractive plant with finely-divided, silvery leaves in basal tufts and slender clusters of small, nodding, yellowish flower heads. Commonly seen on dry slopes of the foothill, montane, and subalpine zones throughout the Rockies. The early settlers used this aromatic plant to make a bitter tea, which was believed to be a tonic and a remedy for mountain (typhoid) fever. Establishes itself on the disturbed soil of roadsides and on overgrazed pasture land.

273. Big Sagebrush, ½X   274. Alpine Mountain-sage, ½X   275. Fringed Mountain Sage, ½X

PRAIRIE SAGEWORT or WESTERN MUGWORT, *A. ludoviciana,* is a variable herbaceous plant (see key), best recognized by a dense tomentum of white, cobwebby hairs. Widely distributed from the Mississippi River westward. Occurs in the mountains on dry open slopes in the foothill and montane zones.

(7) *Aster.* Aster

A large genus of rather similar species and considered to be difficult. The ray flowers range from white to blue, violet, rose, and purple. The pappus is a tuft of soft, white or tawny bristles. In our species, the leaves are nearly always longer than wide; the leaf margins are usually entire, rarely toothed or slightly irregular. There may be one to many heads in the inflorescence; the involucre usually has 2 series of phyllaries in an imbricated arrangement. For a comparison of *Aster, Erigeron,* and *Machaeranthera* involucres, see Plate G, figs. 62, 63, and 64, as these genera are often confused. In general, asters have more heads per stalk, fewer and broader ray flowers, and more definitely imbricated phyllaries than do erigerons. Asters usually bloom in late summer and fall, erigerons in spring and early summer. But such rules-of-thumb are not infallible, and the involucre is a better key.

1. Annuals with inconspicuous rays or the rays lacking.
    2. Involucral bracts linear, acute.
        A. brachyactis
    2. Involucral bracts oblong, obtuse.
        A. frondosus
1. Perennials with conspicuous rays.
    2. Plants with taproot; heads solitary on each stem; rays lavender or purple.
        A. alpigenus ssp.
        haydenii
    2. Plants mostly fibrous-rooted or with rhizomes; heads rarely solitary.
        3. Ray flowers fewer than 15; involucral bracts scarcely herbaceous, more or less strongly keeled; lower stem leaves reduced.
            4. Involucral bracts glabrous or nearly so.
                5. Involucral bracts mostly obtuse at apex; leaves glaucous.
                    A. glaucodes ssp.
                    glaucodes
                    Glaucous aster
                5. Involucral bracts acute; leaves green.
                    A. engelmannii
                    Engelmann Aster
            4. Involucral bracts hairy; flowers deep violet.
                A. perelegans
                Elegant aster
        3. Ray flowers mostly more than 15 (but rarely more than 25); involucral bracts herbaceous, at least at the tips, not keeled; lower stem leaves usually not reduced.
            4. Involucre and peduncle glandular.
                5. Leaves mostly under 10mm wide.
                    6. Leaves mostly linear, much reduced upward.
                        A. pauciflorus
                    6. Leaves broader or not much reduced upward.
                        7. Leaves strongly clasping the stem; plants often over 5dm tall.
                            A. novae-angliae
                            New England aster

7. Leaves barely or not clasping the stem; plants mostly under 4dm tall.

A. campestris
Meadow aster

5. At least some of the leaves 15mm or more wide.
   6. Leaves mostly toothed.

A. conspicuus
Showy aster

   6. Leaves entire or ciliate

A. integrifolius

4. Involucre and peduncle not glandular, pubescent or not.
   5. Plants conspicuously hairy throughout, the hairs stiff and straight.
      6. Plants with well-developed rhizomes.

A. falcatus
Rough white aster

      6. Plants without well-developed rhizomes; either short rhizome or caudex.

A. ericoides
Sand aster

   5. Plants not conspicuously hairy, whether pubescent or not.
      6. Pubescence of stem decurrent from leaf bases (descending in lines).

A. hesperius
Skyblue aster

      6. Pubescence of stem uniform, or the plant glabrate.
         7. At least some of the outer bracts foliaceous, equalling or surpassing the inner bracts in length.
            8. Middle cauline leaves mostly less than 1cm wide, more than 7 times longer than wide.

A. occidentalis
Fremont aster

            8. Middle cauline leaves 1cm wide or more, less than 7 times longer than wide.
               9. Plants about 2.5dm tall, decumbent to ascending; involucral bracts purple-margined and tipped; subalpine to alpine.

A. foliaceus var. *apricus*
Sun-loving aster

               9. Plants mostly over 2.5dm tall; bracts not purple-margined; not alpine.
                  10. Lower leaves often enlarged and persistent; involucral bracts acute to acuminate; the foliaceous bracts linear to lanceolate; wet habitats.

A. foliaceus var. *parryi*
Leafy-bract aster

                  10. Lower leaves seldom enlarged, often deciduous; involucral bracts obtuse to acutish, the foliaceous bracts broadly lanceolate to ovate; drier habitats.

A. foliaceus var. *canbyi*
Leafy-bract aster

         7. Outer bracts neither truly foliaceous nor equalling or surpassing the inner ones.
            8. Outer involucral bracts oblanceolate, spatulate, or obovate-oblong, rounded at apex; rays rose to purple.

A. ascendens
Pacific aster

            8. Outer involucral bracts linear, lance-linear, to oblong, acute at apex.

9. Involucres 4–5mm high; rays white.
A. porteri
Porter aster
9. Involucres 8–9mm high; rays blue.
A. laevis
Smooth aster

276. Sun-loving Aster, ½X

277. Smooth Aster, ½X

SUN-LOVING ASTER, *A. foliaceus* var. *apricus* (fig. 276), is a low-growing, usually tufted plant with violet or bright-rose rays; 1 to few heads on decument or ascending leafy stems. The phyllaries are loose and somewhat leafy, often purple-edged. Alpine and subalpine zones on moist gravel. Other varieties of LEAFY-BRACT ASTER, *A. foliaceus*, rise 3–9dm tall. They are widely distributed in our area from the upper foothills to the subalpine zone. FREMONT ASTER, *A. occidentalis*, is a similar species, with ray flowers that vary from violet to bright rose. Compared with *A. f.* var. *apricus*, it has fewer, taller, more erect, and often reddish stems that are without leaves on their upper portions. Occurs in the upper montane and subalpine zones from Canada south to Colorado.

PACIFIC ASTER, *A. ascendens*, is common, variable, and widely distributed. It has a creeping rootstock, thus forming clumps or beds. The ray flowers are pale, either lavender or pinkish; phyllaries imbricated in 3–4 rows, blunt or bristle-tipped, white at base. Stems usually about 3dm tall, but they may be less or up to 6dm. The plants may be quite smooth or moderately to densely hairy. Grows along roadsides, in marshes, and on hillsides in the foothill, montane, and lower subalpine zones.

SMOOTH ASTER, *A. laevis* (fig. 277). Rays blue; stems 3–9dm tall; one of the few asters having entirely smooth foliage. Lower leaves up to 15cm long, short-petioled; upper leaves noticeably smaller, clasping the stem. This is the common, blue-flowered aster from the Rockies eastward to the Atlantic coast. Found on moist soil of the foothill and montane zones.

SKYBLUE ASTER, *A. hesperius*, is a tall plant of swampy places at lower elevations. It has numerous small heads; rays commonly pale blue but may be whitish or pinkish; stems marked lengthwise with lines of hairs; leaves willow-like.

SHOWY ASTER, *A. conspicuus*, a stout plant 3–9dm tall with large, stiff, sharply-toothed leaves; many heads in a much branched inflorescence; 15–35 violet rays; involucre sticky; phyllaries with recurved tips, imbricated in about 5 rows. Found in open woods of the montane zone, northern Wyoming, Montana, and Idaho, especially in Yellowstone and Glacier National Parks.

HAYDEN ASTER, *A. alpigenus* ssp. *haydenii*. A small plant with spreading, single-headed stalks 8–15cm long; slender basal leaves about 5–10cm long. The heads are comparatively large and showy, with bright purple rays and yellow disks. Occurs in the alpine zone of Wyoming, Idaho, and Montana. This variety was named in honor of Ferdinand V. Hayden, geologist and explorer of the region that now includes Yellowstone National Park, and a director of the U.S. Geological Survey of the Territories. He was one of the small group directly responsible for the congressional act that set the area apart as a national park.

PORTER ASTER, *A. porteri* (fig 278), is a plant with smooth stems and leaves; bright-white rays; stems branched, 20–50cm tall. As the flowers age, the orange disk turns dark-red. Occurs in montane fields and meadows of New Mexico and Colorado.

ROUGH WHITE ASTER, *A. falcatus*, has a creeping underground rootstock forming patches or large clumps. Found along roadsides, on banks, and in meadows of the foothill zone throughout the Rockies.

ENGELMANN ASTER, *A. engelmannii*, is a stout plant 4–12dm tall with numerous white or pinkish, rather ragged flower heads. The leaves are ovate-oblong to broadly lanceolate,

rays white

278. Porter Aster, ½X

5–10cm long; phyllaries are tough and papery, often purple-tinged, imbricated in several rows. Found on moist banks and in woods of the montane and lower subalpine zones from Canada south to Colorado and Nevada.

GLAUCOUS ASTER, *A. glaucodes*, is 25–38cm tall with gray-green, narrow, blunt leaves and pale lavender flowers. Seen in long clumps along mountain roads or among rocks of the montane and subalpine zones of Wyoming, Colorado, Utah, and Arizona.

279. Bahia Chrysanthemum, ½X

SAND ASTER, *Aster ericoides*, is a little, tufted plant, usually not over 13cm tall, with wiry stems and tiny leaves. Each branch bears a single white-rayed and yellow-centered head. Phyllaries imbricated in several rows, pointed, with green midribs and papery margins. Found on sandy soil of the foothills and plains from Wyoming to Texas westward.

(8) *Bahia*. Bahia

One species is likely to be found in our range: BAHIA or FIELD CHRYSANTHEMUM, *Bahia dissecta* (fig 279). Annuals; stems solitary, 3–9dm tall, from a rosette or dark grayish-green, much divided leaves. Inflorescence a corymb of yellow flower heads, each having 12–20 short, broad rays; phyllaries in 2–3 rows, the outer ones 3-nerved; pappus absent. Gravelly soil of the foothills and montane zones from Wyoming to Mexico.

(9) *Balsamorhiza*. Balsamroot

1. Leaves entire or somewhat toothed.

    B. sagittata
    Arrowleaf balsamroot

1. Leaves laciniately dentate to bipinnatifid.
    2. Leaves green, glabrous or sparingly hirsute.

    B. macrophylla
    Large-leaved balsamroot

    2. Leaves canescent or lanate.
        3. Pubescence sericeous, appressed or spreading.

    B. hookeri var. *hispidula*

        3. White tomentum, sometimes tufted.

    B. incana
    Hoary balsamroot

ARROWLEAF BALSAMROOT, *B. sagittata* (fig. 280), is a coarse plant with many leaves from a stout taproot. Basal leaves are 20–38cm long, including the petiole, grayish-velvety, and arrow-shaped. Flower stalks are 20–60cm tall with only a few small leaves; heads yellow, usually solitary, 5–10cm broad; involucre white with matted hairs; phyllaries with long, slender tips. Found on hillsides and open ground in the upper foothill and montane zones throughout the Rockies. Blooming begins in April or May and continues into the summer. Both game and domestic animals feed on this species, and the plants survive well under heavy grazing.

HOARY BALSAMROOT, *B. incana*, has a similar habit of growth, but its leaves are silvery-white and pinnately parted into narrow segments. Its range is from western Wyoming and southwestern Montana westward.

280. Arrowleaf Balsamroot, ²⁄₅ X

Another BALSAMROOT, *B. hookeri* var. *hispidula*, occurs in the western part of our range. Leaves are green and pinnately divided.

(10) *Brickellia*. Thoroughwort or Tasselflower

1. Leaves ovate or cordate-triangular.
    2. Leaves petiolate, teeth not spine-tipped.
        3. Heads 30–50-flowered, drooping.

                                    B. grandiflora
                                    Tasselflower

        3. Heads 10–25-flowered.
            4. Leaves 30–40mm long; involucral bracts erect.

                                      B. californica

            4. Leaves 5–10mm long; involucral bracts spread widely.
                                    B. microphylla var. *scaber*
    2. Leaves subsessile, teeth spine-tipped.

                                      B. atractyloides
1. Leaves oblong to linear, sessile.

                                      B. oblongifolia var. *linifolia*
                                      Narrow-leaved thoroughwort

281. Tasselflower, ½ X

TASSELFLOWER or BRICKLEBUSH, *Brickellia grandiflora* (fig. 281). Heads pale yellowish in nodding clusters, on long stalks from the axils of the upper leaves. Leaves triangular, toothed, rather strongly veined. Found among rocks on dry canyon sides in the foothill and montane zones throughout our range.

(11) *Carduus*. Thistle

One species in our range. NODDING THISTLE, *Carduus nutans*, is a tall, pink to purple, weedy thistle, whose involucral bracts are strongly-reflexed, sharp-pointed, and glabrous. Seen on fields and along roadsides in Montana, but has been spreading from there into other western states. An introduced plant.

(12) *Chaenactis*. False-yarrow

1. Plants annuals.

    C. stevioides

1. Plants perennials.
  2. Heads corymbose, short peduncled.
    3. Low plant, usually under 15cm tall.

    C. douglasii var. *montana*

    3. Plant taller, 15–40cm tall.

    C. douglasii var. *achilleaefolia*
    Dusty maiden

  2. Heads solitary.

    C. alpina
    Alpine dusty maiden

DUSTY MAIDEN, *Chaenactis douglasii* var. *achilleaefolia* (fig. 282), is a grayish plant up to 40cm tall. Leaves pinnately divided with the segments again lobed or divided; flowers white or pinkish; phyllaries of equal length, forming a cylindrical involucre; pappus a crown of 10 narrow, transparent scales, 5 being shorter than the others. Found on shady or rocky ground from the foothills to the subalpine zone, from Montana to Colorado and Nevada. The ALPINE DUSTY MAIDEN, *C. alpina*, is similar but shorter. Occurs at higher elevations.

(13) *Chrysanthemum*. Chrysanthemum

1. Plants annual; rays pale yellow.

    C. coronarium

1. Plants perennial; rays white, if present.

2. Heads usually solitary, on long peduncles.

C. leucanthemum
Ox-eye daisy

2. Heads numerous, inflorescence corymbose; rays very short or absent.

C. balsamita
Mint geranium

OX-EYE DAISY, *Chrysanthemum leucanthemum.* This golden-centered and white-rayed daisy is becoming established along roadsides and ditches in our area. Its ancestors arrived from Europe with early colonists. Their progeny have been establishing themselves in hay meadows and fields of the eastern states, where the plant is considered a troublesome weed by farmers. It has gradually worked its way westward. Narrow leaves with wavy or toothed margins form low, green rosettes. Flower stalks are erect, to 60cm high, almost leafless, usually 1-flowered; phyllaries are pointed, light green lined with brown.

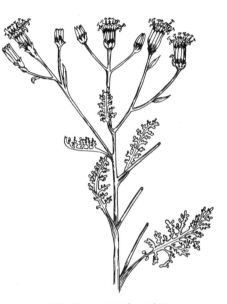

282. Dusty Maiden, ½X

(14) *Chrysothamnus.* Rabbit brush

(Note: The major species within *Chrysothamnus* have numerous subspecies, making it impractical to key them for this general flora.)

1. The entire plant entirely devoid of tomentum.
    2. Plant tall (5–10dm), tufted; woody stems erect, slender, and flexible; the leafy flowering branches white.

C. linifolius

    2. Plant low (5dm or less); freely branched.
        3. Achenes glabrous or nearly so.

C. vaseyi

        3. Achenes densely hairy.
            4. Leaves with only 1 nerve, not over 1.2mm wide.

C. greenei

            4. Leaves often more than 1 nerve, usually over 1.2mm wide.

C. viscidiflorus
Sticky-flowered
rabbitbrush

1. Twigs covered with a feltlike tomentum. (Sometimes appears to be bark. Scrape to be sure.)
    2. Outer involucral bracts prolonged into slender, herbaceous tips; achenes pubescent.

C. parryi
Parry rabbitbrush

2. Outer involucral bracts not herbaceous-tipped; achenes glabrous.

C. nauseosus
Common or golden
rabbit-brush

283. Sticky
Rabbitbrush, ½X

284. Golden Rabbitbrush, ½X

STICKY-FLOWERED RABBITBRUSH, *Chrysothamnus viscidiflorus* (fig. 283), has many forms. A common form is a shrub with several branches from the base, 2.5–5dm tall; leaves narrow and usually twisted; rounded or flat-topped clusters of small yellow flower heads not over 6mm high. Occurs on dry hillsides of the foothill and montane zones from Montana to Colorado. A taller species, GOLDEN RABBITBRUSH, *C. nauseosus* (fig. 284), is also quite variable. A shrub 6–12dm tall; gray or light-green twigs; narrow leaves; large, rounded clusters of golden-yellow heads. A handsome, conspicuous plant when blooming in late summer and fall along ditch and stream banks and on roadsides, in the foothill and montane zones throughout our range.

PARRY RABBITBRUSH, *C. parryi*, a shrub not over 6dm tall, has heads 10–13mm high with slender-tipped phyllaries. Occurs in open forests of the montane and subalpine zones from Wyoming to Utah, Nevada, and Colorado.

(15) *Cichorium*. Chicory

One species in our range: CHICHORY, *Chicorium intybus*, a perennial with taproot; 3–9dm tall, branched; heads numerous, 1–4 together in sessile clusters; flowers mostly bright blue, rarely white. A common roadside weed below 7,000 feet.

(16) *Cirsium*. Thistle

All our thistles have toothed or wavy-margined leaves, and each tooth is tipped with a sharp bristle spine. The phyllaries are also usually spine-tipped. Some thistles have large, showy purple or pinkish heads, but this is a variable character, and even the usually purple-flowered plants sometimes have pale, dingy heads.

1. Upper leaf surface with short, appressed, yellowish prickles.

                                         C. vulgare
                                         Bull thistle

1. Upper leaf surface glabrous to tomentose, but without prickles.
    2. Plants with involucre 1–2cm tall; perennial, with deep creeping roots; forms dense patches; plant glabrous.

                                         C. arvense
                                         Canada thistle

    2. Involucres mostly larger; plants taprooted.
        3. Cauline leaves strongly decurrent, the wing sometimes extending downward to the next node; wings over 15mm long.
            4. Wings on the middle and upper leaves longer than the wings on the lower leaves.
                5. Leaves glabrous or glabrate below and above; heads in terminal clusters involucral bracts cobwebby.

                                         C. eatonii

                5. Leaves white-tomentose below, less so or green above; heads solitary or in loose clusters with 1 terminal; involucral bracts not cobwebby.
                  6. Heads less than 4cm tall.
                    7. Involucral bracts with lacerate fringes near apex.

                                       C. centaureae
                                       American thistle

                  7. Involucral bracts entire near apex.
                    8. Involucre as wide or wider than long.
                                       C. canescens
                  8. Involucre longer than wide.
                                     C. pulcherrium
                6. Heads more than 4cm tall; corollas rose or purple.
                                    C. ochrocentrum

            4. Wings on the lower leaves (25–40mm) longer than wings at apex (5–10mm); heads solitary or in loose clusters with 1 terminal; involucral bracts cobwebbby.

                                       C. subniveum
                                   Jackson Hole thistle

3. Leaves somewhat clasping the stem; if decurrrent, the wings under 12mm long.
    4. Heads in terminal clusters.
        5. Involucral bracts densely cobwebby.

                                     C. scopulorum
                                   Hooker thistle

        5. Involucral bracts not densely cobwebby.
            6. Pappus equaling or exceeding the corolla.

                                     C. foliosum
                                   Elk thistle

            6. Pappus shorter than the corolla.
                7. Corolla 16–24mm long.

                                   C. scariosum
                                   Elk thistle

                7. Corolla 22–30mm long.

                                   C. coloradense

4. Heads solitary or in loose clusters with 1 terminal.
    5. Leaves glabrate above and below; involucral bracts densely
       cobwebby.

C. parryi

    5. Leaves more or less tomentose below, less so above.
      6. Bracts with a conspicuous glutinous dorsal ridge.

C. canovirens
Gray-green thistle

      6. Bracts lack a glutinous dorsal ridge.

C. undulatum
Wavy-leaf thistle

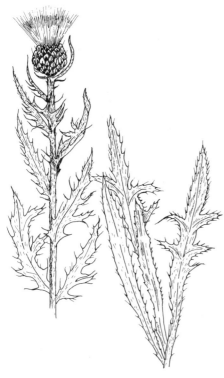

285. Wavy-Leaf Thistle, ²⁄₅ X

WAVY-LEAF THISTLE, *C. undulatum* (fig. 285). Common on the plains and mesas; frequently seen in the foothill and montane zones throughout our range; also on the North Rim of the Grand Canyon.

AMERICAN THISTLE, *C. centaureae,* has pale, dingy, or pink flowers and fringed phyllaries. Occurs in ravines of the montane and subalpine zones, often forming colonies. Wyoming, Colorado, and Utah.

ELK THISTLE, *C. foliosum*, is somewhat similar to Hooker thistle, but it is often stemless, with the heads sitting in a large basal rosette of very spiny leaves. This is the species that saved Truman Everts, a lost explorer in Yellowstone National Park in 1870, from starvation. The peeled root and stems are pleasantly flavored and nourishing, and they are eaten by elk and bears. Grows in meadows, on stream banks, and around ponds of the montane and subalpine zones throughout the Rockies. HOOKER THISTLE, *C. scopulorum*, with stems 20–60cm tall, has a dense cluster of pale heads, which are overtopped by the excessively spiny leaves. A species of rocky slopes in the subalpine and alpine zones from Montana to Colorado and Utah.

In addition to the four native species above, three species introduced from Europe, classed as weeds, are conspicuous and beautiful when in bloom. CANADA or CREEPING THISTLE, *C. arvense*, is one of the most noxious weeds we have. Between its vigorous, deep-seated, creeping rootstock and its efficient seed dispersal system, it can take over whole meadows in a few years and render them useless. It thrives on moist, rich soil and has spread throughout the northern United States. BULL THISTLE, *C. vulgare*, a stout, branched plant 3–18dm tall, with stiff spines and beautiful purple heads, is becoming established in pastures and meadows of our region. NODDING THISTLE, *Carduus nutans*, noted earlier, is increasingly visible on fields and along roadsides in the western states.

(17) Crepis. Hawksbeard

1. Dwarf plants, 2–7cm tall, alpine or subalpine.

C. nana
Dwarf hawksbeard

1. Plants over 8cm tall, usually below alpine.
  2. Herbage glabrous or glaucous, sometimes hispidulous, but never tomentose.
    3. Flowers 20 or more per head.
      4. Leaves somewhat glaucous, the mature involucre glabrous.

C. runcinata ssp. *glauca*

      4. Leaves not glaucous
        5. Involucre pubescent or hirsute.

C. runcinata ssp.
runcinata
Dandelion hawksbeard

        5. Involucre glandular-hispid.

C. runcinata ssp. *riparia*

    3. Flowers 6–12 per head.

C. elegans
Elegant hawksbeard

  2. Herbage tomentose or puberulent.
    3. Involucres glabrous, 5–7-flowered.

C. acuminata
Long-leaved hawksbeard

    3. Involucres not glabrous, obviously hairy.
      4. Heads 5–10-flowered.

C. intermedia
Gray hawksbeard

      4. Heads 10–60-flowered.
        5. Involucre with black, bristly hairs; not glandular.

C. modocensis
Low hawksbeard

        5. Involucre not black-bristly, or if so, the bristles gland-tipped.
          6. Leaf segments mostly linear or lance-linear, entire.

C. atribarba
Slender hawksbeard

          6. Leaf segments lanceolate to deltoid, often toothed; black-bristly involucral hairs gland-tipped, at least some of them.
            7. Involucre definitely glandular.

C. occidentalis
var. *costata*

            7. Involucre not glandular, or sparsely so.

C. occidentalis var.
occidentalis
Western hawksbeard

DANDELION HAWKSBEARD, *Crepis runcinata*, has several bright yellow heads in a flat-topped cluster, and its leaves are usually somewhat like dandelion leaves. A plant of moist meadows and found throughout our range. ALPINE or DWARF HAWKSBEARD, *C. nana,* is closely related and similar, but it has short stems and tufted leaves, which have an enlarged terminal lobe with smaller teeth below. Occurs in moist, rocky situations of the alpine zone.

(18) *Erigeron.* Erigeron

A very large genus containing species not only similar to each other but often confused with asters. Study the information on asters and Plate G, figs. 62 and 63. The majority of erigerons are perennial. In general they have numerous (30–150) and narrow ray flowers, ranging from violet, pinkish, or white to blue and purple. To the west of our range and into the Canadian Rockies there are species with bright-yellow rays. One of them, *Erigeron linearis,* is found as far south as Yellowstone National Park. Many erigerons have only one head to each stem, and most have fewer than five. (Most asters have several to many heads). The phyllaries are of nearly equal length, rarely in more than two rows. The pappus is a tuft of rather brittle bristles, sometimes with an outer ring of shorter bristles or scales.

Erigerons usually bloom in spring and early summer, asters in late summer and fall. Erigerons have been popularly called *fleabane* and *mountain daisies,* but so many different plants in this family have borne these names that confusion results. It is much better to use the generic name, pronounced e-rij'-eron, as a common name.

✓ 1. Leaves palmately or pinnately lobed.
    2. Leaves ternately (once or twice) lobed, not pinnatifid.
        3. Caudex stout, simple or somewhat branched.

                                                              E. compositus
                                                              Cut-leaf erigeron
        3. Caudex diffuse, branched (rhizome-like branches)
                                                              E. vagus
                                                        Wandering erigeron
    ✓ 2. Leaves pinnatifid.

                                                             E. pinnatisectus
                                                          Pinnate-leaf erigeron
1. Leaves entire or minutely toothed.
    2. Ray flowers filiform and erect, or absent.
        3. Inner involucral bracts usually long-attenuate at apex, often glandular;
           cauline leaves lanceolate to broader.
           4. Plants with several to many heads, usually 30cm or taller.

                                                     E. acris var. *asteroides*
                                                     Bitter erigeron
           4. Plants with 1 head to several, usually under 30cm tall.
                                                     E. acris var. *debilis*
        3. Inner involucral bracts sharply acute to acuminate, never glandular; cauline
           leaves usually linear.

                                                   E. lonchophyllus
                                                    Short-rayed erigeron
    2. Ray flowers wider and spreading, conspicuous.
        3. Cauline leaves substantial, lanceolate or wider; plants tall, erect, somewhat
           aster-like.
           4. Involucre woolly-villous.

                                                   E. elatior
                                                   Pink-head erigeron
           4. Involucre not woolly-villous, but often pubescent.

5. Hairs of involucre with black crosswalls near their base (villous-hirsute).

> *E. coulteri*
> Coulter erigeron

5. Hairs of involucre without black crosswalls.
   6. Rays mostly 2–4mm wide; phyllaries attenuate.
      7. Up to 7dm tall; cauline leaves not much reduced; subalpine.

> *E. peregrinus* ssp.
> *callianthemus*
> var. *callianthemus*
> Subalpine erigeron

      7. Under 2dm tall; cauline leaves much reduced (subscapose); alpine.

> *E. peregrinus* ssp.
> *callianthemus* var.
> *scaposus*

   6. Rays 0.5–2mm wide; phyllaries not as above.
      7. Middle leaves mostly as large as or larger than lowermost leaves; upper leaves gradually reduced.
         8. Leaves and stems pubescent throughout.

> *E. subtrinerveris*

         8. Leaves glabrous (margins may be ciliate); stems glabrous below, glabrous or glandular above.
            9. Upper leaves ciliate.
               10. Uppermost leaves lanceolate.

> *E. speciosus* var. *speciosus*
> Showy erigeron

               10. Upper leaves ovate.

> *E. speciosus* var.
> *macranthus*
> Aspen erigeron

            9. Upper leaves not ciliate.

> *E. eximius*
> Pale erigeron

      7. Middle leaves smaller than the lowermost ones; upper leaves abruptly reduced; stems usually curved at the base, with spreading pubescence; involucral bracts glandular.
         8. Stems usually glandular; numerous fibrous roots.
            9. Involucre densely hirsute and glandular; upper leaves not glandular.

> *E. formosissimus* var.
> *formosissimus*
> Beautiful erigeron

            9. Involucre with glandular hairs; uppermost leaves glandular.

> *E. formosissimus* var.
> *viscidus*

         8. Stems not glandular; taproot.

> *E. caespitosus*
> Tufted erigeron

3. Cauline leaves much reduced, linear or oblanceolate; plants low, not aster-like.
   4. Involucre woolly-villous.
      5. Involucral hairs with black or dark purple crosswalls.

> *E. melanocephalus*
> Black-headed erigeron

      5. Involucral hairs with clear crosswalls (or occasionally some bright red-dish-purple).

✓6. Disk corollas 3–3.6mm long; leaves glabrous to somewhat pubescent.

E. simplex
One-flowered erigeron

6. Disk corollas 4–5mm long; leaves strongly pubescent.

E. grandiflorus
Large-flowered erigeron

4. Involucre not woolly-villous; may be glabrous, pubescent, or glandular.

5. Stem pubescence widely spreading.

6. Biennial plants; no definite caudex present.

E. divergens
Branching erigeron

6. Perennial plants with well-developed caudices or taproot.

7. Plant with taproot.

8. Stems with leaves glandular.

E. vetensis
Early blue erigeron

8. Stems usually not glandular.

9. Involucres canescent, with short white hairs.

E. caespitosus
Tufted erigeron

9. Involucres hirsute, the hairs long.

E. pumilus
Early white erigeron

7. Plants without taproot, roots fibrous; involucral bracts glandular; stems usually glandular, curved at base.

E. formosissimus
Beautiful erigeron

5. Stem pubescence appressed, ascending, or absent.

6. Involucre glandular, not hairy; leaves mostly glabrous, oblanceolate or broader.

E. leiomerus
Rock-slide erigeron

6. Involucre hairy, sometimes glandular; leaves more or less hairy, linear to oblanceolate.

7. Involucral hair appressed, strigose.

8. Rays 15–55 to a head, usually white.

E. nematophyllus

8. Rays 125–175 to a head, rarely white.

E. glabellus var. *glabellus*

7. Involucral hairs spreading, hirsute, villous to hispid.

8. Perennials with strong taproots, the caudex more or less branched.

9. Achenes glabrous or nearly so.

E. canus
Hoary erigeron

9. Achenes usually hairy.

10. Some of the lower leaves 4mm or more wide.

E. eatonii
Eaton erigeron

10. Lower leaves less than 4mm wide.

11. Involucral bracts of different lengths.

12. Flowers white to pinkish.

E. pulcherrimus

12. Flowers yellow.

E. linearis

11. Involucral bracts mostly equal in length.

E. engelmannii
Engelmann erigeron

8. Plants without well-developed taproot; annual,
  biennial, or short-lived perennial.
  9. Caudex of slender, rhizomatous branches with
    fibrous roots; perennial.
      E. ursinus
      Bear River erigeron
  9. Plants without caudex; annuals, biennials; rays
    50–100.
    10. Leafy stolons often present.
        E. flagellaris
        Whiplash erigeron
    10. Leafy stolons never present.
        E. strigosus var. *strigosus*

BRANCHING ERIGERON, *E. divergens,* is easily distinguished from other erigerons by its
widely branching habit and its many flower
heads. It may be distinguished from asters
by its numerous and very narrow rays,
which are light purple to pinkish. The plant
is softly hairy; basal leaves are oblanceolate
to spatulate, with entire or somewhat lobed
margins. Occurs on sandy ground and rocky
slopes of the foothills throughout our range.

WHIPLASH ERIGERON, *E. flagellaris* (fig.
286). A plant of foothill and montane meadows. Stems 10–25cm tall; heads always bent
down in bud and appearing to be pink,
though usually white when erect and fully
open; leaves mostly basal and oblanceolate.
Begins to bloom in spring and continues into
summer, when it develops long runners,

286. Whiplash Erigeron, ½ X

which may root at the tips to produce new plants. Occurs throughout our range.

EARLY WHITE ERIGERON, *E. pumilus.* A hairy plant with white rays, found in
tufts growing from a woody caudex. Stems several, 7–25cm tall; leaves narrow
and up to 8cm long. Found on dry mesas and plains from Colorado and
Wyoming south to Arizona. Occurs among oaks and serviceberries in Mesa
Verde National Park and is found in Dinosaur National Monument. Several
similar species are found on rocky slopes and in dry open woods of the
foothills and lower montane zone: *E. engelmannii,* which has almost leafless
stalks and occurs mostly on the western slope of the Continental Divide; and
E. eatonii, which frequently has 3-nerved basal leaves up to 15cm long and is
often found among sagebrush. Both are usually white-flowered, but their rays
may vary to blue or pink.

EARLY BLUE ERIGERON, *E. vetensis,* is a tufted, somewhat sticky plant with several stems, 15–25cm tall, from a heavy, woody caudex. Leaves are mostly basal,
oblanceolate, with a fringe of stiff hairs along their edges. The stalks are leaf-

287. Cut-leaf erigeron, ½X

288. Subalpine Erigeron, ½X

phyllaries covered with pink wool

289. Pink-head Erigeron, ½X

less, but there may be dry leaf bases on the caudex. Occurs in dry woods of the foothill and montane zones from Wyoming to New Mexico.

CUT-LEAF ERIGERON, *E. compositus* (fig. 287), is a little tufted plant with stems 5–15cm tall. Rays are usually white but may be bluish or purple, especially at high altitudes. Leaves mostly basal, once or twice 3-parted into narrow, blunt lobes; the very small stem leaves entire. Sometimes seen is a form of this species called GOLD BUTTONS, which lacks the ray flowers and has only the yellow disk. This species begins to bloom in early spring, coming into flower later at higher elevations. Abundant on openings in the ponderous pine forest in June and found on the tundra in July and August. Widely distributed throughout the Rockies.

✔ SUBALPINE ERIGERON, *E. peregrinus* (fig. 288), varies greatly in height (see key). The heads are large and showy, with comparatively broad, rose-purple or violet rays and orange-yellow disks; phyllaries are attenuated into loose tips; usually only 1 to a stem. Foliage mostly glabrous; basal leaves oblanceolate and petiolate, stem leaves smaller and often clasping. This is one of the conspicuous flowers of subalpine meadows in July and early August and may extend into exposed alpine situations. Occurs in Alaska south to New Mexico and California.

PINK-HEAD ERIGERON, *E. elatior* (fig. 289), is one of the tallest of erigerons, with erect stems to 6dm in height. Heads solitary or 2–3 on one stem, in bud usually bent down; rays numerous, narrow, pink or rose; involucres very woolly with pinkish hairs. This pink effect is produced by red coloring in the crosswalls of the hairs.

PALE ERIGERON, *E. eximius*. Stems from 3–6dm tall; smooth except sometimes a little sticky in the inflorescence; stem leaves more or less 3-nerved, few, far apart; rays light-pinkish, lavender, or white; heads noticeably bent down in bud. Found on stream banks and in moist forest glades in the montane and subalpine zones, from Wyoming and Utah south to Texas and Arizona.

SHOWY ERIGERON, *E. speciosus* var. *speciosus* (fig. 290). Stems 3–6dm tall, each with 1–10 heads; rays many (75–150) and narrow, blue or violet; involucres usually somewhat glandular, with some long, non-glandular hairs; leaves usually 3-nerved, mostly smooth; basal ones oblanceolate, petiolate, stem leaves lanceolate, not noticeably smaller. Found along roadsides, on open slopes, and among rocks throughout our range. The form most commonly seen in Colorado is ASPEN ERIGERON, *E. speciosus* var. *macranthus.* Its stems tend to lean and usually have 3–5 heads, which lack the long hairs on the involucre. Occurs in aspen groves and on montane and subalpine meadows.

290. Showy Erigeron, ½X

E. formosissimus is a similar but more hairy plant, usually growing in clumps having several stems, which are curved at base. Occurs on open ground throughout our range from the foothills to timberline.

COULTER ERIGERON, *E. coulteri*, is a hairy plant 25–50cm tall, with 50–100 white rays. Slightly woolly involucre, though not as dense as those hairs on E. e*latior.* They darken the involucre because there are dark crosswalls near the base of each hair. Lower leaves broadly oblanceolate to oval, petiolate; upper leaves broadly lanceolate, clasping, occasionally dentate. Grows in meadows and at forest edges in the montane and subalpine zones from Wyoming south to New Mexico. In Colorado, it is common on the western slope, rare east of the Continental Divide. Named for John Merle Coulter of the University of Chicago, who, as a young man, had accompanied F. V. Hayden on exploring expeditions and collected plants at high altitudes, in Colorado in particular. He prepared the first manual of Rocky Mountain botany, which was later revised and expanded by Aven Nelson.

Erigerons of the alpine and subalpine zones are usually under 20cm tall. BLACK-HEADED ERIGERON, *E. melanocephalus*, has solitary heads, white rays, and dark purplish and woolly involucres with long hairs. The dark effect is caused by purple pigment in the crosswalls of these hairs. This is a plant of the subalpine and alpine regions, especially around melting snow. Wyoming to Utah and New Mexico. ONE-FLOWERED ERIGERON, *E. simplex* (fig. 291), has leaves mostly basal, oblanceolate, entire; heads solitary; rays blue or purplish; involucre woolly with long white or grayish hairs. Alpine tundra from Montana south to New Mexico and Arizona. ROCK-SLIDE ERIGERON, *E. leiomeris.* Stems scattered or few together from a long taproot or branching caudex extending among loose rocks; basal leaves clustered, spatulate, and petiolate; stem leaves few and much smaller; heads solitary, rays blue or violet. Found among rockslides in the subalpine and alpine zones from Wyoming and Idaho south to New Mexico and Nevada. PINNATE-LEAF ERIGERON, *E. pinnatisectus* (fig. 292), is a tundra plant forming tufts of fern-like leaves, with stems 10–13cm tall; heads about 2.5cm broad with purple rays and bright yellow or orange disks. Alpine regions from Wyoming to New Mexico.

(19) *Eupatorium*. Thoroughwort or boneset

1. Leaves large (6–10cm long), usually in whorls of 3–5.

> *E. maculatum*
> Joe-Pye weed

1. Leaves small (2–5cm long), opposite.

> *E. herbaceum*
> Thoroughwort

JOE-PYE WEED, *Eupatorium maculatum,* is a coarse plant with purple-spotted stems 6–15dm tall; leaves whorled, ovate and lanceolate, short-petioled, and sharply toothed; heads about 10mm high, pink or purple, in large, flat-topped clusters. Widely distributed in North America; found in foothill meadows in the Rocky Mountain region. THOROUGHWORT, *E. herbaceum*, with stems 3–7.5dm tall; leaves opposite, ovate or triangular; flowers whitish. Occurs in foothill canyons and on rocky slopes or southwestern Colorado, Utah, Arizona, and New Mexico.

(20) *Gaillardia*. Blanketflower

1. Leaves deeply and narrowly pinnatifid.

> *G. pinnatifida*
> Pinnate-leaf gaillardia

1. Leaves entire, toothed, or broadly pinnatifid.
   2. Disk flowers red-purple.

> *G. aristata*
> Gaillardia or blanket flower

   2. Disk flowers yellow.

> *G. spathulata*

GAILLARDIA or BLANKETFLOWER, *Gaillardia aristata* (fig. 293). Heads large, 5–7.5cm broad, showy; rays yellow or light orange-yellow, sometimes reddish at base, wedge-shaped, 3-lobed or toothed; disk flowers dark-red when open; corollas of disk flowers bear hairs. Plants with rays partially or entirely red are probably escaped cultivars. Dry slopes and meadows of the foothill and montane zones throughout our range. The PINNATE-LEAF GAILLARDIA, *G. pinnatifida*, occurs in the foothills from southern Colorado southward.

291. One-flowered
Erigeron, ½X

292. Pinnate-leaf Erigeron,
½X

293. Gaillardia, ½X

(21) *Gnaphalium*. Cudweed

1. Heads in leafy-bracted clusters; plants usually under 20cm tall.
    2. Plants loosely and not uniformly tomentose; leaves spatulate or oblanceolate.
                                                  G. palustre
                                               Lowland cudweed
    2. Plants appressed tomentose; leaves linear to narrowly oblanceolate.
        3. Leaves narrowly oblanceolate, some over 3mm wide.
                                                G. uliginosum
                                             Marsh cudweed
        3. Leaves linear, none over 3mm wide.
                                                G. exilifolium

1. Heads not leafy-bracted clusters; plants usually over 20cm tall.
    2. Leaves green and glandular above, tomentose below.

                  G. viscosum
                  Sticky cudweed

    2. Leaves gray-tomentose on both sides.
        3. Leaves slightly decurrent or not at all; involucral bracts white.

                  G. wrightii

        3. Leaves moderately decurrent; involucral bracts straw-colored or yellowish.

                  G. chilense
                  Western cudweed

WESTERN CUDWEED, *Gnaphalium chilense*, is somewhat similar to pearly everlasting but may be easily distinguished upon close observation. Its phyllaries are cream-colored or yellowish and very satiny. Its stems and leaves are coated with cottony hairs and are yellowish-green, rather than grayish as are those of pearly everlasting. Also, cudweed is a sticky plant. Occurs infrequently in the foothill and montane zones of the Rockies.

(22) *Grindelia.* Gumweed

1. Heads discoid, no rays.
    2. Perennials, 50–150cm tall; involucral bracts thickened and leathery at tip; leaves resinous-punctate.
                  G. fastigiata
    2. Annual or biennial plants; involucral bracts only slightly thickened or leathery; upper leaves oblong to oblanceolate, often entire.
                  G. aphanactis
1. Heads with ray flowers.
    2. Biennials.
        3. Cauline leaves oblanceolate, narrowed to a petiole-like base.
            4. Stems 2 or more, 2–4dm tall.
                  G. subalpina var. *subalpina*
                  Gumweed
            4. Stems usually solitary, 4–8dm tall.
                  G. subalpina var. *erecta*
        3. Cauline leaves oblong, sessile, clasping base.
            4. Leaves entire or remotely serrulate; lower leaves may be coarsely and irregularly toothed.
                  G. squarrosa var. *quasiperennis*
            4. Leaves closely and evenly serrulate.
                5. Upper and middle leaves 2–4 times as long as wide.
                  G. squarrosa var. *squarrosa*
                  Gumweed
                5. Upper and middle leaves 5–8 times as long as wide.
                  G. squarrosa var. *serrulata*
    2. Perennials.
        3. Stems and leaves glabrous.
            4. Leaves 2–8cm long, 3–7 times longer than wide, entire to dentate.
                  G. decumbens var. *decumbens*
            4. Leaves 2–6cm long, 6–15 times longer than wide, entire, denticulate to pinnatifid.
                  G. decumbens var. *subincisa*
        3. Stems and leaves crisped-hairy or subscabrous.
                  G. scabra

Gumweed, *Grindelia squarrosa* (fig. 294), is easily recognized by the slender recurved tips of its phyllaries (like little curls) and by its distinctly sticky heads. In bud the disk has a smooth, white, gummy covering. Stems are 20–45cm tall, and leaves are pinnately-lobed or irregularly-toothed. Found on fields and hillsides in the foothill and montane zones, especially in disturbed soil, mainly in Wyoming and Colorado.

(23) *Gutierrezia*. Matchbrush or snakeweed

1. Disk and ray flowers each 3–10 in each head.

G. sarothrae var.
sarothrae
Broom snakeweed

1. Disk and ray flowers each only 1–2 in each head.

G. microcephala
Snakeweed

Broom Snakeweed, *Gutierrezia sarothrae*, is a bushy plant somewhat resembling the goldenrods. Stems 20–60cm tall from a woody base; leaves very narrow; many small, yellow heads clustered at ends of the branches. A plant of dry ground, abundant on the western plains but may be seen in dry situations anywhere in our region below the subalpine zone.

294. Gumweed, ½X

(24) *Haplopappus*. Goldenweed

This genus contains many yellow-flowered species, which vary considerably in size and general appearance. The main characters they have in common are: alternate leaves, phyllaries imbricated in several rows, rays and disk flowers yellow, and pappus of few or many bristles. The species differ so much from each other that some botanists separate them into several distinct genera. These separations are included in the following key to indicate bases for segregation.

1. Leaves with spinulose-tipped teeth.
    2. Involucral bracts papery, with green tips. (*Sideranthus* section)
        3. Annuals; leaves pinnatifid; stem slender, branched.
H. gracilis
Star goldenweed
        3. Perennials; leaves pinnatifid; stems from woody caudex; involucres not glandular.
H. spinulosus
Spiny goldenweed
    2. Involucral bracts more or less foliaceous. (*Pyrrocoma* section)

3. Involucral bracts with broad, obtuse tips.

        H. croceus
        Curlyhead goldenweed

3. Involucral bracts acute or acuminate.
    4. Heads mostly solitary, terminal, rarely 1–2 smaller ones in the upper axils.
        5. Heads large, the disk more than 2cm in diameter.
            6. Stem and involucre villous.

        H. clementis

            6. Stem and involucre glabrate.

        H. integrifolius
        5. Heads smaller, disk less than 2cm in diameter.

        H. uniflorus
        Plantain goldenweed

    4. Heads several to many.

        H. lanceolatus
        Lanceleaf goldenweed

1. Leaves without spinulose-tipped teeth.
  2. Involucral bracts foliaceous, at least in part.
    3. Plants 1dm or more high.
      4. Low shrubs. (*Macronema* section)
        5. Ray flowers absent; twigs tomentose.

        H. macronema
        Macronema

        5. Ray flowers present; twigs not tomentose.

        H. suffruticosus
        Shrubby goldenweed

      4. Plants herbaceous. (*Oreochrysum* section)

        H. parryi
        Parry goldenweed

    3. Plants only a few cm high. (*Tonestus* section)
      4. Stems very short, 1–5cm; involucral bracts obtuse.

        H. pygmaeus
        Pygmy tonestus

      4. Stems slender, 5–15cm; involucral bracts acute.

        H. lyallii
        Alpine tonestus or
        goldenweed

2, Involucral bracts not foliaceous.
    3. Stems leafy; involucral bracts tipped with stiff, sharp point. (*Oonopsis* section)
      4. Plants under 1dm high, somewhat caespitose.
        5.Rays 6–10.

        H. muticaulis

        5. Rayless.

        H. wardii
    3. Stems scapose, at least naked above. (*Stenotus* section)
      4. Involucral bracts mostly acute, usually green throughout.

        H. acaulis
        Stemless goldenweed
      4. Involucral bracts mostly obtuse or rounded, usually green only at tips.

        H. armerioides
        Thrift goldenweed

SPINY GOLDENWEED, *H. spinulosus* (fig. 295). Stems 20–75cm tall, usually several from a woody base; leaves serrate or pinnately divided, teeth or lobes bristle-tipped; head yellow, 12–25mm broad; phyllaries spreading, papery, with green midrib, bristle-tipped. Found on dry ground, in the plains and foothills along the eastern base of the Rockies. STAR GOLDENWEED, *H. gracilis*, is a much-branched, annual plant that is rather similar to the last, with bristle-tipped leaves and phyllaries, and yellow heads. Grows on dry slopes of the foothill and montane zones from western Colorado southward.

MACRONEMA, *H. macronema*, is a small, rounded shrub, 15–38cm tall. It has white-woolly branches, many rayless yellow heads, and phyllaries all about the same length. Occurs in subalpine and alpine zones from Idaho to Colorado and to California.

PARRY GOLDENWEED, *H. parryi* (fig. 296). Stems 15–45cm tall; heads clustered, about 12mm broad, with short, light yellow rays and yellow disk flowers. Found in the montane and subalpine zones, usually in open woods or on the edge of forest, from Wyoming to New Mexico and Arizona. Occurs on the North Rim, Grand Canyon National Park.

LANCELEAF GOLDENWEED, *H. lanceolatus*, is a plant with upward-curving stems 20–45cm tall; many yellow heads 12–25mm broad; leaves lance-shaped, mostly with sharp, spine-tipped teeth, leaves few on upper stems. Grows on moist, alkaline soil in the montane zone. Saskatchewan to British Columbia, south to western Nebraska, south-central Colorado, and northeastern California. PLANTAIN GOLDENWEED, *H. uniflorus*, is similar, but its stems are reddish, 10–43cm tall, often with loose cottony coating; heads usually solitary. Grows on alkaline soil from Montana and Oregon to Colorado, especially on geyser formations in Yelllowstone National Park.

CURLYHEAD GOLDENWEED, *H. croceus* (fig. 297), is a stout plant 25–60cm tall.

295. Spiny Goldenweed, ½X

296. Parry Goldenweed, ½X

297. Curlyhead Goldenweed, ½X

Leaves large, alternate; heads brilliant, solitary, about 7.5cm broad. The orange rays tend to curl backward as the flower heads age; pappus tawny. Grows in montane and subalpine meadows from Wyoming and Utah to New Mexico and Arizona; west of the Front Range in Colorado and occasionally seen on the western slope of Rocky Mountain National Park.

EVERGREEN GOLDENWEED, *H. acaulis*, is a mat-forming plant with a branched, woody base. Leaves basal, narrow, 3-nerved, 1–5cm long; flower stalks 4–11.5cm tall, with few small leaves or none; heads up to 3cm broad, with 6–15 yellow rays. Found on badlands, dry slopes, and rocky ridges of the foothill and montane zones from Montana and Idaho to Colorado and California.

ALPINE GOLDENWEED or TONESTUS, *H. lyalii*. Stems slender, leafy, 5–15cm tall; heads solitary, 12–25mm broad; rays more than 15, yellow; phyllaries lanceolate, tapering to a sharp point, glandular. An alpine plant, found in the high mountains from Alberta and British Columbia to Colorado and Oregon. PYGMY TONESTUS, *H. pygmaeus*, is even smaller than the above, forming low cushions of narrow leaves. The bright yellow flowering heads are scarcely raised above the leaves on short, leafy stems. Phyllaries are leaf-like and blunt. Alpine, from Wyoming to New Mexico.

(25) *Helenium*. Sneezeweed

1. Stem leaves decurrent, forming wings.

> *H. autmnale* var. *montanum*

1. Stem leaves not decurrent.

> *H. hoopesii*
> Orange sneezeweed

ORANGE SNEEZEWEED, *Helenium hoopesii*, is a stout plant, each stem bearing several orange-yellow flower heads in a rounded or flat-topped inflorescence. The heads have a mound-shaped disk nearly 2.5cm broad, and the rays are rather drooping. Lower leaves are oblanceolate or obovate, 12–25cm long, the upper ones progressively smaller and sessile. May be distinguished from curlyhead goldenweed by its clustered heads. Found in meadows and open areas of the upper foothill, montane, and low subalpine zones from Wyoming south to New Mexico. Abundant on the western slope in Colorado, rare on the eastern slope.

(26) *Helianthella*. Little sunflower

1. Involucral bracts ovate, strongly graduated; heads numerous; disks purple or brown; rays 5–6mm long.

> *H. microcephala*

1. Involucral bracts lanceolate or linear-oblong, subequal; heads solitary or few; disks yellow rays 20mm or longer.
   2. Involucral bracts pubescent, not conspicuously ciliate; enlarged basal leaves absent.

> *H. uniflora*
> One-flower helianthella

   2. Involucral bracts conspicuously ciliate, little or no pubescence on their faces; enlarged basal leaves present.

3. Plants tall, stout; leaves up to 50cm long.

         H. quinquenervis
         Little sunflower

3. Plants low, slender; leaves under 10cm long.

         H. parryi
         Parry helianthella

LITTLE SUNFLOWER, *Helianthella quinquenervis* (fig. 298). Stems erect, 3–12dm tall, unbranched, usually with large basal leaves; stem leaves usually 4 pairs, shiny, 5-nerved; heads 8–10cm broad, solitary (or with a few smaller ones below); rays light yellow, disk darker. Found in open, moist woods, montane and subalpine, from Montana to Mexico.

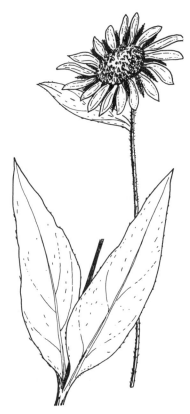

ONE-FLOWER HELIANTHELLA, *H. uniflora*, is similar but lacks the large basal leaves. Both rays and disk are bright-yellow; leaves 3-nerved (although the nerves are not a reliable character on either of these species). One of the conspicuous plants of Yellowstone National Park, it occurs from Montana through Wyoming to northwestern Colorado.

PARRY HELIANTHELLA, *H. parryi*, is a smaller plant, about 3dm tall, with many long-petioled basal leaves. Stem leaves 1–2 pairs, smaller, 3-nerved; heads yellow, about 5cm broad. Occurs in the montane and subalpine zones, especially in ponderosa pine forests, from central Colorado south to New Mexico and Arizona.

298. Five-nerved Helianthella, ½X

(27) *Helianthus.* Sunflower

1. Plants annual.
   2. Lower leaves broadly ovate, somewhat cordate.

         H. annuus
         Kansas sunflower

   2. Leaves ovate-lanceolate to narrowly oblong-lanceolate, all cuneate at base.

         H. petiolaris
         Prairie sunflower

1. Plants perennial.
   2. Lobes of disk flowers red or purple; cauline leaves lanceolate to ovate.

         H. rigidus
         Showy sunflower

   2. Lobes of disk flowers yellow.

3. All the leaves opposite, except for foliar bracts.

H. pumilus
Bush sunflower

3. Some or many of the upper leaves alternate.
    4. Stem and leaves scabrous.

H. maximilianii
Maximilian's sunflower

    4. Stem smooth or nearly so, often glaucous.
        5. Leaves usually coarsely toothed, somewhat pubescent (not scabrous) beneath.

H. grosse-serratus
Sawtooth sunflower

        5. Leaves entire or denticulate, glabrate or scabrous beneath.
            6. Stems mostly 20–30dm tall.

H. nuttallii ssp. *nuttalli*
Nuttall's sunflower

            6. Stems mostly 5–20dm tall.

H. nuttallii ssp. *rydbergii*

299. Bush Sunflower, ½X

BUSH SUNFLOWER, *Helianthus pumilus* (fig. 299). Several stiff stems 3–7.5dm tall from a woody, perennial crown give this plant a bushy appearance. The opposite leaves are ovate to lanceolate, with short petioles, very rough and dull-green; heads 4–8cm broad with yellow rays, disks dingy-yellow or brownish. The name *Helianthus* comes from two Greek words, *helios* (the sun) and *anthos* (the flower). Common on dry slopes of the foothills in Wyoming and Colorado, mostly on the eastern slope.

NUTTALL'S SUNFLOWER, *H. nuttallii*, has erect, unbranched stems up to 3m tall, with several entirely bright yellow heads 5–10cm broad on short branches along their upper portions. Closely related to the Jerusalem artichoke of the central states, *H. tuberosus*, and, like that plant, has tubers on its roots, which are dug by bears and other animals for food. Grows in marshy places and along ditch banks of the foothill and montane zones from Alberta south to New Mexico and Arizona.

KANSAS SUNFLOWER, *H. annuus*, is commonly seen along roadsides, on fields, and on dry plains. Heads 5–10cm broad, with bright yellow rays, the disks dark brown or black; phyllaries bristly-hairy with bristle-fringed margins and with long-pointed tips. PRAIRIE SUNFLOWER, *H. petiolaris*, is very similar, but its phyllaries lack the bristly fringe and the long tips. The two are often found growing together and hybridize freely. Western plains and foothills during late summer from Canada to Mexico. Seeds of all the sunflowers furnish important food for birds and small mammals.

(28) *Heterotheca*. Golden aster

1. Upper part of plant and involucre densely pubescent, hirsute to strigose-sericeous and canescent; usually not glandular.

H. villosa

1. Upper part of plant and involucre less densely pubescent, greenish; glands conspicuous.
    2. Heads somewhat sessile, usually closely subtended by leaves of the peduncle, which are similar to middle and upper stem leaves.

H. fulcrata

    2. Heads appear to have peduncles, subtended by bracts much smaller than middle or upper stem leaves.

H. horrida

GOLDEN ASTER or GOLDENEYE, *Heterotheca villosa* (fig. 300), is a grayish-hairy plant usually having several leafy, ascending stems. Leaves, except the uppermost, petiolate. The yellow, 2.5cm broad heads, surrounded by leafy bracts, are at the ends of the stem or on short branches. Occurs on dry foothill and lower montane slopes throughout our range.

(29) *Hieracium*. Hawkweed

1. Basal and lower cauline leaves small, deciduous early; other leaves more numerous and sessile.
    2. Leaves mostly subentire, narrowly lanceolate.

H. umbellatum
Narrow-leaved hawkweed

    2. Leaves mostly broadly lanceolate or ovate-oblong, acutely dentate.

H. canadensis
Canada hawkweed

1. Basal or lower cauline leaves conspicuously larger than progressively reduced middle and upper leaves.
    2. Achenes tapering from base to summit.

H. fendleri
Fendler hawkweed

    2. Achenes cylindrical.
        3. Flowers white or ochroleucous.

H. albiflorum
White hawkweed

        3. Flowers yellow.
            4. Heads small, black-hairy.

H. gracile
Slender hawkweed

            4. Heads large, white-hairy or glabrate.
                5. Leaves densely long-hirsute.

H. scouleri
Woolly-weed

                5. Leaves smooth or nearly so.

H. cynoglossoides
Hounds-tongue
hawkweed

SLENDER HAWKWEED, *Hieraceum gracile* (fig. 301). Stem usually solitary to 4dm high, with a few, small, dandelion-like yellow heads; involucres densely black-hairy; leaves mostly basal, obovate or oblong-spatulate. Commonly found in moist woods of the montane and subalpine zones, but also in the alpine

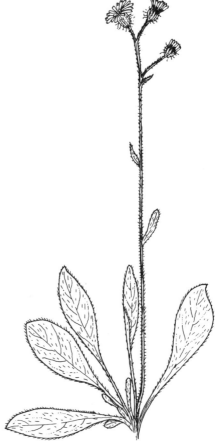

300. Golden-aster, ½ X   301. Slender Hawkweed, ½ X

zone. WHITE HAWKWEED, *H. albiflorum,* has white or creamy flowers; lacks the black hairs on the involucre but has tawny, bristly hairs on the leaves. Occurs in similar situations.

(30) *Hymenopappus.* Hymenopappus

1. Conspicuous white ray flowers.
                                                    H. newberryi
1. Ray flowers absent.
    2. Basal leaf-axils without dense tomentum, glabrous or sparsely tomentose.
        3. Ultimate peduncles 0.5–1.5cm long.
                                                    H. filifolius var. *parvulus*
                                                    Dusty maiden
        3. Ultimate peduncles 2–10cm long.
                                                    H. filifolius var. *nudipes*
    2. Basal leaf-axils with dense tomentum.
        3. Flowers averaging 20 per head; corollas 2–3mm long.
                                                    H. filifolius var. *luteus*
        3. Flowers 20–70 per head; corollas 3–7mm long.

4. Anthers 3–4mm long.
    5. Flowers white.

                    H. filifolius var. *ericopodus*

    5. Flowers yellow.

                    H. filifolius var. *megacephalus*

4. Anthers 2–3mm long.
    5. Heads 1–40 per stem; stem leaves 0–12; Rocky Mountains, usually west of Continental Divide.

                    H. filifolius var. *cinereus*
    5. Heads 5–60 per stem; stem leaves 3–8; plains east of Continental Divide.

                    H. polycephalus

(31) *Hymenoxys.* Hymenoxys

1. Leaves entire, usually linear.
    2. Involucral bracts sparsely hairy to glabrous; leaves glandular-punctate.

                    H. torreyana
    2. Involucral bracts usually densely hairy.
        3. Leaves sericeous, whether sparingly or densely; rays sometimes lacking.

                    H. acaulis var. *acaulis* Actinea
        3. Leaves lanate-villous, whether sparingly or densely, or glabrate; rays always present.
            4. Plant acaulescent; leaves all basal.
                5. Leaves conspicuously glandular-punctate.

                    H. acaulis var. *arizonica*
                5. Leaves not glandular, but may be punctate or epunctate; the alpine form.

                    H. acaulis var. *caespitosa*
             4. Plants caulescent; leaves both basal and cauline.
                H. ivesiana
1. Leaves not entire, some ternately or pinnately divided.
    2. Herbage glabrous or becoming glabrate.
        3. Heads 1–3 per stem.

                    H. richardsonii var. *richardsonii* Colorado rubber plant

        3. Heads often 5 or more per stem.

                    H. richardsonii var. *floribunda*

    2. Woolly alpine; low perennial; large heads.
        3. Involucre short-villous; leaves glabrate.

                    H. brandegei

        3. Involucre woolly; leaves with woolly tufts.

                    H. grandiflora Rydbergia

ACTINEA, *Hymenoxys acaulis* (fig. 302), is an interesting plant, with silvery-silky or woolly leaves and bright-yellow heads. Leaves narrow, 2.5–5cm long, crowded on a caespitose caudex; heads either embedded in the leaves or on slender, leafless stalks 5–38cm tall; ray flowers usually 3-toothed with orange veins. Grows in rocky places of the foothill and alpine zones. Blooms in the foothills in early spring with very short stalks, but with taller ones later in the season. Blooms on the tundra in late June or July, its yellow heads nestled among the silvery leaves. This alpine form is var. *caespitosa.*

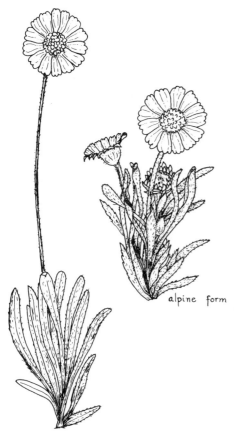

302. Actinea, ½ X

alpine form

RYDBERGIA, *H. grandiflora* (fig. 303). With its large, golden heads and comparatively short stems, this plant is the most showy and easiest to recognize of all our alpine wild flowers. Heads 2.5–5cm broad; rays crowded, 3-lobed; foliage and involucre woolly or cottony; leaves divided into slender segments. Abundant on exposed alpine ridges, from southwestern Montana and east central Idaho to New Mexico and Utah, on Trail Ridge in Rocky Mountain National Park, and on the Beartooth Plateau northeast of Yellowstone. Its heads always face toward the rising sun and remain facing eastward. Named in honor of Per Axel Rydberg, Swedish-born American botanist who greatly contributed to our knowledge of Rocky Mountain flora. *H. brandegei*, of the southern Rockies, is similar.

COLORADO RUBBER PLANT, *H. richardsonii*, if a relative of *H. acaulis* and *H. grandiflora*, is quite different in general appearance. A much-branched plant with several stems from a branched woody caudex, 10–50cm tall; many yellow flower heads 12–20mm broad; leaves divided into 3–7 narrow segments, in masses at base of stems and along stems, with tufts of woolly hairs in the leaf axils; phyllaries in 2 series, the inner ones narrow, the outer broader, keeled and united at base. Found on dry plains and rocky slopes in the foothill and montane zones throughout our range.

(32) *Krigia*. Dwarf dandelion

Only 1 species is likely to be found in our range: *Krigia biflora*, a slender herbaceous perennial with a forking stem. Leaves mostly basal, oblanceolate to spatulate; cauline leaves 1–3; ray flowers only, yellow to orange.

(33) *Kuhnia*. Kuhnia

(This genus is sometimes lumped with *Brickellia*.)
1. Leaves mostly linear, entire or with a pair of basal teeth.
　　　　　　　　　　　　　　　　　　　　　　　K. chlorolepis
1. Leaves linear-lanceolate to oblong, mostly toothed or laciniate.
　　　　　　　　　　　　　　　　　　　　　　　K. eupatorioides var.
　　　　　　　　　　　　　　　　　　　　　　　corymbosa

(34) *Lactuca.* Wild lettuce

1. Pappus brown; achenes beakless.

L. spicata

1. Pappus bright white; achenes with slender beak.
    2. Leaves spiny-margined; midrib and veins may be spiny.
        3. Leaves oblong-lanceolate, sinuate-pinnatifid.

L. serriola
Prickly lettuce

        3. Leaves oblanceolate, only irregularly denticulate.

L. sativa

    2. Leaves not spinose.
        3. Flowers yellow.
            4. Involucre with calyx-like outer set of bracts.

L canadensis

            4. Involucre imbricated.

L. ludoviciana

        3. Flowers blue to purplish.
            4. Plants perennial.

L. tatarica ssp. *pulchella*
Large-flowered blue
lettuce

            4. Plants annual or biennial.

L. graminifolia

(35) *Liatris.* Gayfeather or blazing star

1. Pappus bristles distinctly plumose.

L. punctata
Kansas gayfeather

1. Pappus bristles barbellate, the barbs hardly visible.
    2. Heads 9–15-flowered; inflorescence of many heads.

L. lancefolia

    2. Heads 40–70-flowered; inflorescence of few to several heads.

L. ligulistylis

Clusters of erect, bright rose-purple spikes seen along roadsides and on fields in late summer show that gayfeather is in bloom. KANSAS GAYFEATHER, *L. punctata* (fig. 304), has stems 15–60cm tall, thickly set along the upper portion with bright-purple heads. The stem is covered with narrow, rough-edged leaves, which are gradually reduced in length toward the inflorescence. The pistils have 2 thread-like, twisted, purple appendages at their tips, which give the flower heads a feathery look. Grows in dry situations of the plains, foothills, and lower montane zone along the eastern side of the Continental Divide from South Dakota to Texas and New Mexico.

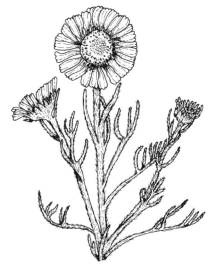

303. Rydbergia, ²⁄₅X

(36) *Lygodesmia*. Skeleton weed

1. Rays 10–12mm long.

L. juncea
Rushpink

1. Rays 1.5–2.5cm long.
   2. Rays 8–20 per head.

L. grandiflora
Milkpink

   2. Rays average 5 per head.

L. arizonica

Milkpink, *Lygodesmia grandiflora*. Stems 15–45cm tall; leaves linear, 5–10cm long, firm and ascending. Heads at the ends of branches appear to be 5–10-petaled pink flowers; actually, each "petal" is a separate flower, as in other members of this family. In dry, gravelly soil of the foothill and montane zones, from Wyoming to Idaho, south to New Mexico and Arizona. Notable in Colorado National Monument.

Rushpink or Skeleton Weed, *L. juncea*, has bare, green stems. Its lower leaves are not over 5cm long, narrow and stiff; upper leaves are reduced to mere scales. Flower heads are pink, similar to *grandiflora* but smaller. On dry foothill slopes in Colorado and New Mexico.

(37) *Machaeranthera*. Tansy-aster

1. Ray flowers lacking.

M. grindelioides var.
grindelioides

1. Ray flowers present.
   2. Leaves, especially the lower, once- or twice-pinnatifid.
      3. Perennials; branched from the base.

M. pinnatifida

      3. Annuals.
         4. Green tips of involucral bracts short, appressed.

M. parviflora

         4. Green tips of involucral bracts long, spreading or recurved.

M. tanacetifolia
Tansy-aster

   2. Leaves entire or merely toothed.
      3. Plants perennial; leaves spinulose-toothed; stems mostly under 10cm tall.

M. coloradoensis
Colorado tansy-aster

      3. Plants annual or biennial.
         4. Leaves broad; heads few, large; phyllaries leafy; plant glandular-hispid, especially above.

M. bigelovii
Tansy-aster

         4. Leaves narrow; heads numerous; phyllaries not leafy, canescent, may be somewhat glandular.

M. canescens

Tansy-Aster, *Machaeranthera bigelovii* (fig. 305), is a widely-branched, annual or biennial, sticky plant with numerous, brilliant reddish-purple flower head; leaves lanceolate or oblanceolate, usually with sharply-toothed or wavy margins; phyllaries sticky, tapering into slender, recurved tips, the best charac-

Rays purple

304. Kansas Gayfeather, ½X   305. Tansy-Aster, ½X

ter by which to distinguish this plant from the true aster (See Plate G, fig. 64). *M. tanacetifolia* is similar, but its leaves are once- or twice-pinnately-divided into narrow segments. The flower heads are distinctive in color. In late summer, a border of bright reddish-purple along the highway is an almost sure sign of the presence of tansy-asters.

COLORADO TANSY-ASTER, *M. coloradoensis*, has violet to rose-purple ray flowers; heads about 2.4cm broad; leaves appressed-hairy with spine-tipped teeth; phyllaries sharp-pointed; up to 13cm tall. Limited to the subalpine zone in central and southwestern Colorado.

(38) *Microseris*. Microseris

One species occurs in our range, and it is one of the false dandelions: MICROSERIS, *Microseria nutans*. Yellow perennials, few to several heads; leaves mostly basal; pappus of plumose bristles; the heads nodding in bud.

(39) *Nothocalais*. False-dandelion

(Sometimes placed in *Microseris* and similar to *Agoseris*.)

1. Pappus of 10–30 slender scales; involucral bracts broadly lanceolate to ovate, dotted black-purple.
                                                                    N. nigrescens
1. Pappus of 30–80 members, some narrow capillary bristles; involucral bracts lance-linear to lanceolate; basal leaves with white-hairy margin.
                                                                    N. cuspidata

306. Golden-shower, ½ X

(40) *Pericome.* Golden shower

One species in our range: GOLDEN SHOWER or TAPERLEAF, *Pericome caudata* (fig. 306). A bushy, branched perennial 6–15dm tall, with bright-green, opposite, very long-tailed, petiolate leaves, cordate-hastate to deltoid; heads numerous, in cymes; flowers golden-yellow. Found on rocky slopes and in canyons of the foothills, southern Colorado, New Mexico, and Arizona.

(41) *Petasites.* Sweet coltsfoot

One species in our range: SWEET COLTSFOOT or ARROWLEAF COLTSFOOT, *P. sagittatus.* An uncommon, rhizomatous plant that grows in bogs or in water. Leaves variable, deltoid-oblong to reniform-hastate, the margins dentate to merely wavy. Flowers white to purplish, with conspicuous tufts of silvery-white pappus in seed. A northern bog plant extending from Alaska southward into Colorado.

(42) *Prenanthes.* Rattlesnake root

1. Basal leaves obovate; involucre loosely hirsute.

P. racemosa

1. Basal leaves triangular-sagittate; involucre glabrous.

P. sagittata

RATTLESNAKE ROOT, *Prenanthes racemosa.* Stems 3–9dm tall, with a narrow, elongated inflorescence. Heads 12–20mm long; flowers pink; phyllaries dark purplish or black, loosely hairy; lower leaves glabrous, oval to oblanceolate; petioles long, winged. Grows on moist ground in ravines and meadows; foothill and montane zones from Alberta to Colorado.

(43) *Ratibida.* Coneflower

1. Ray flowers 7–35mm long; disk in fruit cylindrical, 2–4cm long.

R. columnifera
Prairie coneflower

1. Ray flowers 3–8mm long; disk in fruit oblong, about 1cm long.

R. tagetes

PRAIRIE CONEFLOWER, *Ratibida columnifera*, is a branched plant to 6dm tall. Leaves pinnately divided into oblong or linear segments; heads with a cylindrical or columnar disk, 12–65mm high; rays yellow or purplish, more or less reflexed. On dry plains from Canada south to Colorado and Arizona.

(44) *Rudbeckia*. Coneflower

1. Leaves entire or sparingly serrate.
   2. Rays conspicuous.

   R. hirta
   Black-eyed susan

   2. Rays lacking.

   R. occidentalis
   Western rayless
   coneflower

1. Leaves lobed or divided.
   2. Rays conspicuous.

   R. laciniata var. *ampla*
   Wild golden-glow or tall
   coneflower

   2. Rays lacking.

   R. laciniata var.
   montana
   Colorado rayless
   cone flower

BLACK-EYED SUSAN, *Rudbeckia hirta* (fig. 307), is sometimes confused with gaillardia, but it may be distinguished by its pointed, orange rays and mound-shaped, dark-brown, almost black, disk. Stems 3–6dm tall; heads usually 1 to each stem or long branch; leaves rough-hairy, lanceolate. Widely distributed in North America, on meadows and in aspen groves of the foothill and montane zones in our mountains.

WILD GOLDEN-GLOW or TALL CONEFLOWER, *R. laciniata* var. *ampla* (fig. 308), has stems up to 6dm tall; heads with long, drooping yellow rays; disks cone-shaped, dull yellow or greenish. The large, smooth, long-petioled leaves are much divided. Grows along streams and on wet ground of the foothill and montane zones, from Canada south to New Mexico and Arizona. COLORADO RAYLESS CONEFLOWER, *R. laciniata* var. *montana*, is similar in habit of growth. Its foliage is smooth and grayish; its heads are without ray-flowers and look like stout, dark-greenish thumbs. Found in the mountains of west central Colorado. WESTERN RAYLESS CONEFLOWER, *R. occidentalis*, with stems 6–18dm tall, has similar rayless heads, but its leaves are ovate or ovate-lanceolate, the margins entire or toothed. Occurs from Wyoming and Montana westward.

(45) *Senecio*. Senecio or groundsel

A large genus containing many attractive wild flowers as well as several weedy species. The rays in our species are yellow (except in the case of *S. crocatus*); leaves are alternate; the involucre is made up of one row of narrow, equal phyllaries with a few smaller ones at the base (see Plate G, fig. 69). The

307. Black-eyed Susan, ½X   308. Wild Golden-Glow, ½X

tips of the phyllaries are often black, quite noticeable in some species. A few species have only disk flowers. Some species have nodding heads, meaning that the stalk is bent just below the head so that it faces outward or downward, not upward.

Senecios are either smooth or cobwebby, but they are never rough with stiff hairs as are the sunflowers, or sticky-hairy as are some of the arnicas. Common names for this group are particularly variable in the United States and other countries. Groundsel, ragwort, and butterwort are frequently encountered. The Latin name *Senecio* refers to *old man*—the silvery-white pappus, which is conspicuous as senecios go to seed, suggests white hair. The florists' cinerarias are of this genus.

1. Heads rayless or the rays inconspicuous.
  2. Heads nodding.
    3. Heads 8–12mm high, 8mm wide.

          S. pudicus

    3. Heads 12–20mm high, 15–20mm wide.

          S. bigelovii var. *hallii*
          Bigelow senecio

  2. Heads erect.
    3. Plants annual.

          S. vulgaris

    3. Plants perennial.
      4. Leaves lobed at least halfway to midrib.
        5. Stem leaves with rounded lobes.

          S. debilis

        5. Stem leaves with lobes usually pointed.
          6. Corolla lobes yellow; 6–20 heads.

          S. indecorus

          6. Corolla lobes red or orange; 1–6 heads.
          S. pauciflorus

      4. Leaves sharply and irregularly toothed.

          S. rapifolius
          Turnip-leaf senecio

1. Heads with conspicuous ray-flowers.
  2. Heads nodding, at least in bud.
    3. Leaves pinnatifid; plants tomentose, at least on the lower leaf surfaces.
          S. taraxacoides
          Dandelion-leaf senecio

    3. Leaves smooth, ovate, oblong to roundish, not pinnatifid.
      4. Leaves and stalks reddish-purple.

          S. soldanella
          Alpine senecio

      4. Leaves bright-green.

          S. amplectens var. *holmii*
          Daffodil senecio

  2. Heads erect.
    3. Stems leafy, 6–12dm tall; leaves 5–15cm long.
      4. Leaves triangular, toothed; many heads.

          S. triangularis
          Arrowleaf senecio

      4. Leaves linear or lanceolate, tips pointed, margins saw-toothed.
        5. Heads at center of involucre under 5mm wide.
          S. serra var. *serra*
          Toothed senecio

        5. Heads at center of involucre usually over 5mm wide.
          S. serra var. *admirabilis*
          Toothed senecio

    3. Stems usually under 6dm tall.
      4. Stems leafy, much branched; heads distributed over the rather bushy plant.
        5. Leaves narrow, margins smooth.

          S. spartioides
          Broom senecio

         5. Leaves pinnately divided into unequal segments.
          S. eremophilus var. *kingii*
          Western golden ragwort

      4. Stems not leafy, or leaves much reduced upward; unbranched except in the inflorescence.
         5. Plants tomentose, at least when young.
           6. At least some of the leaves pinnately divided.
             7. Achenes somewhat hirsute or minutely hispid.
          S. plattensis
             7. Achenes glabrous; heads 8–10mm high.
          S. fendleri
          Fendler senecio

6. Leaves not pinnately divided.
   7. Central head larger, on a shorter, stouter stalk than the surrounding heads; leaves yellowish-green.
      8. Involucral bracts conspicuously black-tipped.
         S. integerrimus var. *exaltus*
         Lambstongue
      8. Involucral bracts obscurely or not black-tipped.
         S. integerrimus var. *integerrimus*
   7. Heads all alike; leaves silver-gray or whitish.
      8. Heads numerous; rays 3–5; involucral bracts conspicuously black-tipped.
         S. atratus
         Black-tipped senecio
      8. Heads solitary or few; rays 8–12.
         S. canus
         Pursh senecio
5. Plants glabrous or becoming glabrate at flowering time.
   6. Leaves lobed, at least the upper ones.
      7. Stem leaves well-developed.
         8. Leaves mostly similar throughout.
            9. Stems about 10cm tall.
               S. fremontii var. *fremontii*
               Fremont senecio
            9. Stems 15–50cm tall.
               S. fremontii var. *blitoides*
               Rock senecio
         8. Leaves mostly dimorphic.
            9. Stem leaves lobed no more than halfway to midrib, leaves usually clasping.
               S. dimorphophyllus
            9. Stem leaves pinnatilobate, not clasping.
               S. pauperculus
               Balsam groundsel
      7. Stem leaves progressively smaller than basal leaves; leaves mostly dimorphic.
         8. Heads 1–2 (3); plant with slender rhizomes.
               S. cymbalarioides
          8. Heads usually more than 3; plants with caudex or stout rhizome.
            9. Basal leaves truncate or subcordate, usually toothed.
               10. Rays orange-red.
                   S. crocatus
                   Saffron senecio
               10. Rays yellow.
                   S. pseudaureus
             9. Basal leaves tapering at base, elliptic or obovate, toothed.
               S. streptanthifolius
               Cleft-leaf groundsel
   6. Leaves entire or with small teeth; not dimorphic.
      7. Stem leaves few or wanting.
         8. Leaves greenish; heads 1 to several.
            S. werneriaefolius
         8. Leaves bluish-green; heads many.
            S. wootonii
            Wooton senecio

7. Stems equally leafy to apex; involucral bracts thickened,
with hairy black tips.

S. crassulus
Thick-bracted senecio

LAMBSTONGUE or COMMON SPRING SENECIO, *S. integerrimus* (fig. 309), is an early blooming, rather stout plant. It is somewhat untidy in appearance, with its cobwebby pubescence (which disappears as the plant grows older) and its rather irregularly arranged rays. The upper leaves clasp the stem and taper to a slender point. Found on foothill slopes and open pine forest from Minnesota westward and south to Colorado.

FENDLER SENECIO, *S. fendleri* (fig. 310), is somewhat similar to the last but usually has a more slender stem and more heads (in a more symmetrical, flat-

309. Common Spring Senecio or
Lambstongue Groundsel, ½X

310. Fendler Senecio, ½X

topped cluster). The basal leaves are pinnately divided into folded segments. Grows on dry foothill slopes and open pine forest from Wyoming to New Mexico. Rarely subalpine. Turnip-leaf Senecio, *S. rapifolius*, another plant of the lower altitudes, has obovate, sharply-toothed, petiolate basal leaves; heads numerous, deep yellow, rayless. Occurs on rocky, wooded canyon slopes and ridges from South Dakota to Idaho and south into northern Colorado.

Pursh senecio, *S. canus*, is widely distributed from the foothills to the alpine zone. A silvery-gray plant, usually not over 20cm tall. May have 1-several stems, each with 1-few heads. The leaves are obovate to oblanceolate, entire or more or less lobed. Western Canada south through Montana and Wyoming to Nebraska and Colorado.

Bigelow Senecio, *S. bigelovii*, is one of those senecios found most commonly in meadows and aspen groves of the montane and subalpine zones. A rather stout plant, 3–9dm tall; involucres top-shaped; disk flowers yellow; phyllaries purplish. Found in Colorado, New Mexico, and Arizona; rarely southern Wyoming.

Thick-bracted Senecio, *S. crassulus*, has stems 23–50cm tall; narrow, tapered leaves 8–15cm long, the upper ones sessile. Occurs in the montane and subalpine zones, and in meadows above timberline. From South Dakota to Idaho and south to New Mexico and Utah. *S. crocatus* and *S. dimorphophyllus* are easily confused (check the key), but the former, with orange-red rays, is mostly at lower elevations; the latter, with yellow rays, is mostly subalpine and alpine.

Arrowleaf Senecio, *S. triangularis* (fig. 311), is a tall, lush plant forming large clumps, especially on very wet ground and along streams. Conspicuous in subalpine meadows in Rocky Mountain National Park. Saskatchewan to Alaska and south to New Mexico and California. Toothed Senecio, *S. serra* var. *admirabilis*, is about as tall as the last and with much the same habit, but usually with fewer flower heads. May be distinguished by the narrow,

311. Arrowleaf Senecio, ½X

long-pointed, toothed leaves. From Montana south to southern Colorado. Forms fields of yellow in Yellowstone National Park in July and August.

Several species are found mostly on disturbed soil along roads and trails or on rock-covered slopes: BROOM SENECIO, *S. spartioides,* has narrow leaves that sometimes have 1–2 pairs of short, narrow lobes at the base. Usually seen along highways and on roadsides, covered with yellow flowers in mid-summer or tufts of white pappus later. Nebraska and Wyoming south to Texas and Arizona. WESTERN GOLDEN RAGWORT, *S. eremophilus,* is another plant that brightens the roadsides in July and August. Occurs from Wyoming and Idaho south to New Mexico and Arizona. BLACK-TIPPED SENECIO, *S. atratus,* with its silvery foliage and numerous, small, yellow flower heads, is common along roads and trails of the subalpine zone. Found in Colorado, Utah, and New Mexico, especially along the Trail Ridge Road in Rocky Mountain National Park.

Other senecios are primarily of the alpine rock fields and tundra above timberline: ROCK SENECIO, *S. fremontii* var. *blitoides,* has bright yellow heads that make it easy to see, but it is not always easy to reach, as it flourishes among large, angular boulders or in rock crevices. Leaves oval or spatulate, coarsely toothed. Alpine zone in Colorado and New Mexico. FREMONT SENECIO, *S. fremontii* var. *fremontii,* is similar in appearance and occurs in similar situations in Montana and Wyoming. Found in Glacier and Yellowstone National Parks. DAFFODIL SENECIO, *S. amplectens,* is a handsome plant with long, light yellow rays and a darker disk. Its color and nodding head suggest the name daffodil. In exposed situations it may be very small, growing to 30cm or more in moist, sheltered spots.

ALPINE SENECIO, *S. soldanella,* is a low tufted plant with many basal, roundish, purplish leaves and yellow heads. Found on moist gravel on the high mountains of Colorado. DANDELION-LEAF SENECIO, *S. taraxacoides,* is 5–13cm tall, with leaves more or less coated or patched with white cobwebby hairs. It may be distinguished from true dandelions by the presence of both disk and ray flowers. Occurs on high alpine ridges and peaks of Colorado and New Mexico. WOOTON SENECIO, *S. wootonii,* has stems 25–50cm tall, usually several from a stout rootstock. Most of its smooth, glaucous leaves are basal, spatulate or obovate. The combination of bright yellow flowers and bluish-green leaves makes it especially handsome. Found on moist gravel slopes of the subalpine zone in Colorado and New Mexico. E. O. Wooton, for whom this species was named, was a specialist on the flora of New Mexico.

(46) *Solidago.* Goldenrod

With one exception, the individual heads of goldenrods are very small, but each head is made up of separate disk and ray flowers. The heads are clustered, and an inflorescence usually contains several clusters. Often they are arranged along one side of the branches, which are sometimes arched. The leaves are simple, alternate, and entire or slightly toothed. Because of great similarities among the species, identification can be difficult. Goldenrod pollen is comparatively heavy and is distributed by insects, not by the wind. Consequently, goldenrods are not one of the major causes of hay fever, as is commonly believed.

1. Heads large; involucres 8–12mm long; outer bracts large and foliaceous. (The species
has been sometimes placed in *Haplopappus.*)

> *S. parryi*
> Parry goldenweed or
> goldenrod

1. Heads small; involucres under 8mm long; the bracts small and not foliaceous.
2. Plants with well-developed creeping rhizomes; basal leaves usually not well-
developed.
3. Leaves punctate (sometimes rather obscurely); inflorescence corymbose.
4. Involucre bracts broadly lanceolate, somewhat blunt at tip.

> *S. graminifolia* var. *major*
> Fragrant goldenrod

4. Involucre bracts narrowly lanceolate, somewhat pointed at tip.

> *S. occidentalis*
> Western goldenrod

3. Leaves not punctate; inflorescence more paniculate.
4. Stems glabrous below infloresence.
5. Largest leaves at middle of stem.

> *S. gigantea* ssp. *serotina*
> Smooth goldenrod

5. Largest leaves toward the base. (Varieties poorly defined).

> *S. missouriensis*
> Missouri goldenrod

4. Stems hairy immediately below inflorescence.
5. Rays about 8 to a head.
6. Involucral bracts broadest near middle; leaves moderately hairy.

> *S. mollis*
> Velvety goldenrod

6. Involucral bracts broadest at base; leaves only sparsely hairy.

> *S. sparsiflora*
> Few-flowered goldenrod

5. Rays about 13 per head. (4 or 5 varieties may be found in our
range).

> *S. canadensis*
> Tall goldenrod

2. Plants usually with a short, stout rhizome or a simple caudex; basal leaves
usually well-developed.
3. Leaves glabrous (sometimes with ciliate margins).
4. Achenes glabrous.

> *S speciosa* var. *pallida*

4. Achenes hairy.
5. Rays about 13 per head; petioles on lower leaves ciliate.

> *S. multiradiata* var.
> *scopulorum*
> Northern goldenrod

5. Rays about 8 per head; none of the leaves with ciliate petioles.
6. Plants 0.5–1.5dm tall; basal leaves spatulate or obovate.

> *S. spathulata* var. *nana*
> Dwarf goldenrod

6. Plants 1.5–8dm tall; basal leaves oblanceolate.

> *S. spathulata* var.
> *neomexicana*

3. Leaves pubescent (may be puberulent).
4. Involucral bracts longitudinally striate; achenes glabrous.

> *S. rigida* var. *humilis*
> Hard-leaved goldenrod

4. Involucral bracts not striate; achenes hairy.
5. Disk flowers 5–9 per head, ray flowers same or more.

> *S. nemoralis* var.
> *longipetiolata*
> Gray or field goldenrod

5. Disk flowers 8–16 per head, rays usually fewer.
S. nana
Low goldenrod

SMOOTH or MISSOURI GOLDENROD, *S. missouriensis* (fig. 312), is a glabrous plant 25–40cm tall; leaves 3-nerved; inflorescence open and irregular. Foothill and montane zones throughout our range.

NORTHERN GOLDENROD, *S. multiradiata*, another smooth plant with its head in a narrow panicle (but not 1-sided), has reddish stems 2–5-cm tall. Each head has about 13 ray flowers; leaves obovate, 3-nerved, rounded teeth above, hairy margins at base. Found throughout out range from the upper foothills to the alpine zone. DWARF GOLDENROD, *S. spathulata* var. *nana*, is similar but smaller, 10–15cm tall, with red stems; rounded clusters of small heads; each head with about 8 ray flowers; no hairs along the margins of the leaf bases. Found on rocky slopes in the subalpine and alpine zones. *S. spathulata* var. *neomexicana*, a closely related, taller plant, occurs in the montane and subalpine regions.

LOW or ROUGH GOLDENROD, *S. nana*, is densely covered with short hairs; leaves are indistinctly 3-nerved. Grows in clumps in the foothills and the lower montane areas throughout our range.

312. Smooth Goldenrod, ½X

313. Few-flowered Goldenrod, ½X

FEW-FLOWERED GOLDENROD, *S. sparsiflora* (fig. 313), has finely pubescent foliage. Flower heads in elongated clusters, 1-sided; leaves distinctly 3-nerved. A plant of the foothill and montane zones in Colorado, especially on the western slope.

TALL GOLDENROD, *S. canadensis*, has stems 4.5–12dm tall; leaves 3-nerved, distinctly toothed; inflorescence of 1-sided branches, but more or less flat-topped. Found on wet meadows and along streams of the foothill and montane zones throughout our range.

(47) *Sonchus.* Sow thistle

1. Plants annual; involucre glabrous, not glandular.
    2. Auricles of the leaf bases acute.

> *S. oleraceus*
> Common sow thistle

    2. Auricles of the leaf bases rounded.

> *S. asper*
> Prickly sow thistle

1. Plants perennial; involucres glandular.
    2. Involucre glandular-pubescent, spreading yellow hairs.

> *S. arvensis*
> Field milk thistle

    2. Involucre glabrous but with glandular spots.

> *S. uliginosus*
> Marsh or swamp sow
> thistle

(48) *Stephanomeria.* Wire-lettuce

1. Plants annual.
    2. Leaves entire or sinuate; pappus plumose to the base.

> *S. virgata*

    2. Lower leaves more or less pinnatifid; pappus plumose only about the middle.

> *S. exigua*

1. Plants perennial.
    2. Pappus bristles brownish.

> *S. pauciflora*

    2. Pappus bristles white.
    3. Plants low, 1–3dm tall; larger leaves sharply cleft or pinnatifid.

> *S. runcinata*

        3. Plants taller, 2–7dm tall; leaves linear or filiform, entire or toothed.
        4. Involucre 7–11mm; plant moderately branched.

> *S. tenuifolia* var.
> *tenuifolia*

            4. Involucre 5–8mm; plant slender but branched.

> *S. tenuifolia* var.
> *myrioclada*

(49) *Taraxacum.* Dandelion

1. Introduced weedy, aggressive plants; leaves lobed more than halfway to midrib; outer involucral bracts usually reflexed. (Doubtful that the following two species should be segregated.)
    2. Achenes red, reddish-brown, or purple at maturity.

> *T. laevigatum*
> Red-seeded dandelion

    2. Achenes olive to brown at maturity.

> *T. officinale*
> Common dandelion

1. Native, nonaggressive plants; leaves less dissected; involucral bracts appressed to spreading, but not reflexed.
    2. Achenes red to reddish-brown at maturity; leaves often entire or slightly toothed.

> *T. eriophorum*
> Rocky Mountain
> dandelion

2. Achenes yellowish, olive, brown, or black at maturity (not reddish); leaves often lobed or coarsely toothed.
3. Achenes yellowish or olive to light-brown; horn-shaped swellings at the tips of the inner involucral bracts.

T. ceratophorum
Tundra dandelion

3. Achenes dark brown to blackish; inner involucral bracts rarely horn-shaped.

T. scopulorum
Dwarf alpine dandelion

The COMMON DANDELION, *Taraxacum officinale*, and the RED-SEEDED DANDELION, *T. laevigatum*, are common weeds throughout our area. TUNDRA DANDELION, *T. ceratophorum* (fig. 314), a native species, is frequently seen on grassy alpine slopes. It may be distinguished by the horn-shaped swellings at the tips of its inner phyllaries.

314. Tundra Dandelion, ½X

(50) *Tetradymia*. Horsebrush

1. Plants spiny.
2. Flowers 5–9 per head; branches always tomentose.

T. spinosa
Cotton-horn horsebrush

2. Flowers 4 per head; branches glabrate in age.

T. nuttallii

1. Plants not spiny.

T. canescens
Gray horsebrush

GRAY HORSEBRUSH, *Tetradymia canescens*, is a small, white-felted shrub that grows 20–60cm tall. It has narrow leaves and flower heads in small clusters at the ends of the branches. Each head has 4 whitish phyllaries and 4 yellow flowers. Grows on dry plains and hillsides of the foothill and montane zones from Montana south to New Mexico.

SPINY HORSEBRUSH, *T spinosa*, is a much-branched, very spiny shrub, up to 12dm tall. Its stems and young leaves are covered with a white coating. The heads, containing 5–9 flowers, are arranged along the branches. Occurs on dry plains and hillsides of the foothills from Montana to western Colorado and California.

(51) *Thelesperma*. Greenthread

1. Ray flowers conspicuous.

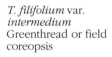

T. filifolium var.
intermedium
Greenthread or field
coreopsis

1. Ray flowers inconspicuous or absent.
    2. Leaves mostly near base.
        3. Leaves glabrous or nearly so.

T. marginatum

        3. Leaves conspicuously hairy.

T. pubescens

    2. Leaves distributed along the stem.

T. megapotamicum

315. Greenthread, ²/₅X

GREENTHREAD or FIELD COREOPSIS, *Thelesperma fili-folium* (fig 315), is a much-branched plant with bright green, smooth stem and leaves. Flower heads on long stalks, large, bright yellow; rays 3-lobed. The phyllaries are in 2 rows, the inner ones broad and united into a cup-shaped involucre, the outer ones slender and spreading. Leaves are pinnately divided into very slender, thread-like segments. A plant of the high plains, mesas, and foothills along the east side of the mountains from Nebraska and Colorado to New Mexico and Texas.

(52) *Townsendia*. Townsendia

1. Involucral bracts with strongly-acuminate points; plants caulescent.
    2. Stems erect and simple, at least below.
        3. Plants perennial; stems glabrate.

T. formosa

        3. Plants usually biennial; stems gray-canescent.
            4. Leaves spatulate (the upper becoming lanceolate).

T. eximia

            4. Leaves in a rosette, obovate-spatulate, tapering into a petiole.

T. parryi
Parry townsendia

    2. Stems branching widely from the base.

T. grandiflora
Showy townsendia

1. Involucral bracts obtuse or acute, but never strongly-acuminate.
  2. Plants caulescent or subcaulescent.
    3. Biennials (or winter annuals).
      4. Diminutive plants with almost filiform stems and roots; heads few.

T. fendleri
Fendler townsendia

      4. Stouter stems and roots; heads several to many.

T. strigosa

    3. Perennials.
      4. Leaves glabrous or sparingly hairy.

T. glabella

      4. Leaves with definite short, grayish hairs.

T. incana

  2. Plants strictly acaulescent.
    3. Plants with long, tangled, woolly hairs.
      4. Involucre 8–16mm wide, 6–10mm long.

T. spathulata

      4. Involucre 17–40mm wide, 8–18mm long.

T. condensata
Cushion townsendia

    3. Plants variously hairy, but not woolly or villous.
      4. Leaves glabrous or soon glabrate; alpine.

T. rothrockii

      4. Leaves more or less permanently hairy; not alpine.
        5. Pappus usually over 8mm long; midveins of leaves conspicuous.

T. exscapa
Easter townsendia

        5. Pappus usually less than 8mm long; midveins of leaves often obscure.
          6. Involucral bracts with tuft of tangled cilia at tip; leaves densely hairy.

T. hookeri
Hooker townsendia

          6. No tuft of tangled cilia at tip of involucral bracts; leaves becoming glabrate.

T. leptotes
Common townsendia

EASTER TOWNSENDIA, *Townsendia exscapa* (fig. 316), is a low, cushion-like plant with white or pinkish flower heads nestled in a dense tuft of narrow, grayish leaves. Heads about 2.5cm broad when fully open; phyllaries in 4–6 series, linear, having irregular, papery margins. HOOKER TOWNSENDIA, *T. hookeri,* is very similar but may be distinguished by the tufts of tangled hairs at the tips of its phyllaries.

316. Easter townsendia, ½X

EASTER TOWNSENDIA, *T. exscapa,* is one of the first plants to bloom. May be found in flower in March, or even earlier, on the mesas and lower foothills, and by April on the fields of the montane zone. Found throughout our range.

SHOWY TOWNSENDIA, *T. grandiflora.* Stems 5–20cm tall, leafy, branching at base; basal leaves oblanceolate, upper ones lanceolate; rays pink or rose-purple; phyllaries in about 3 series, lanceolate with long sharp points and irregular,

317. Salsify, ½X

papery margins. A summer blooming plant found on dry slopes of mesas and foothills of South Dakota, Wyoming, Colorado, and New Mexico. *T. eximia* is similar but taller. Grows in the montane zone of southern Colorado and New Mexico. Its rays are bluish or purple.

PARRY TOWNSENDIA, *T. parryi*. Stems 1–few, 5–25cm tall; heads solitary; rays pale purplish-blue; basal leaves spatulate, entire or 3-toothed, upper leaves reduced. Occurs in the northern part of our range, especially in Teton, Yellowstone, and Glacier National Parks, on dry hillsides and rocky slopes up to timberline. Distinguished from species of aster and erigeron by the large size of the head, and by its pappus, which consists of bristles plus long flat scales.

(53) *Tragopogon*. Salsify

1. Rays purple.

              T. porrifolius
              Salsify

1. Rays yellow.
  2. Involucral bracts shorter than outer ray flowers.

              T. pratensis
              Meadow salsify

  2. Involucral bracts longer than outer ray flowers.
    3. Leaves glabrous, the tips recurved.

              T. miscellus
              Hybrid salsify

    3. Leaves usually hairy, at least in leaf axils, the tips straight.
              T. dubius
              Yellow salsify

YELLOW SALSIFY, *Tragopogon dubius* (fig. 317), is an erect plant 3–9dm tall with long, grass-like leaves that clasp the stem. Flower heads are lemon-yellow with long, pointed phyllaries extending beyond the flowers. The tawny, ball-like seed heads, up to 10cm in diameter, are conspicuous in middle and late summer. Closely related to the vegetable known as oyster plant. The salsifies, aliens from Europe, are now escaped cultivars that have become widely distributed as weeds on roadsides and in disturbed soil throughout our range in the foothill and montane zones.

318. Goldeneye, ½X

(54) *Viguiera*. Goldeneye

1. Leaves narrowly elliptic to lance-ovate.

V. multiflora var. *multiflora*

1. Leaves linear to lance-linear.

V. multiflora var. *nevadensis*

GOLDENEYE, *Viguiera multiflora* (fig. 318), is a much-branched plant, the many yellow flower heads suggesting small sunflowers. The leaves are rough, 2.5–7.5cm long, mostly opposite. Common on dry slopes in the foothill and montane zones throughout our range.

(55) *Wyethia*. Mule's ears

1. Rays pale, yellowish to white.

W. helianthoides
White-head mule's ears

1. Rays bright yellow.
   2. Plant glabrous and smooth throughout.

W. amplexicaulus
Northern mule's ears

   2. Plant hirsutely pubescent or scabrous.
      3. Leaves oblong-lanceolate, tapering to both ends.

W. arizonica
Arizona mule's ears

      3. Leaves broadly linear.
         4. Outer involucral bracts coarsely hirsute, tips more or less erect.

W. scabra var. *scabra*

         4. Outer involucral bracts with fine appressed hairs, tips recurved.

W. scabra var. *canescens*

319. Northern Mule-Ears, ²⁄₅ X

NORTHERN MULE'S EARS, *Wyethia amplexicaulis* (fig. 319), is a coarse plant with large, leathery, elliptic, basal leaves about 3dm long. The leaves have a varnished look and may be somewhat sticky; the smaller upper leaves clasp the stems. Heads 7.5–12.5cm broad, yellow, usually several with the terminal one largest; phyllaries more or less leafy. Grows in moist soil in the montane and subalpine zones from northwestern Colorado to Nevada and Montana, blooming from May to July. ARIZONA MULE'S EARS, *W. arizonica*, is a hairy plant but otherwise similar. Occurs in southwestern Colorado and southward to New Mexico and Arizona. Where the distribution overlaps with *W. amplexicaulis*, hybridization takes place, producing plants intermediate in appearance. Mule's-ears can usurp whole meadows that have been overgrazed.

WHITE MULE'S-EARS, *W. helianthoides*, with white or cream-colored rays, has heads 10–15cm broad. Leaves and phyllaries are finely fringed with marginal hairs. Found on moist soil in Yellowstone National Park and west to Oregon and Washington.

MONOCOTYLEDONOUS PLANTS (MONOCOTS)

Plants that have only one seed-leaf (cotyledon) are called *Monocotyledons*, or *Monocots* for short. They have other characters in common that help us to recognize them: Those with conspicuous flowers have their flower parts in threes. Those with small, chaffy flowers have long, narrow leaves. The leaves of all species have entire margins. All the grass-like plants belong in this group. If the plant you are trying to identify has flowers and leaves that both agree with these characters, you will find it in this section. It may have broad leaves, but if the other characters agree, it will be found here. Even if your plant has narrow leaves and apparently parallel veins, if the flower parts are in fours or fives, look for it under the heading *Dicots*.

KEY 7. DIVISION SPERMATOPHYTA: ANGIOSPERMS. MONOCOTYLEDONS

1. Plants with conspicuous perianth segments (petals and sepals).
  2. Flowers regular.
    3. Pistils numerous, in a head or ring.

                (93) Alismataceae, p. 384
                    Water-plantain Family

    3. Pistils 1, usually 3-celled (3 carpels).
      4. Perianth segments 6, all petal-like.
        5. Ovary superior.

                (106) Liliaceae, p. 400
                    Lily Family

        5. Ovary inferior.

                (107) Iridaceae, p. 411
                    Iris Family

      4. Perianth of 3 petals and 3 sepals.
        5. Petals blue; upper leaf spathe-like.
                (100) Commelinaceae, p. 387
                    Spiderwort Family
        5. Petals not blue.
          6. Leaves 3, broad, just below flower.
                Trillium in Liliaceae, p. 400
          6. Leaves narrow, tapering.

                Calochortus in
                Liliaceae, p. 400
  2. Flowers irregular.
                (108) Orchidaceae, p. 412
                    Orchid Family

1. Plants without conspicuous petals.
  2. Plants aquatic, both submerged and free-floating.
    3. Plants minute; floating fronds; leaves and stems not differentiated.
                (99) Lemnaceae, p. 386
                    Duckweed Family
    3. Plants with distinct stems and leaves.
      4. Flowers perfect; leaves alternate.
        5. Perianth segments 4, distinct.
                (96) Potamogetonaceae, p 385
                    Pondweed Family
        5. Perianth absent.
                (97) Ruppiaceae, p. 386
                    Ditch-grass Family
      4. Flowers imperfect; leaves opposite or whorled.
        5. Sepals 3; petals 3.
                (94) Hydrocharitaceae, p. 384
                    Frogbit Family
        5. Perianth absent.
                (98) Najadaceae, p. 386
                    Water-nymph Family
  2. Plants terrestrial or somewhat aquatic; never free-floating nor completely submerged.
    3. Marsh plants with zigzag stems and floating leaves.
                (104) Sparganicaceae, p. 399
                    Bur-Reed Family
    3. Plants with straight stems; leaves erect.
      4. Stems round (circular in cross-section).
        5. Stems without joints.
          6. Flowers in erect, dense, terminal spikes.
            7. Plants 1–2m high; spikes 7–15mm thick.
                (105) Typhaceae, p. 399
                    Cat-Tail Family

7. Plants less than 10dm high; spikes under 12mm thick.
(95) Juncaginaceae, p. 385
Arrowgrass Family
6. Flowers in clusters, panicles, or heads.
(101) Juncaceae, p. 388
Rush Family
5. Stems jointed and hollow.
(103) Gramineae, p. 393
Grass Family
4. Stems 3-angled (triangular in cross-section).
(102) Cyperaceae, p. 391
Sedge Family

WATER-PLANTAIN FAMILY: ALISMATACEAE

1. Leaves elliptic-ovate; flowers perfect.

Alisma plantago-aquatica var. americanum
Water plantain

1. Leaves sagittate; flowers imperfect.
    2. Beak of achene wanting, or simply an erect tooth in the margin of its wing.
Sagittaria cuneata
Arrowhead
    2. Beak of achene 1/4 to 1/3 as long as the body.
Sagittaria latifolia
Arrowhead or duck-potato

320. Arrowhead, ½X

WATER PLANTAIN, *Alisma plantago-aquatica*, is a plant of marshes and ponds, with basal, sheathing leaves and upright stalks bearing whorls of many, small, white or pinkish flowers. ARROWHEAD or DUCK-POTATO, *Sagittaria latifolia* (fig. 320), varies according to its situation. When completely submerged, its leaves are long and ribbon-like; when protruding from the water, it develops broad, arrowhead-shaped leaves and sends up stalks bearing 3-petalled, white flowers. The plant furnishes food for water birds and muskrats. Its roots produce starchy tubers, which were eaten by Indians and the early settlers.

FROGBIT FAMILY:
HYDROCHARITACEAE
Only one genus and species within our range: ANACHARIS, *Elodea canadensis*, grows submerged in ponds and slow streams. Leaves 1-nerved, transparent,

and opposite on lower stem; upper and middle leaves in whorls of 3. Its flowers are inconspicuous, and it seldom produces seeds; but it grows so rapidly that it can fill ponds and ditches in the foothills. Often used in aquaria.

ARROWGRASS FAMILY: JUNCAGINACEAE

These plants have slender, fleshy basal leaves; leafless, erect stalks thickly set along the upper portions with tiny green flowers; pistils united into a compound ovary. Only 1 genus in our range.

1. Carpels 6; stigmas 6.

> *Triglochin maritima*
> Shore arrowgrass

1. Carpels 3; stigmas 3.

> *T. palustris*
> Marsh arrowgrass

SHORE ARROWGRASS, *T. maritima*, is a stout plant with stems up to 9dm tall. Found on alkaline soil around ponds and in marshes of the high plains and foothills. SWAMP ARROWGRASS, *T. palustris*, is a smaller plant, which occurs in mountain swamps and around lakes. The seeds of both species are valuable food for small mammals and birds. Both grow on the geyser and hot spring formations in Yellowstone National Park.

PONDWEED FAMILY: POTAMOGETONACEAE

Only 1 genus in our range: a group of water plants having both submersed and floating leaves. Perianth segments 4, distinct; stamens 4; anthers nearly sessile. The flowers are very small and are withdrawn under water after blooming, where the seeds ripen. Collectively, these species are one of the most valuable sources of food available in our range for ducks and other wildlife.

1. Some of the leaves floating, the floating leaves elliptic to oblanceolate and petiolate.
  2. Submerged leaves bladeless (narrowly linear).
> *Potamogeton natans*

  2. Submerged leaves with a proper blade.
    3. Blades of two forms, elliptic and lanceolate.
> *P. amplifolius*

    3. Blades of one form, all alike.
      4. Blades linear.
        5. Blades filiform, with stipules adnate to the leaf base.
> *P. diversifolius*

        5. Blades broadest at the base, the stipules free from the rest of the leaf.
> *P. gramineus*

      4. Blades not linear.
        5. Petiole of floating leaves long (3–10cm).
> *P. nodosus*

        5. Petiole of floating leaves short (3–5mm).
          6. Blades elliptic.
> *P. illinoensis*

          6. Blades oblanceolate.
> *P. alpinus*

1. None of the leaves floating.
   2. Leaves with lanceolate or broader blade, amplexicaul.

   P. richardsonii

   2. Leaves linear.
      3. Stipules free from the rest of the leaf; style present.
         4. Leaves not glandular at base.

   P. pusillus

         4. Leaves not glandular at base.

   P. foliosus

      3. Stipules adnate to the leaf base.
         4. Stigma on very short style (0.5mm).

   P. pectinatus

         4. Stigma sessile; no style.

   P. filiformis

DITCH-GRASS FAMILY: RUPPIACEAE

Only 1 genus in this family, and generally only 1 species is held to be good. *Ruppia maritima* is a submerged aquatic herb with long, threadlike, forking stems; slender (almost capillary) alternate leaves, sheathing at the base; perianth absent; flowers perfect, with 2 sessile stamens; 4 small, sessile ovaries, at least usually, but subject to variation.

WATER-NYMPH FAMILY: NAJADACEAE

Only one genus in this family of submerged aquatic herbs. Opposite leaves no more than 1mm wide, minutely serrulate; flowers imperfect, axillary; 1 stamen; 1 pistil, but stigmas 2–4.

1. Leaves acuminate.

   Najas flexilis
   Wavy water-nymph

1. Leaves abruptly acute.

   Najas guadalupensis
   Guadalupe water-nymph

DUCKWEED FAMILY: LEMNACEAE

Minute aquatic plants without differentiated body parts. The flat or disk-like plant-body is called the *thallus*, or sometimes the *frond*. These midget plants occur in great numbers, often forming large, floating colonies. In late summer they are an important duck food. They are perennial plants, yet they seldom produce seed. They survive cold weather by sinking into the mud at the bottom of the pond or stream. Species of duckweed are found throughout North America and almost throughout the world.

1. Thallus 1–3 nerved or nerveless; rootlet solitary.

   Lemna, p. 387
   Duckweed

1. Thallus 7–12 nerved; rootlets several, fascicled.

   Spirodela, p. 387
   Greater duckweed

(1) Lemna. Duckweed

1. Thalli (or fronds) oblong to lanceolate, long-stipitate.

> *L. trisulca*
> Star duckweed

1. Thalli nearly oval, not long-stalked.
    2. Thalli usually gibbous; pale beneath.

> *L. gibba*

    2. Thalli mostly flat.
        3. Thalli 3-nerved, 1.5–4mm wide; green or purplish beneath.

> *L. minor*

        3. Thalli 1-nerved, 0.5–1.5mm wide; green or purplish tinged.

> *L. minuta*

(2) Spirodela. Greater duckweed

One species in our range: GREATER DUCKWEED, *Spirodela polyrhiza*. Fronds suborbicular; rootlets fascicled in the center of the frond.

SPIDERWORT FAMILY: COMMELINACEAE

The plants of this family are somewhat succulent and have a slimy juice. The leaves have sheathing bases and long, tapering blades. Flowers 3-petaled, blue or purplish, lasting only 1 day.

1.Flowers irregular; 3 of the 6 stamens anther-bearing.

> *Commelina*, p. 387
> Dayflower

1. Flowers regular; stamens all anther-bearing.

> *Tradescantia*, p. 387
> Spiderwort

(1) *Commelina*. Dayflower

1. Spathe 3–6cm long; not connate at base.

> *C. dianthifolia*
> Dayflower

1. Spathe less than 3cm long, connate at base.

> *C. erecta var. angustifolia*

DAYFLOWER, *Commelina dianthifolia*, has an irregular flower borne in a green, funnel-shaped spathe. Occurs in the foothills and occasionally in the montane regions from Mexico as far north as central Colorado.

(2) *Tradescantia*. Spiderwort

1. Plant bright green or yellowish-green, never glaucous; sepals with some long, lax, glandular hairs.

> *T. bracteata*

1. Plants glaucous; sepals with glandular-puberulent hairs.

> *T. occidentalis var. scopulorum*
> Western spiderwort

321. Western Spiderwort, ½X

WESTERN SPIDERWORT, *Tradescantia occidentalis* (fig. 321). A very fragile blossom on a coarse plant. The flower has 3 similar, blue or purplish petals and 6 stamens. The filaments are hairy, and this character is said to account for the name spiderwort. The long leaves have sheathing bases and stick out at awkward angles to the stem, so that the plant begins to appear very weedy as it ages. Grows commonly along the base of the foothills from Arizona to Montana, in Zion Canyon, and in several of the southwestern national monuments. The plants are showy and attractive during the mornings in June and early July. A closely related species, *T. pinetorum*, may be found in Arizona. The tubers on its roots were used for food by the Indians.

RUSH FAMILY: JUNCACEAE

This group is very closely related to the lily family. Its flower pattern is the same as that of lilies, but its sepals and petals in most cases, have become reduced to small brownish scales. As do the grasses and sedges, these plants depend on wind-pollination. Its leaves, when present, are long and slender. For these reasons, the rushes are often perceived as grass-like plants. But they may be distinguished from both sedges and grasses by the pattern of the brown flowers, which have 6 similar perianth segments, representing both calyx and corolla; 3–6 stamens; 1 pistil with 3 stigmas, which ripens into a capsule containing 3 to many seeds. Only 2 genera are commonly found in our area.

1. Leaves soft, flat; stems hollow; capsule 1-celled and 3-seeded.
> *Luzula,* p. 390
> Woodrush
1. Leaves stiff, terete or flat; stems usually with a spongy pith; capsule 3- or 1-celled, many-seeded.
> *Juncus,* p. 388
> Rush

(1) *Juncus.* Rush

1. Leaves septate (divided by internal cross-partitions, giving a knotted appearance); inflorescence terminal.
> 2. Leaves compressed (flattened on two opposite sides).
>> 3. Stamens 3.
>>> *J. ensifolius*
>> 3. Stamens 6.

4. Heads 2 or more.

J. saximontanus

    4. Heads solitary.

J. mertensianus
Subalpine rush

2. Leaves terete or only slightly compressed.
    3. Stamens 3.

J. tweedyi

    3. Stamens 6.
        4. Perianth segments and capsule usually obtuse.

J. alpinus

        4. Perianth segments and capsule usually acute or acuminate.
            5. Inner segments longer than the outer.

J. nodosus

            5. Outer segments longer than the inner.
                6. Heads pale.

J. torreyi

                6. Heads dark-brown.

J. nevadensis

1. Leaves not septate.
    2. Inflorescence terminal.
        3. Flowers capitate.
            4. Leaves cylindrical and hollow.
                5. Leaves flattened upward.

J. triglumis

                5. Leaves not flattened upward, but channeled at base.

J. castaneus

            4. Leaves flat, not cylindrical.

J. longistylis

        3. Flowers solitary, in panicles.
            4. Stems simple, naked.
                5. Leaf blades cylindrical.

J. vaseyi

                5. Leaf blades flat.
                    6. Perianth segments pale.

J. tenuis

                    6. Perianth segments brown with green midrib.

J. confusus

            4. Stems diffusely branched, leafy.

J. bufonius

    2. Inflorescence appearing lateral and sessile.
        3. Flowers few, in panicles simple or nearly so.
            4. Stems leafless.

J. drummondii

            4. Stems somewhat leafy.
                5. Perianth segments white-margined; the capsule blunted and
                    indented.

J. hallii

                5. Perianth segments green, the capsule pointed.

J. parryi

        3. Flowers many, in more or less compound panicles.
            4. Perianth segments with a brown stripe on either side of the midrib.

J. balticus
Arctic rush

            4. Perianth pale green throughout.

J. filiformis

The rushes have stiff, pithy, green stems that occur in clumps. Their leaves are often reduced to mere sheaths. They can be recognized among other vegetation because they are a much darker green than most plants. Found in moist

322. Arctic Rush, ½X

323. Subalpine
Rush, ½X

324. Common Woodrush, ½X

or wet places. ARCTIC RUSH, *J. balticus* (fig. 322), is one of the most common. It will be seen forming clumps or zones of very dark green around seepage spots from the plains to timberline. SUBALPINE RUSH, *J. mertensianus* (fig. 323), forms clumps of smooth stems 15–38cm tall, each topped with a round head of small dark flowers. Each flower has 6 sharply-pointed perianth segments. At flowering time the pink stigmas show up against the brownish-black perianth. Found in moist places of the subalpine zone throughout our range.

(2) *Luzula*. Woodrush

1. Flowers in an open panicle, solitary or paired at the ends of the branches.
    2. Perianth 3–3.5mm long.
    L. glabrata
    2. Perianth 1.5–2.5mm long.

3. Stem leaves usually 2–3, rarely over 5mm wide.
L. wahlenbergii
3. Stem leaves usually more than 3, or else over 5mm wide.
L. parviflora
Common woodrush
1. Flowers congested, capitate or spicate.
  2. Inflorescence nodding, usually a single spike.
L. spicata
Spiked woodrush
  2. Inflorescence erect, in 2 to several subglobose or oblong heads.
L. multiflora

Woodrushes have flat, soft, grass-like leaves, hollow stems, and capsules containing only 3 seeds. COMMON WOODRUSH, *L. parviflora* (fig. 324), usually grows in a tuft 3–6dm tall; inflorescence drooping, open, many-flowered. Found widely distributed over northern North America and in our mountains from 8,000 feet to timberline. SPIKED WOODRUSH, *L. spicata*, is a smaller, more compact plant of mountainous regions and occurs in our subalpine and alpine zones. It has a small, nodding cluster of spikes. There are thin, chaffy bractlets between them, which give a frosted appearance to the inflorescence.

SEDGE FAMILY: CYPERACEAE

The sedges constitute a large family of grass-like plants, which many people do not distinguish from the true grasses. They grow in cold, wet places, especially in arctic and alpine regions, and make up a large proportion of the plants found at timberline. Most of them have triangular stalks without joints, but a few may be quadrangular, flattened, or terete. The stem leaves, if present, are 3-ranked, each folded and its base enclosing the stalk. In most cases, all the leaves rise from the rootstock. The perianth is absent or of hypogynous bristles; stamens 1-3; styles 2-3. The flowers are arranged in spikelets, one in the axil of each scale (glume, bract); the spikelets are solitary or in spicate clusters. They depend upon the wind for pollination, and the seeds are achenes. Sedges may be tufted or have creeping, underground stems, which sometimes form tubers called *ground nuts*. These are hunted by animals and waterfowl for food.

Given the large number of difficult species in this family, it is practical to provide only a key to the genera in our range. Once you know where your plant belongs, check that genus in the standard flora for your state or region. Several representative species are described and illustrated following the key.

1. Flowers perfect; the spikes capitate or umbellate.
  2. Spikelets flattened, with the scales in 2 ranks; perianth absent.
Cyperus
Galingale
  2. Spikelets not flattened, the imbricated scales all around; perianth (in the form of bristles) usually present.
    3. Stamens mostly 3; perianth of 1 to several bristles.
      4. Bristles few and short, at the base of the achene.
        5. Leaves (at least the basal) well-developed.
Scirpus
Bulrush
        5. Leaves reduced to only sheathing base.
Eleocharis
Spike rush

4. Bristles many, long-exserted.

Eriophorum
Cottonsedge

3. Stamens 1; perianth absent.

Hemicarpha
Dwarf bulrush

1. Flowers imperfect; spikelets solitary, spicate, or paniculate.
　2. Achenes naked.

Kobresia
Kobresia

　2. Achene enclosed in an enveloping scale (perigynium), which is closed except at
　　the apex.

Carex
Sedge

Some members of BULRUSH, *Scirpus*, are giants. They grow in water around the edges of ponds and reservoirs or in marshes and can reach a height of 3.6m. The species most likely to be found in the mountains, *S. maritimus* var. *paludosus*, rarely reaches more than 1.5m and may be much less. The flowers are in clusters of several fat spikes at the top of the main stalk. Two long, unequal leaves extend above them.

COTTONSEDGE, *Eriophorum*, is a conspicuous plant in swamps and along borders of ponds in the high mountains during late summer, as its flower heads become tufts of white or tawny, silky bristles. The NARROW-LEAVED COTTONSEDGE,

325. Mertens
Sedge, ½X

326. Fish-scale Sedge, ½X

E. polystachion, is the most common species in the central Rockies. *E. chamissonis* is more northern, occurring from Yellowstone National Park northward to Alaska.

SEDGE, *Carex*, is a very large and difficult genus. In general, the inflorescence is made up of a few compact spikes 6mm–5cm long. In several of the high altitude species the spikes are quite conspicuous, black or very dark brown. They often make a lush-looking growth along subalpine streams and in meadows near timberline. MERTENS SEDGE, *C. mertensii* (fig. 325), has light-colored heads, which are usually drooping, and its leaf blades are often more than 6mm wide. Grows 3–9dm tall. Found in Montana, especially in Glacier National Park. FISH-SCALE SEDGE, *C. heteroneura* var. *chalciolepis* (fig. 326), is a plant 20–60cm tall; heads 2.5cm long, black, heavy, erect or drooping. Common in wet places of the subalpine zone, but it occurs from 8,000 to 13,000 feet from Wyoming through Colorado into Utah and Arizona.

ROCKY MOUNTAIN SEDGE, *C. scopulorum*, is one of the most abundant sedges of the subalpine and alpines zones. Grows 25–38cm tall; the new leaves come up through tufts of coarse, old, dry leaves. But in wet places it shows up as patches of bright green in the otherwise brownish landscape. Much dry tundra is covered with the short, curly leaves of *C. rupestris*. A common sedge of the mesas and foothills, SUN SEDGE, *C. heliophila*, grows in small clumps 10–25cm tall, with noticeable, yellowish blooms in early spring. Its yellowish-green leaves are somewhat curved. Similar and closely related to *C. pensylvanica*, which is very common throughout the eastern United States.

KOBRESIA, *Kobresia bellardii*, a close relative of the sedges, is a densely-tufted plant with slender stalks and almost thread-like leaves. Grows in the arctic and on high ridges and peaks of the Rockies in snow-free areas. It is the dominant plant on mature alpine tundra. Its uniform stands of short, close, grass-like growth may be recognized in late summer and fall by an orange-gold color.

FEW-FLOWERED SPIKERUSH, *Eleocharis pauciflora*, is another member of this family found on margins of ponds and in bogs of the subalpine zone. Grows from slender, underground stems and has no obvious leaves. Its green stalks, 10–20cm tall, are tipped by small, compact flower spikes.

GRASS FAMILY: GRAMINEAE

When looked at closely, grasses in bloom are found to be beautiful, but most people fail to see them, and few think of them as wild flowers. The grasses lack showy flower parts and attractive scents. They cannot, therefore, depend upon insects but must be wind pollinated. This is a very large plant family and one of the most important to man. They provide much of the pasture, hay, and grain that support domestic animals, becoming indirectly the source of our meat. Our cereals and flour are derived from the seeds of grasses.

Some grasses grow in wet situations, but generally the plants of this family are adapted to regions of low rainfall. In dry areas, notably our high plains and foothill slopes where the annual precipitation is only 8 to 15 inches, plant communities made up predominantly of several kinds of grasses form the natural vegetative cover. Often there is insufficient moisture to support the growth of trees.

The grass cover is very important in soil conservation. Grass roots penetrate deeply into the ground. They not only help keep the soil from washing away, but, as they die and are renewed, they add humus to the soil. Most grasses are perennial, but much of their root system and all of the leaves and stalks die and are replaced annually. That process alone adds a great amount of organic material to the soil every season. (Significantly, the 6 genera important as food crops are all annuals, that is, shallow-rooted grasses replanted every year: rice, corn, wheat, rye, oats, and barley). If grassland is moderately grazed, it is constantly being improved by the growth of the grass. There is a direct relationship between the amount of leaf growth and the amount of root growth. If more than half the bulk of the grass plant is eaten by stock or game on a pasture, the plants cannot make enough roots to support vigorous growth. Both the pasturage and the soil will become poorer each year. But, if grazing is carefully regulated so that at least half of the leaves and stems of each plant are left, the forage value will improve each year.

If classified only according to their habit of growth, grasses are of two types: *sod formers* and *bunch grasses*. In the first group, the stems spread horizontally, either underground or at the surface, sending leaves up and roots down from the nodes. They form a continuous, interwoven, sod groundcover. Bunch grasses grow singly in tufts and never form a tight sod. Both types have round, jointed stems, which are usually hollow except at the joints (called *nodes*). The leaves are flat and narrow, jointed to a sheath-like base that surrounds the stalk.

The grasses are only difficult in that the terms used for the floral parts are different from those used for other blooming plants. Identification is usually based on the structure of the spikelet, the parts of which are often minute. As a rule, the two lowest scales (or bracts) of the spikelet are *glumes*, which do not subtend flowers. The outer (and usually larger) of the pair is the *lemma*, the inner (and usually smaller) is the *palea*. There may be 1 to many pairs.

The flowers enclosed between the lemma and the palea (the *florets*) are small and generally perfect (a few are unisexual). The inflorescence may be a spike, a raceme, or a panicle of spikelets. The perianth consists of 2–3 tiny scales called *lodicules*; stamens usually 3; 2 styles and 2 stigmas, the stigmas plumose; ovary superior. The term *ligule*, when used with grasses, refers to the thin appendage on the inside (top) of a leaf at the junction of the blade and sheath.

Given the large number of difficult species in this family, it is practical to provide only a key to the genera in our range. (For more about grasses, see James P. Smith, Jr., *A Key to the Genera of Grasses of Conterminous United States.*) Once you know where your plant belongs, check that genus in the standard flora for your state or region. Only a few of the best known and most easily recognized species are described or illustrated following the key to the genera.

1. Spikelets borne in an open or spike-like panicle or raceme, the spikelets with distinct pedicels, whether long or short.
  2. Spikelets 1-flowered (1 fertile floret in each spikelet).
    3. Glumes 4; palea 1-nerved.
      4. Uppermost floret perfect, the others empty.
          Phalaris
          Canary-grass
      4. Uppermost floret perfect, the others staminate.
          Hierochloë
          Sweetgrass.

3. Glumes 2, rarely 1; palea 2-nerved (except in *Cinna*).
    4. Lemma with a long terminal awn, and closely embracing the grain.
        5. Fruiting lemma thin and membranous.

> *Muhlenbergia*
> Muhly

        5. Fruiting lemma firm and hardened.
           6. Awns 3-branched.

> *Aristida*
> Three-awn

           6. Awns simple.
               7. Awns twisted, persistent on the lemma.

> *Stipa*
> Needlegrass

               7. Awns straight, deciduous from the lemma.

> *Oryzopsis*
> Ricegrass

    4. Lemma short-awned or awnless, and loosely surrounding the grain.
        5. Inflorescence a dense spike.
           6. Spikelets persistent; lemma short-awned or awnless; spikes cylindrical.

> *Phleum*
> Timothy

           6. Spikelets early deciduous; lemma with a dorsal awn; spikes cylindrical, ovoid, or capitate.

> *Alopecurus*
> Foxtail

        5. Inflorescence a loose panicle.
           6. Pericarp discharging seed at maturity; nerves of lemma not pilose.

> *Sporobolus*
> Dropseed

           6. Pericarp permanently surrounding the seed.
               7. Palea 1-nerved; stamen 1.

> *Cinna*
> Woodreed

               7. Palea 2-nerved; stamens 3.
                  8. Lemma naked at the base.

> *Agrostis*
> Bentgrass

                  8. Lemma with tuft of long hairs at the base.

> *Calamagrostis*
> Reedgrass

2. Spikelets 2–many-flowered.
    3. Lemma usually shorter than the glumes; the awns usually bent.
        4. Awns of the lemma dorsal.
           5. Lemma erose-truncate.

> *Deschampsia*
> Hairgrass

           5. Lemma 2-toothed, with awn twisted and bent.

> *Trisetum*
> Trisetum

        4. Awn of the lemma terminal, between the teeth.

> *Danthonia*
> Oatgrass

3. Lemma usually longer than the glumes; the awn terminal and straight (rarely dorsal in *Bromus*) or none.
    4. Rachilla (the axis of the spikelet) long-hirsute.

> *Phragmites*
> Common reed

    4. Rachilla glabrous or with short hairs.
        5. Lemma 3-nerved or rarely 1.
           6. Inflorescence spike-like.

> *Koeleria*
> Junegrass

6. Inflorescence an open panicle.
>*Catabrosa*
>Brookgrass

5. Lemma 5-nerved or more.

6. Spikelets with upper florets sterile and folded about each other.
>*Melica*
>Melic grass

6. Spikelets with the upper floret perfect, or narrow and abortive.

7. Stigmas arising below the apex of the ovary.
>*Bromus*
>Bromegrass

7. Stigmas arising at the apex of the ovary.

8. Lemma compressed and keeled.

9. Awn pointed.
>*Dactylis*
>Orchard-grass

9. Awn pointless; glumes 1–3-nerved.
>*Poa*
>Bluegrass

8. Lemma convex or rounded on the back.

9. Lemma acute or awned.
>*Festuca*
>Fescue

9. Lemma obtuse and scarious at apex.

10. Lemma prominently 5–7-nerved.
>*Glyceria*
>Mannagrass

10. Lemma obscurely 5-nerved.
>*Puccinellia*
>Alkali grass

1. Spikelets in 2 rows, sessile or nearly so.

2. Spikelets on one side of the continuous axis (forming 1-sided spikes).

3. Spikelets with perfect flowers.

4. Spikelets deciduous as a whole, the articulation below the glumes; the glumes unequal.
>*Beckmannia*
>Sloughgrass

4. Spikelets in part persistent, the articulation above some of the glumes; spikes 1–4.
>*Bouteloua*
>Grama

3. Spikelets with imperfect flowers, the pistillate very different from the staminate, and on very short culms (stems).
>*Buchloë*
>Buffalo grass

2. Spikelets alternately on opposite sides of axis, which is often articulated (jointed).

3. Spikelets mostly solitary at each joint of the rachis.
>*Agropyron*
>Wheatgrass

3. Spikelets 2 or more at each joint of the rachis.

4. Spikelets 1-flowered.
>*Hordeum*
>Barley

4. Spikelets 2–many-flowered.

5. Rachis continuous.
>*Elymus*
>Wild rye

5. Rachis readily separating into joints.
>*Sitanion*
>Squirreltail

BUFFALO GRASS, *Buchloë dactyloides,* is a short grass of the high plains and low hills that is about 1dm tall. Its stems spread on the surface of the ground, rooting

327. Parry Oatgrass, ½ X   328. Blue Gramma, ½ X

at the nodes, forming a dense, tough sod. The staminate flowers are held above the leaves; the pistillate flowers will be found tucked down close to the ground.

PARRY OATGRASS, *Danthonia parryi* (fig. 327), is strictly a mountain grass, 20–60cm tall. Heads few, large, sometimes drooping; leaves slender, tufted, 10–24cm long. Each spikelet is enclosed in papery glumes 2.5cm long. A few bent and twisted awns stick out from between these glumes. Each awn is attached to one of the lemmas of the enclosed spikelet. Found in open woods or on rocky hillsides through the mountains from Alberta to New Mexico. In Colorado, it occurs between 6 and 10,000 feet.

SWEETGRASS, *Hierochloë odorata*, is common in swampy mountain meadows. It grows about 30–45cm tall, and the ripe spikelets are a rich golden-brown. Also common in the eastern United States, where its sweet-scented stems were used by Indians to make baskets.

BLUE GRAMMA, *Bouteloua gracilis* (fig. 328), is a grass of medium height, 15–30cm tall. The spikelets are arranged along one side of the rachis; the

329. Alpine
Timothy, ½X

330. Squirrel-Tail, ½X

spikes, 2.5cm long, 1–3 per stalk, are set at an angle and look like purple flags. Common in the foothills and on open mountain slopes up to 8,500 feet. SIDE OATS GRAMMA, *B. curtipendula*, is taller and has many short spikes along one side of the erect slender stalk.

TIMOTHY, *Phleum pratense*, is a tall, introduced grass. Spike 5–10cm long, compact, cylindrical. Found on good moist soil in meadows and along roads or trails where hay has been carried. The native ALPINE TIMOTHY, *P. alpinum* (fig. 329), is similar but not as tall; spike shorter, usually dark purplish. Grows in subalpine meadows.

BLUEGRASS, *Poa*. Several species occur naturally in our range, and some European species have become established here. There are both sod-forming and bunch grasses in this genus. The common KENTUCKY BLUEGRASS used in lawns is a sod-forming kind. Some of the little, tufted grasses found above timberline are bunch grass poas. Most kinds provide good forage, and some are called mutton-grass.

SPIKE FESCUE, *Leucopoa kingii*, is a stout, conspicuous bunch grass of the pine forests, with stalks 4–8dm tall and dense panicles 7–18cm long. Its bluish-green leaves are flat, tough, striated, and about 7mm wide.

SQUIRRELTAIL, *Sitanion hystrix* (fig. 330). The bristly, brush-like inflorescence of squirreltail is made up of groups of florets that have long, spreading awns. When these florets ripen, the central axis of the spike disarticulates, and the seeds, with their attached awns, are then free to be moved about by wind or other agents. The long, sharp awns also catch in the hair of animals, further assuring wide distribution of the seeds. The axis of the more compact head of FOXTAIL BARLEY, *Hordeum jubatum*, breaks up in much the same way. Its florets also have sharp, rigid, but more slender awns about 5cm long, making the inflorescence very bristly.

NEEDLE-AND-THREAD, *Stipa comata,* has a slender stem about 6–9dm tall, with a narrow, sparse, and drooping inflorescence. Each floret has an awn that may be more than 15cm long, twisted and bent. This long awn is the *thread*, and the opposite end of the grain, which is very sharp, is the *needle*. Other species of needlegrass in our area are similar but do not have quite such long awns.

COMMON REED, *Phragmites australis*, is the largest grass in our region. It grows up to 3.6m in height; leaves 12–25mm wide. Grows in very wet places on the plains and lower foothills. Inflorescence a large, fluffy panicle. Worldwide in distribution. In Mexico and in our Southwest it was used for thatching, arrow shafts, weaving rods, mats, cords, and nets.

BUR-REED FAMILY: SPARGANIACEAE

This family has only one genus. Rootstocks creeping and stoloniferous; roots fibrous; flowers imperfect, the upper heads staminate, the lower pistillate; stamens usually 5; the ovary superior.

1. Fruits sessile, angular.
    2. Emergent and erect.
    Sparganium eurycarpum
1. Fruits stalked, terete or fusiform (thick but tapering at each end).
    S. emersum
    2. Submersed or floating.
        3. Inflorescence branched.
        S. angustifolium
        3. Inflorescence simple.
        S. minimum

BUR-REED, *Sparganium angustifolium*, is the most likely of these water plants to be found. It has zigzag stems and long, narrow leaves arising from the mud beneath the water; most of the leaves bend and become floating at the surface. All of these species provide good food for muskrats, water birds, and deer. Found in shallow water around the margins of ponds and lakes.

CATTAIL FAMILY: TYPHACEAE

There is one genus in this family. The cattails are tall plants of marshes and ditch borders. They have long, strap-like leaves; cylindrical spikes of tiny, crowded flowers, which are of two kinds: straw-colored staminate flowers above dark brown pistillate ones. The staminate flowers soon wither and fall away leaving the long, bare axis exposed.

1. Leaves 1–2cm broad; plant 1–2m tall; spike 1–3dm long, the upper half staminate, the lower half pistillate; pistillate flowers without bractlets.
    Typha latifolia

1. Leaves narrower, mostly about 5mm wide; plants 1–1.5m tall; the staminate and pistillate portions of the spike distinct; pistillate flowers with bractlets.

T. angustifolia

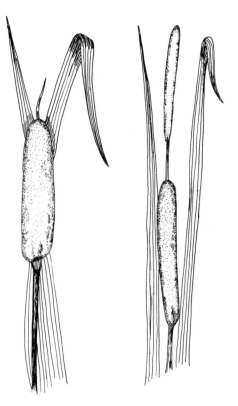

331. Broad-Leaved
Cattail, ½X

332. Narrow-Leaved
Cattail, ½X

Of the two, BROAD-LEAVED CATTAIL, *Typha latifolia* (fig. 331), is the more conspicuous and the more common in most places. The two often grow together in wet places of the foothills and mountain valleys. The smaller NARROW-LEAVED CATTAIL, *T. angustifolia* (fig. 332), has 2.5–5cm of bare stalk between the staminate and pistillate flowers. Cattails have been valuable to man in several ways. Indians obtained starch from the root, peeled and roasted the lower part of the stem and the young flower spikes for food, and used the down from the ripened spikes for padding cradle boards and dressing wounds. The Apache used the yellow pollen as a sacred offering. The long, flat leaves were woven into mats by the American pioneers and used to make seats for the old-fashioned rush-bottomed chairs. A cattail marsh is a good place for wildlife, especially muskrats, geese, red-wing blackbirds, and marsh wrens.

LILY FAMILY: LILIACEAE

All the plants in this family have their flowers built on the numerical plan of 3 and 6. In our species, the sepals and petals are virtually alike, except in *Calochortus* and *Trillium*. Leaves are mostly linear, usually not differentiated into distinct blade and petiole; leaf margins are entire; the root is usually a bulb, corm, or a fleshy rootstock. In some genera the bulbs are edible and were used by Indians and pioneers as food; in other genera they are poisonous.

1. Sepals and petals alike, forming a 6-parted perianth.
    2. Plants coarse; flower stalks stout, 60cm–2m tall.
        3. Leaves along the stem, large and pleated.

Veratrum, p. 408
Corn lily

3. Leaves mostly basal, in large tufts.
    4. Leaves stiff and sharp-pointed; flowers about 5cm long.
                            Yucca, p. 409
                            Yucca
    4. Leaves flexible, reduced in size upward; flowers small.
                            Xerophyllum, p. 408
                            Beargrass
2. Plants with smooth and soft-textured leaves; stalks under 60cm tall.
    3. Stems leafy, basal leaves few or absent.
        4. Plants with some leaves whorled.
            5. Flowers erect, 5–10cm long in our range, and orange-red.
                            Lilium, p. 406
                            Lily
            5. Flowers pendant, not over 2.5cm long in our range, greenish-yellow to yellow to purplish.
                            Fritillaria, p. 405
                            Fritillary
        4. Leaves always alternate, never more that 1 at a node.
            5. Flowers or red berries pendant from underside of stem.
                            Streptopus, p. 407
                            Twisted stalk
            5. Flowers at end of leafy stem.
               6. Flowers usually 1; berry 3-lobed, red.
                            Disporum, p. 405
                            Fairybells
               6. Flowers several or many; berries round, greenish.
                            Smilacina, p. 407
                            False Solomon's seal
3. Leaves all or mostly basal.
    4. Flowers 2.5cm or more broad.
        5. Leaves 2–5, at least 1/3 as broad as long.
            6. Flowers white, leaves usually 5.
                            Clintonia, p. 405
                            Bead lily
            6. Flowers yellow; leaves 2.
                            Erythronium, p. 405
                            Snow lily
        5. Leaves many, grass-like; flowers white and stemless.
                          Leucocrinum, p. 406
                          Sand lily
    4. Flowers less than 2.5cm broad.
        5. Flowers solitary or in umbels.
            6. Flowers blue or purplish; leaves without onion odor.
                      Brodiaea, p. 403
                      Wild Hyacinth
            6. Flowers not blue.
               7. Flowers pink, rose, or whitish; leaves with onion odor.
                      Allium, p. 402
                      Wild onion or garlic
               7. Flowers creamy white with dark veins, usually solitary; leaves without onion odor.
                      Lloydia, p. 406
                      Alpine lily
        5. Flowers in racemes.
            6. Flowers blue or purplish.
                      Camassia, p. 404
                      Camas
            6. Flowers white or cream to greenish-white.

7. Anthers 2-celled, oblong or ovate.
>> *Tofieldia,* p. 408
>> False asphodel
7. Anthers cofluently 1-celled (blended into 1), cordate or reniform.
>> *Zigadenus,* p. 410
>> Death camus
1. Sepals and petals distinct, 3 each, the sepals narrow and green.
>> 2. Leaves broadly ovate, in a whorl of 3.
>>>> *Trillium,* p. 408
>>>> Trillium or wake-robin
>> 2. Leaves few, alternate, long and narrow.
>>>> *Calochortus,* p. 403
>>>> Mariposa lily

(1) *Allium.* Wild onion or garlic.

1. Leaves hollow, terete.
>> *A. schoenoprasum*
>> Wild chives
1. Leaves not hollow, either flat or terete.
>> 2. Bulbs coated with a conspicuous network of coarse fibers (reticulate).
>>>> 3. Involucral bracts mostly 1-nerved.
>>>>>> 4. Leaves 3 or more per scape; tepals usually pink.
>>>>>>>> 5. Flowers normal, 10–25 per umbel.
>>>>>>>>>> *A. geyeri var. geyeri*
>>>>>>>> 5. Fewer than 10 normal flowers per umbel, the rest replaced by bulblets.
>>>>>>>>>> *A. geyeri var. tenerum*
>>>>>> 4. Leaves usually 2 per scape; tepals usually white.
>>>>>>>> *A. textile*
>>>>>>>> Wild onion
>>>> 3. Involucral bracts 3- to 5-nerved.
>>>>>> *A. macropetalum*
>> 2. Bulb coats thin membranous, either nonfibrous or nonreticulate.
>>>> 3. Umbel nodding.
>>>>>> *A. cernuum*
>>>>>> Nodding onion.
>>>> 3. Umbel erect.
>>>>>> 4. Bulb elongate, with a stout primary rhizome below.
>>>>>>>> *A. brevistylum* ✓
>>>>>>>> Shirt-style onion
>>>>>> 4. Bulb small, subglobose, no rhizome below.
>>>>>>>> 5. Leaves longer than the short scape.
>>>>>>>>>> *A. brandegei*
>>>>>>>> 5. Leaves shorter than the scape.
>>>>>>>>>> *A. acuminatum*

All the WILD ONIONS have bulbs, and their leaves have a distinct onion odor. All were used as food by the Indians, the trappers, and the early settlers. The small flowers are in umbels, carried by slender, leafless stalks. A white-flowered species, *A. textile* (fig. 333), is abundant on the plains and dry foothills, where it forms clumps. The Latin word *textile* was used to describe the net-like coat of fibres, which resembles a coarse, woven cloth covering the bulb. The species most frequently found at high altitudes is GEYER ONION, *A. geyeri.* It also has a netted bulb coat, but its blossoms are usually deep pink. The NODDING ONION, *A. cernuum* (fig. 334), with pale pink flowers, can be recognized by its bent umbel. Blooms in fields and meadows of the montane zone in early summer. WILD CHIVES, *A. schoenoprasum,* is the only species with hollow leaves.

(2) *Brodiaea.* Wild hyacinth

Only 1 species occurs in our range: *Brodiaea douglasii* has a cluster of funnel-shaped, blue flowers at the top of a smooth scape, 3–6dm tall, erect and stout, which grows from an underground corm. It occurs on plains and foothills in western Montana, southward into Utah and Idaho.

(3) *Calochortus.* Mariposa lily

MARIPOSA, the Spanish name for these lovely flowers, means *butterfly.* The goblet-shaped blossoms have 3 narrow, greenish sepals, 3 broad petals, each with a darker spot at its base, the spots fringed with yellow hairs. Basal leaves 1–2, slender, grayish; stem leaves 1–2. Grows on open, grassy or sagebrush-covered slopes and in aspen groves of the pine belt.

333. Wild Onion, ½X

334. Nodding Onion, ½X

1. Ovaries and fruits orbicular to oblong; basal leaf broad and flat.
    2. Fruits nodding; petals usually clawed.
                C. elegans
                Pussy-ears
    2. Fruits erect; petals rounded or acute at apex.
                C. eurycarpus
1. Ovaries and fruits linear; basal leaf linear, flat or channeled.
    2. Sepals much longer than the petals; anthers linear.
                C. macrocarpus
    2. Sepals rarely longer than the petals.
        3. Hairs on face of petals branched and gland-tipped.
                C. gunnisonii
                Gunnison Mariposa
        3. Hairs on face of petals not branched nor gland-tipped.
                C. nuttallii
                Sego lily

GUNNISON MARIPOSA, *Calochortus gunnisonii* (fig. 335), is 25–45cm tall, with white, pinkish, or lavender flowers. It is the common mariposa on the eastern side of the Continental Divide. Occurs from South Dakota to New Mexico.

SEGO LILY or NUTTALL MARIPOSA, *C. nuttalli*, is the state flower of Utah. Its bulbs were an important food for Indians and pioneers. Frequently found on the western slope in Colorado, where its flower is usually ivory-white with a conspicuous dark splotch at the base of each petal. Abundant in Utah and found on both rims of the Grand Canyon. In Dinosaur National Monument a brilliant pink form may be found; yellow and orange forms are found in New Mexico and Arizona.

335. Gunnison Mariposa, ²∕₅X   336. Camas, ½X

PUSSY-EARS, *C. elegans*, has smaller, creamy-white flowers; petals hairy, borne singly or 2–3 together. It has one long leaf, which overtops the flower. Found in grassy, partially-shaded meadows and on slopes of the pine and aspen belts of Montana and westward into Idaho and Oregon. Frequent in Glacier National Park, occasional in Yellowstone. The tall *C. macrocarpus*, with large purple blossoms in which each petal has a green stripe, grows around Flathead Lake in Montana.

(4) *Camassia*. Camas

1. Pedicels usually tightly appressed in fruit; tepals mostly spreading as they wither.
   C. quamash var. quamash
1. Pedicels erect and incurving in fruit; tepals twisting together to cover ovary as they wither.
   C. quamash var. utahensis

Camas, *Camassia quamash* (fig. 336). Its bulb was an important food for the Indians of Utah, Idaho, and Montana, where it grows abundantly on the plains and hillsides up to 8,000 feet. The flower stalks, 3–6dm tall, rise from a cluster of strap-shaped, bright green leaves and hold open racemes of blue flowers.

(5) *Clintonia*. Bead lily

Only 1 species in our range: Queen's Cup, *Clintonia uniflora*, is a plant of the northern woods; occurs only in Glacier National Park and in the mountains of Idaho. Broad leaves 2–3; usually 1-flowered, white, bell-shaped; plant widely rhizomatous; berry bright metallic-blue.

(6) *Disporum*. Fairybells

One species in our range: Fairybells, *D. trachycarpum*. Leaves ovate to oblong-oblance-olate, acute; flowers 1–2, creamy white, on pendulous, pubescent pedicels; berries globose, orange to red, at the ends of the stem; plant suberect, pubescent, sparsely branched.

(7) *Erythronium*. Snow lily or dogtooth violet

Snow Lily, *Erythronium grandiflorum* (fig. 337), is the only species in our range. Flowers 1–3, bright yellow, the tepals turned back, hanging from a slender stalk 15–38cm tall; anthers white, yellow, red, or purple; leaves only 2, flat, shining, sheathing the base of the naked stem.

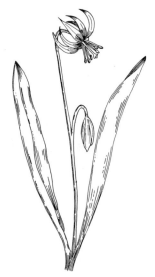

337. Snow-Lily, ²/₅ X

338. Yellow Bells, ½ X

(8) *Fritillaria*. Fritillary

1. Flowers yellow or orange; style 1.

F. pudica
Yellow bells

1. Flowers purple and mottled; styles 3.

F. atropurpurea
Purple fritillary

Yellow Bells, *F. pudica* (fig. 338), grows 10–30cm tall, with a hanging bloom that turns from green to yellow to orange-brown. Grows on grassy plains, mesas, and hillsides of Montana, Wyoming, and northwestern Colorado.

339. Sandlily, ½X

340. Western Wood-Lily, ½X

Common in Yellowstone National Park. PURPLE FRITILLARY, *F. atropurpurea*, is taller and has 1–4 hanging flowers of a mottled purplish color. Its range is similar to that of yellow bells.

(9) *Leucocrinum*. Sand lily

One species only: *Leucocrinum montanum*, SAND LILY or STAR LILY (fig. 339), is one of the earliest spring flowers, found at the lower elevations in April. Stemless; numerous fleshy roots from a short rootstock; pedicels and linear leaves rise from the rootstock; flowers few to many, white, more than 2.5cm broad.

(10) *Lilium*. Lily

One species in our range: WESTERN WOOD LILY, *Lilium philadelphicum* var. *andium* (fig. 340), is one of the most brilliant and rarest of our mountain wild flowers. Flowers are large and showy, 1–3, stiffly erect, tepals orange-red to brick-red, spotted near the base; anthers purple. Once frequently seen in mountain meadows and open woods from New Mexico northward into Canada, but picking has reduced its numbers to near extinction in most places. Protect it carefully!

(11) *Lloydia*. Alpine Lily

One species in North America. ALP LILY, *Lloydia serotina* (fig. 341), is a slender plant 5–15cm tall. Stem leafy, the leaves nearly filiform; usually 1-flowered, creamy white, veined with purple. Grows from small bulbs along a creeping, under-

341. Alplily, ½X

342. Star Solomon's Plume, ⅖X

ground stem. In the Rockies it is found in the alpine zone, often on the summits of exposed ridges and high peaks to 13,000 feet; sometimes at lower altitudes in very cold, exposed situations. An arctic plant.

(12) *Smilacina*. False Solomon's seal

1. Flowers paniculate.

S. racemosa
Claspleaf Solomon's plume

1. Flowers racemose.

S. stellata
Star Solomon's plume

These plants have terminal clusters of small white flowers that produce berries. The STAR SOLOMON'S PLUME, *S. stellata* (fig. 342), has flowers like 6-pointed stars. In the foothills, it is one of the very early spring flowers in canyons and along streams. Blooms later at higher elevations on open slopes, in meadows, and occasionally on dry banks. The CLASPLEAF SOLOMON'S PLUME, *S. racemosa* (fig. 340), is less common. Its flowers are smaller and more crowded. Occurs in shady, moist locations.

343. Claspleaf Solomon Plume, ⅖X

(13) *Streptopus*. Twisted stalk

1. Stems coarsely pubescent below the first branches.

S. amplexifolius var. americana

1. Stems glabrous.

S. amplexifolius var. chalazatus

ˇ TWISTED STALK, *Streptopus amplexifolius* (fig. 344), has greenish or yellowish-white flowers and bright red berries, both hanging from the undersides of long, arching branches. Found in the spruce and aspen zones from New Mexico to Montana.

344. Twisted Stalk, ²⁄₅X

(14) *Tofieldia*. False asphodel

One species in our range: FALSE ASPHODEL, *Tofieldia glutinosa* var. *montana*, a tufted plant with fibrous roots; simple stems leafy only at base, viscid-pubescent above, 3–6dm tall; few-flowered, flowers greenish, white, or yellow. Grows in wet places of the montane regions in northern Wyoming, Montana, and northern Idaho. Found in Teton, Yellowstone, and Glacier National Parks.

(15) *Trillium*. Trillium or wake-robin

One species in our range: WESTERN TRILLIUM, *Trillium ovatum*. Its 3 ovate leaves and erect flower with 3 white petals make this plant easy to recognize. Leaves in a whorl near the top of stem; flower solitary; sepals green. Petals turn reddish or purplish with age.

√(16) *Veratrum*. Corn lily or skunk cabbage

One species generally recognized in our range: CORN LILY, *Veratrum californicum* var. *californicum*. Because of its large, coarse, pleated, and strongly-veined leaves this plant is sometimes called skunk cabbage in our area, where the true skunk cabbage does not grow. It is a tall perennial, 5–15dm high, with broad, strongly-veined leaves, the upper leaves lanceolate; large flowers in a terminal pubescent panicle, whitish with a greenish base. It grows on marshy ground of the aspen and spruce belts; its stout shoots push up soon after the snow disappears. In Colorado, it is found much more commonly on the western than on the eastern slope. Usually abundant wherever it grows.

(17) *Xerophyllum*. Beargrass

One species in our range: NORTHERN BEARGRASS, *Xerophyllum tenax*, is very conspicuous when in bloom, with its tall clusters of creamy-white flowers on stems 6–10dm tall. The plant tufted has needle-shaped leaves 5–8dm long, the upper leaves reduced to bristle-like bracts. Grows on hillsides of Glacier National Park, and in Montana, Idaho, and into northern California. (In the Southwest the name *beargrass* is used for the genus *Nolina*, a different plant).

(18) *Yucca*. Yucca or soapweed

1. Fruit an erect, dry capsule.
    2. Style green; leaves long, narrow, rigid.
    Y. glauca
    2. Style white; leaves short, lanceolate-spatulate.
    Y. harrimaniae
1. Fruit pendulous, whether fleshy or dry.
    2. Fruit fleshy and edible.
    Y. baccata
    Datil
    2. Fruit dry.
    Y. baileyi

YUCCA or SOAPWEED, *Yucca glauca* (fig. 345), is one of the most conspicuous and interesting plants of the foothills of eastern Colorado and New Mexico. Its tufts of dagger-like, evergreen leaves give a light green color to many otherwise barren slopes. In May or June, each clump may send up flower spikes 6–15dm tall, bearing numerous creamy-white blossoms. The flowers hang down and are partially closed during daylight hours. In darkness the 6 thick petals open.

The stout, greenish pistil has 6 white stamens around it. The pollen in the anthers is very heavy and sticky, not at all powdery, and cannot be scattered by wind or wandering insects. But a small, white night-flying moth of the genus *Pronuba*, under 2.5cm long, brings about pollination in the following way: The female yucca moth flies to a yucca flower, takes a ball of

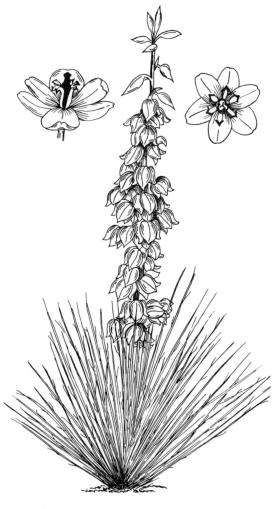

345. Yucca, ⅕ X

346. Wand-Lily, ½X

pollen from one of its anthers, and carrys it to another flower. There she pierces the ovary wall and deposits an egg inside an ovule. Then she crawls up on the pistil and forces the little ball of pollen into the depression in the top of the stigma. She will repeat the process several times. By placing the pollen on the stigma, she assures fertilization of the yucca ovules and development of the seeds. Thus, when the tiny grubs hatch, they find themselves surrounded with food (the developing seeds), and they eat their way out of the yucca seed pod. If you look closely at almost any ripened pod, you will be able to discover the tiny hole where the grub came out. Without the help of *Pronuba*, there would be no seeds.

SPANISH BAYONET is another name for our yucca. Its roots were used by Indians and pioneers in place of soap, and the fibers of the leaves were used by the Indians to make a coarse rope. These coarse fibers can often be seen fraying out and forming decorative curls along the leaf edges, particularly in DATIL, *Y. baccata*, a species of southern Colorado, New Mexico, and Arizona. The leaves of datil are broader and thicker than those of *Y. glauca*, and its fleshy, berry-like fruits are edible. Datils may be seen in Mesa Verde National Park.

(19) *Zigadenus*. Death camas

1. Tepals 8–11mm long, not clawed; stamens perigynous.

Z. elegans var. elegans
Wand lily

1. Tepals under 7mm long, at least the inner series clawed; stamens hypogynous.
   2. Inflorescence usually paniculate; stamens mostly 1–2mm longer than the perianth.

Z. paniculata

   2. Inflorescence usually racemose; stamens mostly about equal to the perianth, rarely up to 1mm longer.

Z. venenosus var. gramineus
Death camas

The common name of this genus indicates its poisonous character and the fact that it must be distinguished from the true *Camas* described above. No parts of any plants of *Zigadenus* should even be tasted. The flowers of all species are greenish or creamy-white, so the plant in bloom is easily distinguished from the edible, blue-flowered *Camas*. The leaves and habit of growth are somewhat similar. In one of the most poisonous species, DEATH CAMAS, *Z. veneno-sus*, the cream-colored flowers are small and arranged in a compact raceme. This plant can be a serious danger to cattle and sheep because the cluster of bright-green leaves appears in early spring before there is much other growth to feed on.

WAND LILY, *Z. elegans* (fig. 346), which is less poisonous, grows in the mountain meadows and into the alpine zone. It has tufts of narrow leaves from which there may be several flowering stalks 20–45cm tall. The individual blossoms are about 12–18mm across, in a star-pointed saucer shape, and each tepal has a greenish spot at its base.

347. Rocky Mountain
Iris, ⅖X

IRIS FAMILY: IRIDACEAE

The plants of this family grow from tough, underground stems. Stalks erect; leaves long and slender; flower parts in threes. The stamens are on the bases of the sepals, and iris differ from lilies in having an inferior ovary.

1. Style branches large and petaloid.

Iris, p. 411

1. Style branches filiform.

Sisyrinchium, p. 411
Blue-eyed grass

(1) *Iris.* Iris or flag

Only one species in our range: ROCKY MOUNTAIN IRIS, *Iris missouriensis* (fig. 347). Stem slender; the leaves few, mostly basal, shorter than the stem; flowers 1–2, light or purplish-blue. The 3 petals and 3 petal-like style branches are erect; the 3 petal-like sepals curve downward. Grows throughout the Rocky Mountain states, especially in wet meadows, but may be found on hillsides or other apparently dry places if spring moisture has been abundant.

(2) *Sisyrinchium.* Blue-eyed grass

1. Bracts very unequal, the outer (lower) often surpassing the inflorescence.
    2. Outer bracts usually clasping at the base for more than 4mm; tepals narrowly elliptic to oblanceolate.

S. idahoense var.
occidentale

    2. Outer bracts usually clasping at the base less than 4mm; tepals broadly elliptic.

S. montanum

1. Bracts moderately unequal (subequal), neither usually surpassing the inflorescence.
S. demissum

348. Blue-Eyed
Grass, ⅖ X

BLUE-EYED GRASS, *Sisyrinchium*. The plants in this genus have small, blue flowers, usually less than 12mm broad; 3 petals and 3 sepals alike. They grow in meadows and are rather inconspicuous because of their tufted, grass-like leaves, and because their flowers open only in bright sunshine. Found in nearly all areas of the United States except deserts, and they occur from 4,000 to 10,000 feet in the Rockies. *S. montanum* (fig. 348), is the common species in Colorado and Wyoming; *S. demissum* in New Mexico and Arizona; *S. idahoense* in Yellowstone National Park.

ORCHID FAMILY: ORCHIDACEAE

This is the largest of all plant families. Most of its members live in tropical and subtropical regions, but about 30 species are found in the Rocky Mountains. Many of them have small, inconspicuous flowers, but, small as they are, each has the distinctive characters of the orchid blossom: corolla irregular (bilaterally symmetrical), the lower petal larger, sometimes spurred or developed into a sac or pouch; the stamens and pistil grown together (the style united with the stamens to form a column); the ovary inferior. The flower structure is highly specialized for insect pollination. Some of our species are wholly or partially saprophytic, growing in moist leaf-mold or rotted wood soil under forest trees.

1. Plants without green leaves, yellowish to brownish-red.
Corallorhiza, p. 414
Coral root
1. Plants with green leaves.
   2. Base of the lip forming a spur; bracts subtending the flowers.
      3. Apex of the lip entire; flowers greenish or white.
         4. Leaves persisting until the maturity of the fruit.
Plantanthera, p. 416
Bog Orchid
         4. Leaves near base withering at or before anthesis.
Piperia, p. 416
Wood orchid
      3. Apex of the lip divided into 2 lateral lobes and a small central lobe.
Coeloglossum, p. 413
Rein orchid
   2. Base of the lip not forming a spur.
      3. Lip inflated, saccate, or pouch-like.
         4. Plants with a single basal leaf; lip bearded; stem from a fleshy corm.
Calypso, p. 413
Fairy slipper

4. Plants with 2 or more leaves; lip not bearded; stem from fleshy or fibrous roots.
    5. Lip inflated, pouch-like; leaves generally 2, cauline, green.
        Cypripedium, p. 414
        Lady's slipper
    5. Lip saccate; leaves more than 2.
      6. Leaves basal and rosulate, dark green often mottled with white.
        Goodyera, p. 415
        Rattlesnake plantain
      6. Leaves cauline, green.
        Epipaetis, p. 415
        Helleborine
3. Lip not inflated, flat to concave.
  4. Leaves 1–2; flowers greenish or greenish-yellow.
    5. Leaves 1 (2), basal; flowers greenish-yellow.
        Malaxis, p. 416
        Addersmouth
    5. Leaves 2, near the middle of the stem; flowers greenish.
        Listera, p. 416
        Twayblade
  4. Leaves more than 2, mostly basal; flowers white in a twisted spike.
        Spiranthes, p. 418
        Ladies' tresses

(1) *Calypso.* Fairy slipper

Only one species in the genus: FAIRY SLIPPER, *Calypso bulbosa* (fig. 349). Stems 6–12cm high with 2–3 brownish-green sheathing bracts and a linear bract at the summit. Single radical leaf, ovate; flower drooping, light rose, the lip brownish-pink mottled with purple. Found in cool, moist, coniferous forests of North America; extends south along our high mountains, where such forests are present, as far as Arizona. In favorable locations, often on north-facing slopes where snow collects in winter, it sometimes occurs in great numbers. Because it blooms in late May, before many people visit our mountain forests, it has not suffered the fate of its larger cousin, the lady's slipper. It should be protected even so, as it is a very fragile plant. Picking or transplanting usually destroys the root.

(2) *Coeloglossum.* Rein orchid

Only one species in our range: LONG-BRACTED ORCHID, *C. viride* var. *virescens.* Flowers greenish, in a long, leafy-bracted spike 2–5dm tall; leaves sheathing, reduced and narrowed above; plants glabrous with fleshy roots.

349. Fairy Slipper, ½X

(3) *Corallorhiza*. Coralroot

1. Sepals and petals pinkish, conspicuously striped with reddish-brown or purple; spur absent; the lip entire.

> *C. striata*
> Striped coralroot

1. Sepals and petals yellow or pink to dark red, not conspicuously striped, though they may be veined; spur often present; lip often lobed.
    2. Sepals more or less 5mm long, yellow to greenish to whitish; lip somewhat 3-lobed by lateral clefts.

> *C. trifida*
> Yellow-stem coralroot

    2. Sepals 6–13mm long, often reddish, usually 3-nerved.
        3. Lip neither lobed nor toothed.

> *C. wisteriana*
> Spring coralroot

        3. Lip usually lobed or toothed.
            4. Slender column, 6–7mm long; spur present; lip with 1–2 spots.

> *C. mertensiana*
> Western coralroot

            4. Stout column, 3.5–5mm long; spur much reduced, adnate to ovary; lip with several spots.

> *C. maculata*
> Spotted coralroot

SPOTTED CORALROOT, *Corollorhiza maculata* (fig. 350). A few stalks of coralroot, standing in a spot of sunlight on the brown pine needle floor of a mountain forest, sometimes catch the eye of a hiker. Their stems are a translucent reddish-brown, and the small flowers show white tongues spotted with purple. These plants are without chlorophyl and absorb their nourishment from rotted wood in the forest soil. Stems 15–45cm tall; flowers about 13mm long. Frequently found in spruce and pine forests over the United States.

(4) *Cypripedium*. Lady's slipper

1. Leaves 2, opposite or nearly so; flowers 2–4, clustered immediately above the leaves; lip purplish.

> *C. fasciculatum*
> Brownie lady's slipper

1. Leaves several, flowers 1–3, not clustered; lip white or yellow, lightly purple tinged.
    2. Lip yellow, often purplish-dotted; flower usually 1.

> *C. calceolus var. pubescens*
> Yellow lady's slipper

    2. Lip white to purplish-tinged, but not dotted; flowers usually 2.

> *C. montanum*
> Mountain lady's slipper

YELLOW LADY'S SLIPPER, *Cypripedium calceolus* (fig. 351), is easily recognized by the large, yellow, inflated lip, but it is one of the rarest of Rocky Mountain wildflowers and is considered an endangered species. Its leafy stem grows 20–60cm tall. A few plants still persist in isolated areas, and they should never be picked. Grows in moist forest openings and at edges of meadows from New Mexico and Arizona north into Canada, and throughout most of the United States east of the Rockies. Our variety, *C. c. pubescens*, has greenish-yellow lateral petals. BROWNIE LADY'S *Slipper*, *C. fasciculatum*, usually has several stems, each with 2–3 yellowish, brown, or purplish flowers 2.5–4cm long.

350. Spotted
Coral-Root, ½ X

351. Yellow Ladyslipper, ½ X

(5) *Epipactis*. Helleborine

One species in our range: GIANT HELLEBORINE, *Epipactis gigantea*. Stems 1 to many, to 12dm tall; leaves numerous, sheathing especially below, elliptic-lance-olate, 3–15cm long; flowers 3–15, pink, purplish, or coppery-green, subtended by large, leafy bracts; lip greenish-yellow. Grows in wet meadows, especially near seeping springs at the base of cliffs. It is uncommon but found in isolated locations in the Arbuckle Mountains of Oklahoma, and in Texas, the Grand Canyon of Arizona, Wyoming, California, and rarely in central and western Colorado, often in association with maidenhair fern, *Adiantum*.

(6) *Goodyera*. Rattlesnake plantain

1. Lip scarcely saccate; leaves usually white-mottled or with white midrib.
                         G. oblongifolia
                         Rattlesnake orchid

1. Lip plainly saccate; leaves neither white-mottled nor with white midrib.
> *G. repens.*
> Rattlesnake orchid

RATTLESNAKE ORCHID: *Goodyera repens* (fig. 352), with flowers in a 1-sided raceme, is rare. *G. oblongifolia*, with its flowers in a loose spiral, is more common. Their leaves are in rosettes at the base of the 15–20cm stems, which bear small, whitish flowers.

(7) *Listera*. Twayblade

1. Lip cleft to about half its length into 2 somewhat divergent lobes; leaves subcordate.
> *L. cordata*
> Twayblade
1. Lip entire or only slightly notched with rounded lobes.
> 2. Lip petal narrowed toward its base.
>> 3. Lip petal with 2 narrow teeth at base, the petal margins glabrous.
>>> *L. caurina*
>>> Northwest twayblade
>> 3. Lip petal lacking narrow teeth at base, the petal margins ciliolate.
>>> *L. convallarioides*
>>> Broad-leaved twayblade
> 2. Lip petal hardly narrowed toward its base (or not at all).
>> *L. borealis*
>> Northern twayblade

TWAYBLADE, *Listera cordata* (fig. 353), and BROAD-LEAVED TWAYBLADE, *L. convallarioides*, delicate plants 7–20cm tall, having a single pair of leaves at the middle of the stalk, are found in permanently damp, shaded situations where the soil contains much humus. The inflorescence is a few-flowered spike.

(8) *Malaxis*. Adder's mouth

1. Inflorescence spicate; flowers greenish-yellow; perianth parts oblong to ovate.
> *M. mascrostachya*
> Mountain malaxis
1. Inflorescence racemose; flowers purplish; perianth parts linear to lanceolate.
> *M. ehrenbergii*

In our range, the two species of Adder's mouth are found only in New Mexico and Arizona. Not common.

(9) *Piperia*. Wood orchid

One species in our range: *Piperia unalascensis*, formerly placed in the genus *Habernaria*. WOOD ORCHID, found in damp woods, has a strict and slender stem 3–7dm tall, leafy below only, the leaves withering early, lanceolate to oblanceolate; flowers greenish, the petals sometimes whitish.

(10) *Plantanthera*. Bog orchid

1. Leaves 1 (2), borne low on stem; lip linear; spur and lip equal in length; perianth greenish or yellowish-green.
> *P. obtusata*
> One-leaf orchid
1. Leaves 2 or more (if 2, usually basal).

2. Leaves 2, broad, basal, and opposite; floral scape with a bract; spur about 2cm long; pale to deep greenish-white.

P. orbiculata var. orbiculata
Two-leaved wood orchid

2. Leaves 2 or more, nearly basal or borne on the stem.
  3. Flowers white; lip conspicuously dilated at base.
    4. Spur as long as, or shorter than, the lip.
      5. Nectary (the organ secreting nectar) slender, equal to or shorter than the lip.

P. dilitata var. dilitata
White bog orchid

    5. Nectary stout or club-shaped, shorter than the lip.

P. dilitata var. albiflora

    4. Spur longer than the lip.

P. dilitata var. leucostachys

353. Twayblade,
½ X

352. Rattlesnake Orchid, ½ X

354. Green Bog-
Orchid, ½ X

355. One-Leaf
Orchid, ½ X

    3. Flowers green to yellow-green or whitish-green; lip inconspicuously dilated
      at base.
      4. Spur about equal to lip in length; plant very slender, 30–60cm tall.
                                       P. hyperborea var. gracilis
                                        Green bog orchid
      4. Spur either much shorter or much longer than lip.
        5. Spike loosely to densely flowered; lip tapered from a widened base.
                                         P. hyperborea var.
                                        purpurascens
        5. Spike slender and remotely flowered; lip linear to elliptical.
                                        P. stricta
                                        Green bog orchid

These bog orchids may sometimes be found in *Habernaria*. The White Bog Orchid, *P. dilitata*, has a slender spike of pure white, fragrant flowers. Found in wet, grassy spots on stream banks, beside trails, or in boggy ground around ponds and lakes. The Green Bog Orchids, *P. hyperborea* (fig. 354, preceeding page), and *P. stricta,* are similar except for the green color of the flowers, sometimes tinged with purple, and their lack of fragrance. They occur in similar boggy situations. The flowers of *H. hyperborea* are more closely arranged than are those of *P. stricta.*

The Two-Leaved Wood Orchid, *P. orbiculata*, which grows in Glacier National Park and vicinity, has greenish-white flowers 2.5cm or more long. Its 2 rounded leaves, 15–20cm long, spread flat on the ground. The One-Leaf Orchid, *P. obtusata* (fig. 355, preceeding page), is a small plant with greenish flowers, usually found in permanently wet, shaded situations where the soil contains humus.

(11) *Spiranthes.* Ladies' tresses

One species in our range: Ladies' Tresses, *Spiranthes romanzoffiana* var. *romanzoffiana*, with stems more or less leafy, to 6d tall but frequently only 2dm, is found in grassy meadows and on stream banks. Its fragrant, white flowers are closely set in about 3 ranks, and the stalk is spirally twisted. Blooms in meadows of the montane zone in middle and late summer.

GLOSSARY

Acaulescent	Stemless.
Achene	A small, hard. one-seeded, dry fruit.
Acuminate	Taper-pointed. Plate C 27.
Acute	Sharp-pointed or ending in a point that is less than a right angle. Plate C 26.
Adnate	Attached to, especially used of unlike parts; united in growth.
Adventive	Introduced but not yet naturalized.
Alternate	(Used of leaves, buds, or branches). Occurring singly at the nodes. Plate A 3.
Ament	Same as catkin.
Amplexicaul	(Of leaves.) Clasping the axis by the base.
Anther	The essential part of the stamen, which contains the pollen.
Anthesis	The opening of the flower.
Apiculate	Terminated by a short, sharp, flexible point.
Appressed	Lying close and flat.
Arachnoid	Cobwebby.
Articulation	Joint or place where separation naturally takes place.
Auriculate	Furnished with auricles, or ear-like appendages.
Awl-shaped	Sharp-pointed from a broader base.
Axil	The upper angle between a leaf and the stem. Plate A 3.
Axillary	Occurring in the axils.
Axis	The central line of any body; the organ around which others are attached.
Banner	Or standard. The uppermost petal of a papilionaceous corolla.
Barbellate	With barbs along the main axis.
Beak	A substantial point, prolonged pistils and fruits.
Bifid	Two-cleft.
Bilabiate	Two-lipped.
Biseriate	In two rows or in two series.
Blade	The flat, expanded portion of a leaf or petal.
Bract	A much reduced leaf without a petiole and usually close below (subtending) a flower.
Bracteate	Bearing bracts.

Bulb	A round, underground bud of fleshy, overlapping scales (leaf bases) attached to a short, flattened stem, as in onions.
Caespitose	Growing in a compact mat or tuft.
Calyx	The outer circle of perianth segments, made up of sepals, which may be either separate or joined.
Campanulate	Bell-shaped.
Canescent	Grayish-white; the surface covered with fine white hairs.
Capillary	Very slender or hairlike.
Capitate	Headed; in heads; a dense or compact cluster.
Capsule	A dry fruit composed of more than one carpel, which splits open when ripe.
Carpel	The unit of structure of the pistil, which may consist of one or more carpels.
Catkin	A scale spike of inconspicuous flowers, as in willows. An *ament*.
Caudex	The persistent, woody base of an otherwise herbaceous stem.
Caulescent	Stemmed or stem-bearing.
Cauline	Pertaining to a stem or axis.
Ciliate	With a fringe of hairs on the margin.
Ciliolate	Slightly ciliate.
Circumscissile	Opening (dehiscing) by a transverse line around the fruit or anther, the valve usually coming off as a lid.
Clavate	Club-shaped, the body thickened toward the top.
Commissure	The plane of junction of two carpels.
Comose	Tufted; bearing a tuft of hairs.
Compound	Made up of 2 to many similar parts—as a compound ovary or a compound leaf.
Confluent	Merging or blending together.
Conifer	A cone-bearing tree.
Connate	Like or similar structures united or joined.
Connivent	Converging, but not fused.
Cordate	Heart-shaped.
Coriaceous	Of leathery texture.
Corolla	The inner circle of perianth segments, a collective name for the petals. Commonly used when the petals are united.

Corona	Crown. Any appendage standing between the corolla and stamens, or on the corolla.
Corymb	A flat or convex flower cluster, with branches arising from different levels. Plate F 53.
Cotyledon	The first or seed leaves of a plant, usually differing from the later leaves.
Crenate	With rounded teeth. Used of margins of leaves or petals.
Culm	The stem of grasses.
Cuneate	Wedge-shaped.
Cuspidate	Tapering to an elongated point, concavely constricted on the sides.
Cyme	Like a corymb, but with the central or terminal flowers in the cluster opening first.
Deciduous	Used of leaves that fall off at the end of one season of growth, or of petals or sepals that fall early.
Decompound	More than one compound.
Decumbent	Bent at the base of the stem and more or less leaning or lying on the ground, the tip tending to rise.
Decurrent	Extending downward and adnate to the stem.
Deflexed	Reflexed.
Dehiscent	Opening by some method of dehiscence: the natural opening of a closed vessel, as of an anther or a pod.
Deltoid	Triangular.
Dentate	Toothed. Having sharp teeth pointing straight out.
Denticulate	Finely dentate.
Diadelphous	In 2 sets or clusters.
Digitate	Handlike.
Dimorphous	Of two forms.
Dioecious	Having staminate and pistillate flowers on different plants of the same species.
Diphyllous	With two leaves.
Disk	The face of any flat body, especially the central region of a head of flowers, as in the Composite Family.
Dissected	Cut deeply into many lobes or divisions.
Divaricate	Spreading far apart: extremely divergent.
Divergent	Spreading broadly.
Dolabriform	Attached at the middle, the two ends free.
Dorsal	Relating to the back or the outer surface.

Drupe	A stone fruit.
Ecology	The study of plants in relation to other living organisms and to their environment.
Elliptical	Oval in outline.
Emarginate	With a shallow notch at the apex.
Entire	The margin not at all toothed, notched, or divided.
Epigynous	Upon the ovary. The ovary inferior.
Erose	Irregularly eroded, as if gnawed.
Exserted	Protruding from, as the stamens may protrude from the corolla.
Falcate	Sickle-shaped.
Farinose	Covered with a mealy powder.
Fascicle	A close cluster.
Fertile	Capable of producing seed, pollen, or sporangia.
Filament	The stalk of a stamen; or any slender, thread-like structure. Plate D 29.
Filiform	Thread-shaped.
Fistulose	Cylindrical and hollow.
Floret	A small flower, especially applied to the individual flowers of grasses or composites.
Foliaceous	Leaf-like.
-foliolate	With leaflets.
Follicle	A simple pod, opening only down the inner structure.
Frond	The leaf of a fern.
Fusiform	Spindle-shaped; narrowed both ways from a swollen middle.
Galea	A helmet or hood.
Genus	(plural, *genera*) A group of plants made up of closely related species.
Gibbous	Swollen on one side.
Glabrate	Almost glabrous, or becoming glabrous with age.
Glabrous	Having no hairs, bristles, or other pubescence.
Gland	A secreting structure or surface.
Glandular	Bearing secreting glands.
Glaucescent	Slightly glaucous.
Glaucous	Covered with fine, white powder, giving a bluish color to foliage.
Globose	Shaped like a globe.

Glumes	(Used in describing grasses). The outer husks or bracts of each spikelet.
Glutinous	Sticky.
Granulate, *Granular,* *Granulose*	Covered with very small grains.
Habitat	The situation in which a plant grows in its wild state.
Hastate	Arrowhead-shaped, the basal lobes extending outward instead of downward.
Head	A short, compact inflorescence.
Herbaceous	Applied to plants of soft texture, of plants whose stems die back to the ground at the end of the growing season.
Hirsute	Hairy with stiffish or beard-like hairs.
Hispid	Bristly: beset with stiff hairs.
Hyaline	Thin, whitish, transmitting light.
Hypantheum	In epigynous and perigynous flowers only: the tube from the base or the tip of the ovary to the point where the sepals, petals, and stamens are inserted.
Hypogynous	Inserted under the pistil. The ovary superior.
Imbricated	Applied to leaves or flower parts arranged in an overlapping pattern, as shingles on a roof.
Imperfect	Unisexual. Flowers that lack either stamens or pistils. Unisexual.
Incised	Cut, or slashed irregularly.
Indusium	(plural, *indusia*) The very small covering over the *sorus* on ferns.
Inferior	Applied to the ovary when the stamens, or the perianth, are inserted on its top. Epigyny.
Inflorescence	The arrangement of flowers on the stem; the flower cluster as a whole.
Insectivorous	Used of plants that have organs developed to digest insects.
Internode	That part of an axis between two nodes.
Involucel	A secondary involucre.
Involucre	A whorl or set of bracts surrounding a flower, umbel, or head.
Involute	Rolled inward.
Irregular	Used to describe a calyx or corolla in which the parts are not all alike, i.e., the calyx or corolla is not radially symmetrical.

Keel	A projecting ridge on a surface, like the keel of a boat. Used especially to describe the structure formed by the fusing of the two lower petals in flowers of the pea family.
Lacerate	Torn or irregularly cut.
Laciniate	Slashed; cut into deep, narrow lobes.
Lanate	Woolly; covered with long, soft, entangled hairs.
Lanceolate	Lance-shaped. Plate C 18.
Leaflet	A segment of a compound leaf.
Legume	The fruit (pod) of a member of the pea family; also a member of that family.
Lemma	The outer bract of the grass floret.
Lenticels	Wartlike, usually light-colored spots on the bark of trees or shrubs.
Lenticular	Lentil-shaped (convex on both sides).
Ligulate	Tongue or strap-shaped.
Ligule	A strap-shaped corolla as in the ray flowers of the composites. Also, a projection from the top of the sheath in grasses.
Linear	Narrow and flat, the margins parallel. Plate C 17.
Lobe	A division of a leaf, especially a rounded one; also divisions of a united corolla.
Locule	A compartment or cell.
Lodicule	One of 2 or 3 scales appressed to the base of the ovary in grasses.
Loment	The fruit of a legume, contracted between the seeds.
Lunate	Crescent-shaped.
Marcescent	Withering, but the dried parts persisting.
-merous	Refers to the number of flower parts, i.e., a flower having 5 sepals, 5 petals, 5 carpels, and 5 stamens is 5-merous.
Monodelphous	Stamens united into one group, their filiments connate.
Monocotyledonous	Used of plants having only one seed leaf.
Monoecious	With stamens and pistils in separate flowers on the same plant.
Mucronate	Terminating abruptly in a sharp spur.
Muricate	Beset with short and hard points.
Nectary	An organ that secretes nectar.
Node	A point on a stem from which one or more leaves arise.
Nodding	Used of flowers or buds that hang down.

Ob-	A prefix indicating inversion.
Oblanceolate	Lance-shaped with the tapering point toward the base. Plate C 19.
Obovate	The reverse of ovate, the broad end upward. Plate C 23.
Obtuse	Blunt, rounded.
Ochroleucous	Cream colored.
Opposite	Applied to leaves and branches when an opposing pair occurs at each node. Plate A 4.
Orbicular	Circular.
Ovary	The part of the pistil that contains the ovules and ripens into the "seedpod." Plate D 59.
Ovate	Shaped like a section through a hen's egg with the broader end toward the base.
Palea	The inner bract of the grass floret.
Palmate	Applied to a leaf whose leaflets, divisions, or main ribs all spread from the apex of the petiole, like a hand with outspread fingers. Plate B 10-11.
Palmatifid	Cut about halfway down in palmate form.
Panicle	A type of repeatedly-branched inflorescence with flowers on pedicels. Plate F57.
Paniculate	Resembling a panicle.
Papilionaceous	Butterfly-shaped; flowers as in the pea family, with standard, wings, and keel.
Pappus	The bristles, scales, or awns at the tip of a "seed," in composites.
Parasitic	An organism that depends on living tissue of another living organism for a source of food.
Parietal	Attached to the walls, as of the ovary.
Pectinate	Comb-like, with close narrow divisions.
Pedicel	The stalk of each individual flower of a cluster. Plate D 29.
Pedicellate	Having pedicels.
Peduncle	A flower stalk, whether of a single flower or of a cluster.
Peltate	With a stalk to its center rather than at its margin; shield-shaped.
Perfect	Bisexual. Flowers with both pistils and stamens.
Perfoliate	Applied to a leaf through whose base the stem appears to pass.
Perianth	A collective name for the sepals and petals.
Pericarp	The wall of a ripened ovary, i.e., the wall of a fruit.

Perigynium	The sac that encloses the ovary and fruit in *Carex* and *Kobresia*.
Perigynous	Borne or arising from around the ovary and not beneath it. Ovary superior.
Petal	One of the inner perianth segments.
Petaloid	Petal-like, resembling or colored like petals.
Petiolate	Having petioles.
Petiole	The stalk of an individual leaf. Plate B 8.
Phyllaries	Bracts, especially those making up the involucre of composites. Plate G 62-71.
Phyllodium	Leaf-like petiole with no blade.
Pilose	Hairy; clothed with soft slender hairs.
Pinna	(plural, *pinnea*) A primary division of a fern frond.
Pinnate	Used of leaves having leaflets disposed along the main axis of the leaf; also a type of leaf-veining where the secondary veins arise from a midrib. Plate B 13.
Pinnatifid	Pinnately-cleft.
Pistil	The female, or seed-bearing, organ of the flower, made up of ovary, style, and stigma. Plate D 29.
Pistillate	Having pistils but no functional stamens; female.
Placenta	The surface of the ovary to which the ovules are attached.
Plumose	Plumed or feathery.
Pollen	The grains or spores in the anther that contain the male element.
Polygamous	Bearing bisexual and unisexual flowers on the same plant.
Polygamodioecious	A dioecious species that has a few flowers of the opposite sex or a few bisexual flowers on all plants.
Pome	A fleshy fruit such as the apple or haw.
Pruinose	Having a bluish-white bloom on the surface.
Pseudoscape	A false, naked scape, usually underground, between the roots and the leaves.
Puberulent	Covered with fine, almost imperceptible, down.
Pubescent	A covering of fine soft hairs.
Punctate	With translucent or colored dots, or pits.
Pustulate	Blistery, usually minutely so.
Raceme	An elongated inflorescence in which the stalked (pedicellate) flowers are arranged singly along a central axis. Plate F 55.

Rachilla	A secondary axis or rachis. The axis that bears the floret in the grasses and sedges.
Rachis	The axis (to which other parts are attached) of a compound leaf or of the inflorescence; the stalk of a fern frond.
Radiate	Used of composites that have ray flowers.
Radical	Belonging to, or coming from, the root.
Ray	The marginal flower of a head or cluster when different from the rest.
Receptacle	The more or less enlarged top of the stalk to which other flower parts are attached. Plate D 29.
Recurved	Bent downward or backward.
Reflexed	Abruptly recurved.
Regular	Radially symmetrical; a calyx or corolla where the parts are all similar. Plate E 44-48.
Reniform	Kidney-shaped.
Reticulate	Netted.
Retrorse	Directed downward or backward.
Revolute	The margins rolled toward the lower side.
Rhizome	Underground stem.
Rhomboidal	Rhomboid-shaped.
Rootstalk	Underground stems, rhizome.
Rosette	Cluster of leaves radiating from a stem, usually close to the ground.
Rosulate	In rosettes.
Rotate	Wheel-shaped, applied to flat, open corollas. Plate E 45.
Rugose	Wrinkled.
Saccate	Bag-shaped or pouchy.
Sagittate	Arrowhead-shaped, the basal lobes pointing downward.
Salverform	Corolla with a slender tube and an abruptly expanded limb. Plate E 46.
Samara	Indehiscent winged fruit, as in the maple.
Saprophytic	Describes plants that live on decayed vegetable matter such as rotted wood or leaf mold. No green parts.
Scabrous	Rough or harsh to the touch.
Scale	A reduced, leaf-like body that is not green.
Scape	Leafless peduncle rising from the ground.

Schizocarp	A dry fruit that splits into two indehiscent parts, as in most umbels.
Scurfy	Covered with minute, scaly particles.
Scorpioid	Coiled at the apex like the tail of a scorpion.
Secund	One-sided, or borne on one side.
Seed	A ripened ovule; it contains an embryo.
Sepal	One of the separate parts of the calyx, usually green but may be petaloid. Plate E 40-43.
Septate	Divided by partitions.
Septum	A partition, as of a pod.
Sericeous	Silky.
Serrate	With margin cut into teeth pointing forward. Plate C 21.
Serrulate	Minutely serrate.
Sessile	Without any stalk, as a leaf without a petiole, a flower without a pedicel, or an anther without a filament.
Setose	Covered with bristles.
Sigmoid	S-shaped.
Silicle	A short fruit in the Cruciferae, usually not more than 3 times as long as wide.
Silique	An elongate fruit in the Cruciferae, more than 3 times as long as wide.
Simple	Said of a leaf when not compound, or of a stem when not branched.
Sinuate	Wavy-margined.
Sinus	The space between two lobes.
Sorus (plural, *sori*)	The fruit dots usually found on the underside of fruiting fern fronds.
Spathe	The bract or leaf surrounding or subtending a flower cluster.
Spatulate	Gradually narrowed downwards from a rounded summit. Plate C 22.
Spicate	Spike-like.
Spike	An inflorescence in which the flowers are sessile on a more or less elongated, common axis. Plate F 54.
Spikelet	(in grasses and sedges). The smallest flower cluster in an inflorescence, usually forming a distinct and compact unit.
Spinulose	With small spines over the surface.
Spore	A simple reproductive body, usually nearly microscopic in size.

Sporangia	The structure that contains spores.
Sporocarp	A receptacle containing sporangia.
Stamen	The pollen-bearing organ, made up of the filament and the anther. Plate D 29.
Staminate	Having stamens but no pistil. Male.
Staminode	A sterile stamen.
Stellate	Star-like.
Sterile	Lacking functional sex organs.
Stigma	The part of the pistil that receives the pollen. Plate D 29.
Stipe	A stalk.
Stipitate	Having a stipe.
Stipules	Appendages at the base of the petiole of a leaf. Plate B 8.
Stolons	Horizontal shoots at ground surface that take root to form new plants.
Striate	With fine, parallel lines, grooves, or streaks.
Strict	Straight and upright.
Strigose	With appressed, rigid bristles or hairs.
Style	If the stigma is raised above the ovary, the connecting portion is the style. Plate D 29.
Stylopodium	An enlargement at the base of the style, especially in Umbelliferae.
Sub-	A prefix indicating almost, nearly, somewhat, or below.
Submersed	Said of plants or their parts when growing under water.
Subtend	To stand below, but close to.
Subulate	Awl-shaped.
Succulent	Used of plants, or their parts, that are fleshy and juicy, usually thickened.
Taproot	The primary root along the main axis of the plant.
Tendril	A slender, clasping or twining outgrowth of stems and leaves.
Tepals	Segments of those perianths where the sepals and petals are not clearly differentiated.
Terete	Cylindrical.
Ternate	In threes.
Thallus	The flat, leaf-like organ in some cryptogams, the stem and leaves not differentiated.
Thyrse	A compact and pyramidal panicle.
Tomentose	Covered with matted, woolly hairs (*tomentum*).

Tomentulose	Somewhat tomentose.
Trifid	Cut into three segments.
Triquetrous	Three-angled in cross section.
Truncate	Horizontal, as if cut off.
Tuber	A short, thickened structure, usually part of a stem and underground.
Tubercle	A small, rounded, protruding body.
Turbinate	Top-shaped.
Umbel	A type of inflorescence in which all the rays (branches) originate at the same point, umbrella-like. Plate F 52.
Unisexual	Imperfect. Having only either female or male flowers.
Utricle	A small bladder.
Valvate	Opening by valves or pertaining to valves.
Valve	Segment of fruit that splits at maturity.
Venation	The type of arrangement of veins, usually in leaves. Plate B.
Verrucose	With a wart-like or nodular surface.
Verticillate	Arranged in whorls.
Villous	Shaggy with long and soft hairs.
Viscid	Sticky, glutinous.
Whorl	A circle or ring of organs, especially a leaf arrangement where three or more arise from the same node. Plate A 6.
Wing	A thin, usually dry extension of an organ or structure; the lateral petals of a flower in the pea family.
Woody	Of firm texture; applied to plant parts above ground that remain alive from season to season.
Woolly	Lanate; clothed with long and entangled, soft hairs.

USEFUL REFERENCES

(1) State Floras

Arizona: Thomas H. Kearney and Robert H. Peebles, *Arizona Flora*. Berkeley & Los Angeles: University of California Press, 2nd ed., 1964.

Colorado: H. D. Harrington, *Manual of the Plants of Colorado*. Chicago: Sage Books, 2nd ed., 1964.

Idaho: Ray J Davis, *Flora of Idaho*. Dubuque: W.C. Brown, 1952.

Montana: Robert D. Dorn, *Vascular Plants of Montana*. Cheyenne: Mountain West Publishing, 1984.

New Mexico: William C. Martin and Charles R. Hutchins, *A Flora of New Mexico*, 2 vols. Vaduz: J. Cramer, 1980-1981.

Utah: S. L. Welsh, N. Duane Atwood, Sherel Goodrich, and Larry C. Higgins, *A Utah Flora*. Provo: Brigham Young University, 1987.

Wyoming: Robert D. Dorn, *Vascular Plants of Wyoming*. Cheyenne: Mountain West Publishing, 1988.

(2) Regional Floras

C. Leo Hitchcock, Arthur Cronquist, Marion Ownbey, and J. W. Thompson, *Vascular Plants of the Pacific Northwest*, 5 vols. Seattle and London: University of Washington Press, 1955-1969. For an abridged, more concise edition see Hitchcock and Cronquist, *Flora of the Pacific Northwest*. Ibid., 1973.

Ronald L. McGregor, et al., *Flora of the Great Plains*. Lawrence: University of Kansas Press, 1986.

Arthur Cronquist, Arthur Holmgren, Noel Holmgren, James Reveal, and Patricia Holmgren, *Intermountain Flora*, 3 vols. (incomplete). Published for the New York Botanical Garden, 1972-1984.

INDEX